MOLECULAR
BIOLOGY

MOLECULAR BIOLOGY

EDITED BY
T.A. BROWN
Department of Biochemistry and Applied Molecular Biology,
UMIST, Manchester M60 1QD, UK

βIOS
SCIENTIFIC
PUBLISHERS

ACADEMIC PRESS

© BIOS Scientific Publishers Limited, 1991

All rights reserved by the publisher. No part of this book may be reproduced or transmitted, in any form or by any means, without permission in writing from the publisher.

First published in the United Kingdom 1991 by
BIOS Scientific Publishers Limited,
St Thomas House, Becket Street, Oxford OX1 1SJ

ISBN 1 872748 00 7

A CIP catalogue record for this book is available from the British Library.

This Edition published jointly in the United States of America by Academic Press, Inc. and BIOS Scientific Publishers Ltd.

Distributed in the United States, its territories and dependencies, and Canada exclusively by Academic Press, Inc., 1250 Sixth Avenue, San Diego, California 92101 pursuant to agreement with BIOS Scientific Publishers Limited, St Thomas House, Becket Street, Oxford OX1 1SJ, UK.

Designed and typeset by the Opus Design Group Ltd, Oxford, UK
Printed by Information Press Ltd, Oxford, UK

The cover illustration shows, with permission, the pGEM-11Zf(−) vector available from Promega Corporation, 2800 Woods Hollow Road, Madison, WI 53711, USA

The information contained within this book was obtained by BIOS Scientific Publishers Ltd from sources believed to be reliable. However, while every effort has been made to ensure its accuracy, no responsibility for loss or injury occasioned to any person acting or refraining from action as a result of the information contained herein can be accepted by the publishers, authors or editors.

PREFACE

There is nothing new under the sun and so it would be foolhardy to suggest that *Molecular Biology Labfax* is an entirely new departure in scientific publishing. It is, however, different from the existing cloning manuals in that it is designed as a companion rather than a guide for molecular biology research. *Molecular Biology Labfax* does not contain procedures or methodology but instead is a detailed compendium of the essential information — on genotypes, reagents, enzymes, reaction conditions, cloning vectors and suchlike — that is needed to plan and carry out molecular biology research. Some of this information is already available in cloning manuals, catalogs and possibly on pieces of paper kept somewhere safe, but tracking down exactly what you need to know takes time and can be a frustrating experience. With molecular biology becoming an increasingly sophisticated science, an acute need has arisen for a databook to complement the traditional cloning manuals. *Molecular Biology Labfax* is intended to meet this need.

To be useful, the coverage of *Molecular Biology Labfax* has to be right. The scope of the book is of necessity a compromise between a desire to include everything and a need to keep within a reasonable size limit. The reader will expect to find extensive details of *Escherichia coli* genotypes and genetic markers, restriction enzymes, DNA and RNA modifying enzymes, chemicals and reagents, cloning vectors, restriction fragment patterns and suchlike. These topics are covered in as comprehensive a way as possible, so for instance in Chapter 4 all the known restriction enzymes are described along with reaction conditions for all the commercially available ones. A few items that might be expected are not included on the basis that they are of specialist interest, an example being cloning vectors for eukaryotes. Topics such as these will be covered by future editions in the Labfax series. Although experimental protocols are not given, certain key information is provided for subjects such as the growth of *E. coli* strains, use of radiochemicals, electrophoresis of nucleic acids, and hybridization analysis. These topics are of widespread importance in molecular biology procedures and so warrant sections of their own. Readers using other standard techniques will find their needs met by the data presented throughout the book. For instance, the DNA sequencer will find data on dideoxynucleotides and Maxam–Gilbert reagents (Chapter 2), radionucleotides and detection methods (Chapter 3), enzymes for chain termination sequencing (Chapter 5), M13 cloning vectors (Chapter 6), and electrophoresis systems (Chapter 8), as well as details of the genetic code and codon usages for interpretation of sequence information (Chapter 7).

A second essential requirement is accuracy. Wherever possible I have double-checked items that I have had doubts about, going back to the original publications if necessary. In a few cases the literature contains annoying contradictions that I have been unable to resolve, with *E. coli* genotypes providing some of the biggest headaches. The relevant entries carry a footnote or other warning to alert the reader and I welcome enlightenment if anyone knows any of the answers.

Without the help of a number of people *Molecular Biology Labfax* would never have been completed. I am very grateful to Rich Roberts, Toshimichi Ikemura and Michael McClelland for their contributions, as well as GIBCO-BRL, Pharmacia, Promega, USB, Clontech, Stratagene and FMC for providing artwork for the cloning vectors and electrophoresis sections. I would like to give a general thank-you to the various colleagues and friends who helped me out with points here and there. Half-way through the enterprise it became clear that you need to be slightly unbalanced to compile a databook: I became even more worried on the occasions when I thought I was actually enjoying the experience. Because of this I must thank my wife, Keri, who made sure I survived to tell the tale.

T.A. Brown

CONTENTS

3. RADIOCHEMICALS   79

4. RESTRICTION AND METHYLATION   93

5. DNA AND RNA MODIFYING ENZYMES   139

6. CLONING VECTORS 193

10. CENTRIFUGATION

11. SAFETY

SAFETY NOTE

Safety is a critical aspect of laboratory work and all molecular biologists should be aware of the precautions needed to ensure personal safety and the safety of colleagues. National and local safety precautions must be followed at all times. All reputable suppliers of laboratory chemicals provide risk and safety information with those products that present a potential hazard, but several risk and safety classification schemes exist. This book cites risk and safety data for products where appropriate: full details of the classification schemes used, together with information on protection from microbiological and radiochemical hazards, are given in Chapter 11.

CONTRIBUTORS

T.A. BROWN
Department of Biochemistry and Applied Molecular Biology, UMIST, Manchester M60 1QD, UK

T. IKEMURA
National Institute of Genetics, Mishima, Shizuoka-ken 411, Japan

M. McCLELLAND
California Institute of Biological Research, 11099 North Torrey Pines Road, La Jolla, CA 92037, USA

R.J. ROBERTS
Cold Spring Harbor Laboratory, Cold Spring Harbor, NY 11724, USA

ABBREVIATIONS

ATP	adenosine triphosphate	m.p.	melting point
Bq	becquerel	mRNA	messenger ribonucleic acid
bp	base pair	min	minute
b.p.	boiling point	mol	mole
BSA	bovine serum albumin	mol. wt	molecular weight
c.p.m.	counts per minute	nt	nucleotide
cccDNA	covalently-closed-circular DNA	ORF	open reading frame
Ci	Curie	pH	hydrogen-ion exponent
DNase	deoxyribonuclease	RNase	ribonuclease
DNA	deoxyribonucleic acid	sec	second
dNTP	deoxyribonucleotide	ssDNA	single-stranded DNA
d.p.m.	disintegrations per minute	ssRNA	single-stranded RNA
DTT	dithiothreitol	T_m	melting temperature
dsDNA	double-stranded DNA	tRNA	transfer ribonucleic acid
dsRNA	double-stranded RNA	u.v.	ultraviolet
eV	electron volt	(v/v)	volume/volume
f.p.	flash point	(w/v)	weight/volume
g	acceleration due to gravity	(w/w)	weight/weight
h	hour	X-gal	5-bromo-4-chloro-3-indolyl-β-D-galactopyranoside
kb	kilobase		
kd	kilodalton		

CHAPTER 1
BACTERIA AND BACTERIOPHAGES

1. *E. COLI* STRAINS USED IN RECOMBINANT DNA EXPERIMENTS

The *E. coli* strains routinely used in recombinant DNA experiments are listed with their genotypes in *Table 1*. The genotypes are described in accordance with the standard nomenclature as defined below.

1.1. Individual genes

(i) Each mutant locus is described by a three-letter abbreviation (e.g. *ara* = arabinose utilization). The abbreviations are defined in *Table 2*.

(ii) The capital letter following the locus refers to the individual gene that is mutated (e.g. *araD* = L-ribulosephosphate 4-epimerase). The genes are also described in *Table 2*.

(iii) Numbers following the gene designation refer to the specific allele involved (e.g. *araD139*).

(iv) A superscript '–' is generally not used, since, by convention, only mutated genes are listed in the genotype. A superscript '+' may be used to emphasize a locus or gene that is wild-type (e.g. *lac*$^+$ = no mutations in the genes involved in lactose utilization).

(v) A superscript 'q' indicates a constitutive mutation (e.g. *lacI*^{q} = constitutive mutant for the *lac* repressor).

(vi) An amber mutation is denoted by 'am' following the gene designation (e.g. *malBam*).

(vii) If an antibiotic response is listed in the genotype then a superscript 'r' or 's' is used to denote resistance or sensitivity respectively (e.g. *kan*^{r} = kanamycin resistant).

1.2. Deletions

Deletions are denoted by 'Δ' with the deleted gene or genes listed in brackets, possibly followed by an allele designation outside of the brackets [e.g. Δ*(gal–uvrB)40* = deletion of the region from *gal* to *uvrB*].

1.3. Fusions

(i) A fusion is denoted in the same way as a deletion, except that the symbol 'Φ' is used.

(ii) A prime (') is used to designate that the fused gene is incomplete (e.g. '*lacZ* indicates that the *lacZ* gene involved in the fusion is deleted in the 5' region; *lacZ*' indicates a deletion in the 3' region).

(iii) A superscript '+' (e.g. *lacZ*$^+$) denotes that the fusion involves an operon rather than a single gene.

1.4. Insertions

(i) An insertion is denoted by '::', preceded by the position of the insertion and followed by the inserted DNA (e.g. *trpC22*::Tn*10* = insertion of Tn*10* into the *trpC* gene, allele 22).

(ii) If the insertion does not occur within a known gene then the map position is denoted by a three-letter code. The first letter is always *z*, followed by *a–g* to indicate a 10 min interval, and *a–i* to indicate a 1 min interval (e.g. *zgi* = 79 min).

1.5. Phages and plasmids

A plasmid or lysogenic phage carried by the bacterium is listed at the end of the genotype in brackets [e.g. (pMC9) = carries plasmid pMC9].

1.6. Fertility status

(i) Strains are assumed to be F⁻ unless the status is given.
(ii) F⁺ and Hfr strains are denoted by the relevant symbol at the start of the genotype.
(iii) When the strain is listed as F', the genes carried on the episome are listed in square brackets. The F' status is usually placed at the end of the genotype.

The full restriction and modification status of individual strains is not given in *Table 1*. The genotypes are correct for *hsdR*, *hsdS* and *hsdM*, but *mcrA*, *mcrB* and *mrr* are not included. This is because for many strains the *mcrA*, *mcrB* and *mrr* genotypes have not yet been determined. Full descriptions of the restriction and modification status, as far as is known, of important strains are given in *Table 3*. *Table 4* classifies strains according to their specific application(s) in recombinant DNA experiments.

Table 1. Genotypes of *E. coli* strains used in recombinant DNA experiments

Strain	Genotype	References
594	*rpsL*	1
1101	F⁺ *supE*	2
71/18	*supE thi* Δ*(lac–proAB)* F'[*proAB⁺ lacI^q lacZ*ΔM15]	3–6
χ1776[a]	*tonA53 dapD8 minA1 supE42* Δ*(gal–uvrB)40 minB2 rfb-2 gyrA25 thyA142 oms-2 metC65 oms-1* Δ*(bioH–asd)29 tte-1 cycB2 cycA1 hsdR2*	7, 8
AG1[b]	*recA1 endA1 gyrA96 thi hsdR17 supE44 relA1*	9
BB4	*supF58 supE44 hsdR514 galK2 galT22 trpR55 metB1 tonA* Δ*(lac)U169* F'[*proAB⁺ lacI^q lacZ*ΔM15 Tn*10(tet^r)*]	1
BHB2600	*supE supF* (λCH616)	10
BHB2688	N205 *recA* (λ*imm434 c*Its *b2 red3 Eam4 Sam7)/*λ	11, 12
BHB2690	N205 *recA* (λ*imm434 c*Its *b2 red3 Dam15 Sam7)/*λ	11, 12
BJ5183	*endA sbcB recBC galK met str^r thi-1 bioT hsdR*	13
BL21(DE3)[c]	*hsdS gal* (λ*c*Its857 *ind*1 Sam7 *nin*5 *lacUV5*-T7 gene 1)	14
BNN93	see C600	
BNN102[d]	*supE44 hsdR thi-1 thr-1 leuB6 lacY1 tonA21 hflA150*[chr::Tn*10(tet^r)*]	15, 16
C-1A	wild-type	17, 18
C600[e]	*supE44 hsdR? thi-1 thr-1 leuB6 lacY1 tonA21*	13, 15, 16, 19–21
C600*galK*	see C600	
C600*hflA*	see BNN102	
CES200	*sbcB15 recB21 recC22 hsdR*	23
CES201	*recA sbcB15 recB21 recC22 hsdR*	24
CJ236	*dut1 ung1 thi-1 relA1* (pCJ105[*cam^r*F'])	25
CR34	see C600	

Table 1. Continued

Strain	Genotype	References
CSH18	*supE thi* Δ*(lac–pro)* F'[*proAB*$^+$ *lacZ*$^-$]	26, 27
D1210	*supE44 hsdS20 recA13 ara-14 proA2 lacI*^{q} *galK2 rpsL20 xyl-5 mtl-1*	28
DH1	*supE44 hsdR17 recA1 endA1 gyrA96 thi-1 relA1*	13, 29
DH5	*supE44 hsdR17 recA1 endA1 gyrA96 thi-1 relA1*	13, 29
DH5α	*supE44* Δ*lacU169* (φ80 *lacZ*Δ*M15*) *hsdR17 recA1 endA1 gyrA96 thi-1 relA1*	13
DH20	*supE44 hsdR17 recA1 endA1 gyrA96 thi-1 relA1* F'[*lacI*^{q} *lacZ*$^+$ *proAB*$^+$]	13
DH21	*supE44 hsdR17 recA1 endA1 gyrA96 thi-1 relA1* F'[*lacI*^{q} *lacZ*$^+$ *proAB*$^+$]	13
DP50*supF*[f,g]	*supE44 supF58 hsdS3 dapD8 lacY1 glnV4* Δ*(gal–uvrB)47 tyrT58 gyrA29 tonA53* Δ*(thyA)57*	30
ED8654	*supE supF hsdR metB lacY gal trpR*	18, 31
ED8767[g]	*supE44 supF58 hsdS3 recA56 galK2 galT22 metB1*	31
ER1398[h,i]	*supE44 hsdR endA1 thi*	20
ER1451[j]	*supE44 endA1 hsdR17 gyrA96 relA1 thi* Δ*(lac–proAB)* F'[*traD36 proAB*$^+$ *lacI*^{q} *lacZ*Δ*M15*]	20
ER1647*	*serB28* Δ*(mrr–hsdRMS–mcrB) mcrA1272 recD*	32
ER1648*	*serB28* Δ*(mrr–hsdRMS–mcrB) mcrA1272*	32
GM48	*thr leu thi lacY galK galT ara tonA tsx dam dcm supE44*	3
GM2163	*hsdR dam dcm supE*	20, 33
HB101[g,k]	*supE44 hsdS20 recA13 ara-14 proA2 lacY1 galK2 rpsL20 xyl-5 mtl-1*	13, 20, 34, 35
HMS174	*recA1 hsdR rif*^{r}	14, 36
JC7623	*recB1 recC22 sbcB15*	37
JC9956	*recA99 thr-1 leu-6 thi-1 lacY1 galK2 ara-14 xyl-5 mtl-1 proA2 his-4 argE3 str-31 tsx-33*	38
JM83	*ara* Δ*(lac–proAB) strA thi-1* (φ80 *lacZ*Δ*M15*)	3, 39
JM101	*supE thi-1* Δ*(lac–proAB)* F'[*traD36 proAB*$^+$ *lacI*^{q} *lacZ*Δ*M15*]	3, 20, 40
JM103[l]	*supE thi-1 endA1 hsdR4 sbcB15 strA* Δ*(lac–proAB)* F'[*traD36 proAB*$^+$ *lacI*^{q} *lacZ*Δ*M15*]	13, 41
JM103Y	see JM103	
JM105[m]	*endA sbcB15 hsdR4 rpsL thi* Δ*(lac–proAB)* F'[*traD36 proAB*$^+$ *lacI*^{q} *lacZ*Δ*M15*]	3
JM106	*supE44 endA1 hsdR17 gyrA96 relA1 thi* Δ*(lac–proAB)*	3
JM107	*supE44 endA1 hsdR17 gyrA96 relA1 thi* Δ*(lac–proAB)* F'[*traD36 proAB*$^+$ *lacI*^{q} *lacZ*Δ*M15*]	3, 20
JM108	*recA1 supE44 endA1 hsdR17 gyrA96 relA1 thi* Δ*(lac–proAB)*	3
JM109	*recA1 supE44 endA1 hsdR17 gyrA96 relA1 thi* Δ*(lac–proAB)* F'[*traD36 proAB*$^+$ *lacI*^{q} *lacZ*Δ*M15*]	3

* See note on page 24.

Table 1. Continued

Strain	Genotype	References
JM109(DE3)[c]	*recA1 supE44 endA1 hsdR17 gyrA96 relA1 thi* Δ*(lac–proAB)* F′[*traD36 proAB⁺ lacI^q lacZΔM15*] (λ*cIts857 ind*1 *Sam*7 *nin*5 *lacUV5*-T7 gene 1)	43
JM110[n]	*dam dcm supE44 thi leu rpsL lacY galK galT ara tonA thr tsx* Δ*(lac–proAB)* F′[*traD36 proAB⁺ lacI^q lacZΔM15*]	3
K802	*supE hsdR gal metB*	20, 44
KK2186	see JM103	45
KLF41	*leuB6 hisG1 recA1 argG6 metB1 lacY1 gal-6 xyl-7 mtl-2 malA1 rpsL104 tonA tsx supE44* F′141	1
LE30	*mutD5 rpsL azi galU95*	1
LE292	HfrH *argEam rpoB galT*::[λΔ*(int–*FII)]	1
LE392	*supE44 supF58 hsdR514 galK2 galT22 metB1 trpR55 lacY1*	13, 18, 20, 31
LE392.23	*supE44 supF58 hsdR514 galK2 galT22 metB1 trpR55 lacY1* Δ*(argF–lac)U169*	1
LG90	Δ*(lac–proAB)*	46
M5219	*lacZ trpA rpsL* (λ*bio252 cIts857 H1*)	47, 48
MAL103	Δ*(gpt–proAB–argF–lac)XIII rpsL* [Mud1 *(lac,Ap)*] (Mu*cts62*)	1
MB100	Δ*(argF–lac)U169 rpsL150 relA1 flbB5301 deoC1 ptsF25 rbsR leuABCD*::Tn*10*	1
MB101	*araCam araD* Δ*(argF–lac)U169 trpam malBam rpsL relA thi supF* Φ*(araBA′–lacZ⁺)101* [λ*p1(209)*]	1
MBM7007	*araCam araD* Δ*(argF–lac)U169 trpam malBam rpsL relA thi*	1
MBM7014	*araCam araD* Δ*(argF–lac)U169 trpam malBam rpsL relA thi supF*	1
MBM7014.5	*hsdR2 zjj202*::Tn*10(tet^r) araD139 araCU25am* Δ*(lac)U169*	20
MBM7060	*araCam araD* Δ*(argF–lac)U169 trpam malBam rpsL relA thi supF* (λ*p1048*)	1
MC1000	*araD139* Δ*(araABC–leu)7679 galU galK* Δ*(lac)X74 rpsL thi*	1
MC1061	*hsdR araD139* Δ*(araABC–leu)7679* Δ*(lac)X74 galU galK rpsL thi*	16, 49, 50
MC4100	*araD139* Δ*(argF–lac)U169 rpsL150 relA1 flbB3501 deoC1 ptsF25 rbsR*	1
MH225	*araD139* Δ*(argF–lac)U169 rpsL150 relA1 flbB3501 deoC1 ptsF25 rbsR* Δ*(ompC′– lacZ⁺)10–25* [λ*p1(209)*]	1
MH513	Δ*(argF–lac)U169 rpsL150 relA1 flbB3501 deoC1 ptsF25 rbsR* Φ*(ompF′–lacZ⁺)16–23* [λ*p1(209)*]	1

Table 1. Continued

Strain	Genotype	References
MH760	*araD139 Δ(argF–lac)U169 rpsL150 relA1 flbB3501 deoC1 ptsF25 rbsR ompR472*	1
MH1160	*araD139 Δ(argF–lac)U169 rpsL150 relA1 flbB3501 deoC1 ptsF25 rbsR ompR101*	1
MH1471	*araD139 Δ(argF–lac)U169 rpsL150 relA1 flbB3501 deoC1 ptsF25 rbsR envZ473*	1
MH2101	*araD139 Δ(argF–lac)U169 rpsL150 relA1 flbB3501 deoC1 ptsF25 rbsR ompR101 Φ(ompC'–lacZ⁺)10–25 [λp1(209)]*	1
MH2472	*araD139 Δ(argF–lac)U169 rpsL150 relA1 flbB3501 deoC1 ptsF25 rbsR ompR472 Φ(ompC'–lacZ⁺)10–25 [λp1(209)]*	1
MH5101	*Δ(argF–lac)U169 rpsL150 relA1 flbB3501 deoC1 ptsF25 rbsR ompR101 Φ(ompF'–lacZ⁺)16–23 [λp1(209)]*	1
MH5473	*Δ(argF–lac)U169 rpsL150 relA1 flbB3501 deoC1 ptsF25 rbsR envZ473 Φ(ompF'–lacZ⁺)16–23 [λp1(209)]*	1
MM294[i]	*supE44 hsdR endA1 thi*	13, 20
MV1184[o]	*ara Δ(lac–proAB) rpsL thi (φ80 lacZΔM15) Δ(srl–recA)306::Tn10(tetʳ) F'[traD36 proAB⁺ lacIq lacZΔM15]*	39
MV1193	*Δ(lac–proAB) rpsL thi endA sbcB15 hsdR4 Δ(srl–recA)306::Tn10(tetʳ) F'[traD36 proAB⁺ lacIq lacZΔM15]*	51
MZ-1	*galK 8attL BamN₇N₅₃cIts857 H1 his ilv bio N⁺*	52
N3098	*lig7ts supF*	1
N4830	see N4830-1	
N4830-1[p]	*su his ilvA galK8 Δ(chlD–pgl) [λBam N⁺ cI857]*	53
NK5486	*thyA rha strA lacZam*	54
NM519	*hsdR recB21 recC22 sbcA23*	55
NM522	*supE thi Δ(lac–proAB) hsd5 F'[proAB⁺ lacIq lacZΔM15]*	56
NM531	*supE supF hsdR trpR lacY recA13 metB gal*	55
NM538	*supF hsdR trpR lacY*	57
NM539	*supF hsdR lacY (P2cox)*	57
NM554	*recA1 araD139 Δ(ara–leu)7696 Δ(lacI)7A galK u hsr hsmt strA*	90
P2392	P2 lysogen of LE392	90
PLK-17	*hsdR lac supE gal*	91
PLK-A	*recA lac hsdR gal supE*	91
PLK-F'	*recA lac hsdR gal supE F'[proAB lacIq lacZΔM15 Tn10(tetʳ)]*	91
Q358	*supE hsdR φ80ʳ*	58
Q359	*supE hsdR φ80ʳ (P2)*	58

Table 1. Continued

Strain	Genotype	References
R594	*galK2 galT22 rpsL179 lac*	59
RB791	W3110 *lacI^qL8*	60
RR1^g,k	*supE44 hsdS20 ara-14 proA2 lacY1 galK2 rpsL20 xyl-5 mtl-1*	13, 20, 34 35, 61, 62
RT3	*araD139 Δ(argF–lac)U169 rpsL150 relA1 flbB3501 deoC1 ptsF25 rbsR envZ3*	1
RT203	*araD139 Δ(argF–lac)U169 rpsL150 relA1 flbB3501 deoC1 ptsF25 rbsR envZ3 Φ(ompC'–lacZ^+)10–25* [λp1(209)]	1
SE3001	*araD139 Δ(argF–lac)U169 rpsL150 relA1 flbB3501 deoC1 ptsF25 rbsR Δ(malK–lamB)1*	1
SE5000	*araD139 Δ(argF–lac)U169 rpsL150 relA1 flbB3501 deoC1 ptsF25 rbsR recA56*	1
SG263	*araCam araD Δ(argF–lac)U169 trpam malBam rpsL relA thi supF malPQ::Tn10*	1
SG265	*Δ(gpt–proAB–argF–lac)XIII ara argEam gyrA rpoB thi supP* (P1*cry*)	1
SG404	*araD139 Δ(argF–lac)U169 rpsL150 relA1 flbB3501 deoC1 ptsF25 rbsR asd* (P1*cam*) F'141	1
SG480	*araD139 Δ(argF–lac)U169 rpsL150 relA1 flbB3501 deoC1 ptsF25 rbsR Δ(malPQ–bioH– ompB)61*	1
SG608	*araD139 Δ(argF–lac)U169 rpsL150 relA1 flbB3501 deoC1 ptsF25 rbsR Φ(ompC'–lacZ^+)10–25* [λpRT2.3]	1
SG624	*araD139 Δ(argF–lac)U169 rpsL150 relA1 flbB3501 deoC1 ptsF25 rbsR envZ22 Φ(ompC'–lacZ^+)10–25* [λp1(209)]	1
SG626	*araD139 Δ(argF–lac)U169 rpsL150 relA1 flbB3501 deoC1 ptsF25 rbsR aroB Φ(ompC'–lacZ^+)10–25* [λp1(209)]	1
SK1590	*gal thi sbcB15 endA hsdR4*	63
SK1592	*thi supE endA sbcB15 hsdR4*	3
SK2267	*endA1 hsdR4 supE44 thi-1 lacZ4* or *lac-61 gal-44 ton58 [rfa] recA1 sbcB15*	13
SL10	HfrH *thi sup^o Δ(lac–proAB) galE Δ(pgl–bio)*	3
SMR10^q	*su* (λ*cos2 ΔB xis*1 *red*3 *gamam*210 *c*Its857 *nin*5 *Sam*7)/λ	64, 65
SURE^r	*recB recJ sbcC201 uvrC umuC*::Tn5*(kan^r) lac Δ(hsdRMS) endA1 gyrA96 thi relA1 supE44* F'[*proAB^+ lacI^q lacZ*ΔM15 Tn10*(tet^r)*]	65
SV101	*araD139 Δ(argF–lac)U169 rpsL150 relA1 flbB3501 deoC1 ptsF25 rbsR malPQ*::Tn10	1
SW101	*araD139 Δ(araABC–leu)7679 zab*::Tn10 *Δ(argF–lac)U169 rpsL150 relA1 flbB5301 deoC1 ptsF25 rbsR*	1

Table 1. Continued

Bacteria/Phages

Strain	Genotype	References
TAP90[s]	supE44 supF58 hsdR pro leuB6 thi-1 rpsL lacY1 tonA1 recD1903::mini-tet	66
TD1	araD139 Δ(argF–lac)U169 rpsL150 relA1 flbB3501 deoC1 ptsF25 rbsR recA56 srlC300::Tn10	3
TG1	supE hsdΔ5 thi Δ(lac–proAB) F'[traD36 proAB+ lacI^q lacZΔM15]	67
TG2	supE hsdΔ5 thi Δ(lac–proAB) Δ(srl–recA)306::Tn10(tet^r) F'[traD36 proAB+ lacI^q lacZΔM15]	68
TK821	araD139 Δ(argF–lac)U169 rpsL150 relA1 flbB3501 deoC1 ptsF25 rbsR ompR331::Tn10	1
TK827	Δ(argF–lac)U169 rpsL150 relA1 flbB3501 deoC1 ptsF25 rbsR Φ(ompF'-lacZ+)16–23 ompR331::Tn10 [λp1(209)]	1
W5449	hsdR18 supE44 recB21 recC22 sbcB15 tonB56 tsx-33 ara14 argE3 galK2 his4 lacY1 leuB6 mtl-1 proA2 rpsL31 xyl-5 trpB9579 thi-1	13
XL1-Blue[r]	supE44 hsdR17 recA1 endA1 gyrA46 thi relA1 lac F'[proAB+ lacI^q lacZΔM15 Tn10(tet^r)]	9
XS101	recA1 hsdR rpoB331 F'[kan]	69
XS127	gyrA thi rpoB331 Δ(lac–proAB) argE F'[proAB+ lacI^q lacZΔM15]	69
Y1088[t]	Δ(lac)U169 supE supF hsdR metB trpR tonA21 proC::Tn5 (pMC9)	16, 20, 70
Y1089[t]	araD139 Δ(lac)U169 Δ(lon) rpsL hflA150[chr::Tn10(tet^r)] (pMC9)	15, 16
Y1090[t]	araD139 Δ(lac)U169 Δ(lon) rpsL supF trpC22::Tn10(tet^r) (pMC9)	16, 20, 70
Y1090r[– t,u]	araD139 Δ(lac)U169 Δ(lon) rpsL supF hsdR trpC22::Tn10(tet^r) (pMC9)	15, 16, 22
YK537	supE44 hsdR hsdM recA1 phoA8 leuB6 thi lacY rpsL20 galK2 ara-14 xyl-5 mtl-1	71

[a] The nutritional requirements and detergent sensitivity of χ1776 means it can be used for high-containment experiments. It has a high transformation frequency.

[b] AG1 carries an uncharacterized mutation that improves transformation efficiency.

[c] The λDE3 lysogen carries the gene for T7 RNA polymerase.

[d] Also called C600hflA.

[e] The genotype of C600 is confused. The strains C600 and BNN93 are considered the same by some sources, though C600 is in fact hsdR+ whereas the hsdR– version should be called BNN93. There is also a C600galK strain (22) and C600hflA, usually called BNN102. C600, BNN93 and BNN102 are also different with respect to mcrB (Table 3). C600 is the same as CR34.

[f] DP50supF was originally used for high-containment experiments with λ vectors. It has now been superceded by strains that are easier to grow.

[g] Derived from E. coli B.

[h] ER1398 is an mcrB– version of MM294.

[i] The literature is confusing as to whether these strains are *hsdR2* or *hsdR17*. Ref. 68 states that they are *pro⁻*, but this is contrary to the original publications (13, 20).

[j] ER1451 is an *mcrB⁻* version of JM107.

[k] HB101 and RR1 are isogenic except that HB101 is *recA⁻* and RR1 is *recA⁺*.

[l] Some stocks of JM103 have lost *hsdR4* and become P1 lysogenic (42). KK2186 (= JM103Y) is genetically identical to the authentic JM103.

[m] Ref. 68 states *supE44* but this is not given in the original publication (3).

[n] Ref. 68 states *hsdR17* but this is not given in the original publication (3).

[o] Some stocks of MV1184 do not carry the F' episome.

[p] N4830-1 is a P1 transductant of N4830.

[q] SMR10 is derived from *E. coli* C.

[r] Stratagene, who market these strains, state that they carry an 'uncharacterized' mutation which makes plaques and colonies more intensely blue when plated on X-gal or equivalent.

[s] The mini-*tet* insertion in *recD* of TAP90 improves the yield of Spi⁻ phages.

[t] pMC9 is pBR322 carrying *lacIq*.

[u] Y1090r⁻ is also called Y1090*hsdR*.

Table 2. Phenotypes of loci and genes relevant to recombinant DNA experiments (72–74)

Locus	Gene	Phenotype
amp		ampicillin sensitivity or resistance
ara		arabinose utilization
	araB	ribulokinase (EC 2.7.1.16)
	araC	regulatory gene (activator–repressor protein)
	araD	L-ribulosephosphate 4-epimerase (EC 5.1.3.4)
arg		arginine biosynthesis
	argE	acetylornithine deacetylase (EC 3.5.1.16)
	argF	ornithine carbamoyltransferase (EC 2.1.3.3)
	argG	argininosuccinate synthetase (EC 6.3.4.5)
aro		aromatic amino acid biosynthesis
	aroB	dehydroquinate synthetase
asd		aspartate semialdehyde dehydrogenase (EC 1.2.1.11)
azi		azide sensitivity or resistance
bio		biotin biosynthesis
	bioH	block before pimeloyl CoA
	bioT	not in the literature; possibly a misprint in the original reference
cam		chloramphenicol sensitivity or resistance
chl		chlorate resistance
	chlD	molybdenum-containing factor
cyc	*cycA*	transport of D-alanine, D-serine and glycine
dam[a]		DNA adenine methylase
dap		diaminopimelate biosynthesis
	dapD	succinyl-diaminopimelate aminotransferase

Table 2. Continued

Locus	Gene	Phenotype
dcm[b]		DNA cytosine methylase
deo		deoxyribose biosynthesis
	deoC	deoxyribose-phosphate aldolase (EC 4.1.2.4)
dut		deoxyuridinetriphosphatase (EC 3.6.1.23)
end	endA[c]	DNA-specific endonuclease I
env		cell envelope
	envZ	regulation of outer membrane protein biosynthesis
flb	flbB	flagella synthesis
gal		galactose utilization
	galE	UDP-galactose 4-epimerase (EC 2.7.7.12)
	galK	galactokinase (EC 2.7.1.6)
	galT	galactose-1-phosphate uridylyltransferase (EC 2.7.7.10)
	galU	glucose-1-phosphate uridylyltransferase (EC 2.7.7.9)
gln		glutamine biosynthesis and activation
	glnV	glutamine-tRNA2
gpt		guanine-hypoxanthine phosphoribosyltransferase (EC 2.4.2.8)
gyr		DNA gyrase
	gyrA	subunit A, resistance to nalidixic acid
hfl	hflA	high frequency of lysogeny by phage λ
his		histidine biosynthesis
	hisG	ATP phosphoribosyltransferase (EC 2.4.2.17)
hsd		host specific restriction and/or modification
	hsdM[d]	DNA methylase M
	hsdR[e]	endonuclease R
	hsdS[f]	specificity determinant for hsdM and hsdR
htp	htpR	regulatory gene for proteins induced at high temperatures
ilv		isoleucine and valine biosynthesis
kan		kanamycin resistance and sensitivity
lac[g]		lactose biosynthesis
	lacI[h]	repressor protein
	lacY	galactoside permease
	lacZ[i]	β-galactosidase (EC 3.2.1.23)
lam	lamB	phage receptor protein (maltose uptake)
leu		leucine biosynthesis
	leuA	α-isopropylmalate synthase (EC 4.1.3.12)
	leuB	β-isopropylmalate dehydrogenase

Table 2. Continued

Locus	Gene	Phenotype
	leuC	α-isopropylmalate isomerase subunit
	leuD	α-isopropylmalate isomerase subunit
lig		DNA ligase
lon[j]		ATP-dependent protease La
mal		maltose utilization
	malA	maltose uptake (= *malPQT*)
	malB	maltose catabolism (= *malEFGK*)
	malE	periplasmic maltose binding protein
	malF	maltose transport
	malG	maltose active transport
	malK	maltose permeation
	malP	maltodextrin phosphorylase (EC 2.4.1.1)
	malQ	amylomaltase (EC 2.4.1.25)
	malT	positive regulatory gene for *mal* regulon
met		methionine biosynthesis
	metB	cystathionine γ-synthase (EC 4.2.99.9)
	metC	cystathionine γ-lyase (EC 4.4.1.1)
min	*minA* } *minB* }	Formation of mini-cells containing no DNA
mtl		mannitol utilization
mut		high mutation rate
	mutD	generalized high mutability
omp		outer membrane protein biosynthesis
	ompB	= *ompR*
	ompC	outer membrane protein 1b
	ompF	outer membrane protein 1a
	ompR	positive regulatory gene for *ompC* and *ompF*
oms		osmotic sensitivity
pgl		6-phosphogluconolactonase (EC 3.1.1.31)
pho		phosphate utilization
	phoA	alkaline phosphatase (EC 3.1.3.1)
pnp[k]		polynucleotide phosphorylase (EC 2.7.7.8)
pro[g]		proline biosynthesis
	proA	γ-glutamyl phosphate reductase (EC 1.2.1.41)
	proB	γ-glutamyl kinase (EC 2.7.2.11)
pts		phosphotransferase system
	ptsF	fructosephosphotransferase enzyme II
rbs	*rbsR*	ribose utilization

Table 2. Continued

Locus	Gene	Phenotype
rec		general recombination and radiation repair
	recA[l]	major recombination gene, also involved in λ induction
	recB[m]	exonuclease V subunit, component of RecBCD pathway
	recC[m]	exonuclease V subunit, component of RecBCD pathway
	recD	component of RecBCD pathway
	recJ[n]	component of RecE and RecF pathways
rel		regulation of RNA synthesis
	relA	ATP:GTP 3'-pyrophosphotransferase
rfb		cell wall biosynthesis
rha		rhamnose utilization
rif		rifampicin sensitivity or resistance
rna		ribonuclease I
rpo		RNA polymerase (EC 2.7.7.6)
	rpoB	β-subunit
rps		small ribosomal protein
	rpsL	30S ribosomal subunit protein S12, may confer resistance to streptomycin
rrn		rRNA
	rrnD	rRNA operon at 72 min
	rrnE	rRNA operon at 90 min
sbc		suppressor of *recBC*
	sbcA[o]	uncharacterized, involved in RecE pathway
	sbcB[m]	exonuclease I, involved in RecF pathway
	sbcC[n]	uncharacterized, involved with *sbcB* in RecF pathway
srl		sorbitol utilization
	srlA	D-glucitol-specific enzyme II of phosphotransferase
	srlC	regulatory gene
str		streptomycin sensitivity or resistance
sup		suppressor
	supE	amber (UAG) suppressor, inserts glutamine
	supF	amber (UAG) suppressor, inserts tyrosine
tet		tetracycline sensitivity or resistance
thi		thiamine biosynthesis
thr		threonine biosynthesis
thy	*thyA*	thymidylate synthase (EC 2.1.1.45)
ton		phage T1 resistance
	tonA	outer membrane protein receptor ($= fhuA$)
tra	*traD*	suppresses conjugal transfer of F plasmids

Table 2. Continued

Locus	Gene	Phenotype
trp		tryptophan biosynthesis
	trpR	aporepressor protein
tsx		phage T6 and colicin K resistance
tyr		tyrosine biosynthesis and activation
	tyrT	tyrosine tRNA1
umu	*umuC*[n]	sensitivity to u.v.
ung		uracil-DNA glycosylase
uvr		repair of u.v. radiation damage to DNA
	uvrB	excision nuclease
	uvrC[n]	excision nuclease
xyl		xylose utilization

[a] DNA adenine methylase (Chapter 4, Section 1.3) adds a methyl group to the N^6 position of adenines present in the target sequence 5'-GATC-3' (75). DNA cloned in a *dam*⁻ strain is subject to a relatively high level of ssDNA breakage.

[b] DNA cytosine methylase (Chapter 4, Section 1.3) adds a methyl group to the C^5 position of cytosines in the target sequences 5'-CCAGG-3' and 5'-CCTGG-3' (76, 77).

[c] The *endA1* mutation improves plasmid yield, especially from mini-preps.

[d] DNA methylase M is the modification subunit of the *Eco*K restriction–modification system. An *hsdM* strain is $r_k^+ m_k^-$.

[e] Endonuclease R is the restriction subunit of the *Eco*K restriction–modification system. An *hsdR* strain is $r_k^- m_k^+$.

[f] The HsdS gene product is the site recognition subunit of the *Eco*K restriction–modification system. An *hsdS* strain is $r_k^- m_k^-$.

[g] The Δ*(lac–proAB)* deletion removes the *lac* operon and surrounding region, including the two genes involved in proline biosynthesis. Most Δ*(lac–proAB)* strains carry the deleted genes on an F' episome. When maintained on minimal media the F'[*proAB*] genes provide positive selection for cells that retain their F' status.

[h] Most strains used with vectors employing *lac* selection are *lacI^q*. This mutation results in over-production of the *lac* repressor (78), minimizing the low level of *lac* expression that occurs in uninduced wild-type cells.

[i] The *lacZΔM15* deletion removes the amino-terminal α-peptide (amino acids 11–41) of β-galactosidase. Most cloning vectors that employ *lac* selection carry a gene that codes for the α-peptide and rescues the *lacZΔM15* mutation by α-complementation (79).

[j] The *lon* mutation results in increased stability of fusion proteins but results in a mucoid phenotype. Some plasmids are unstable in *lon* strains.

[k] The *pnp* mutation results in increased stability of eukaryotic mRNA sequences, improving the expression of some eukaryotic genes cloned in *E. coli*. Some plasmids are unstable in *pnp* strains.

[l] A *recA* strain is recombination deficient and has a reduced ability to multimerize plasmids and rearrange cloned sequences (3, 80). However, *recA* strains grow relatively slowly and have low transformation frequencies. Strains that are used to propagate *red gam* λ vectors must be *recA*⁺.

[m] Strains that are *recBC sbcB* display decreased rearrangement of cloned palindromic sequences. However, these strains are relatively slow growing (though better than *recBC* mutants: refs 81 and 82) and have low transformation frequencies. Some plasmids are unstable in *recBC sbcB* strains.

[n] The genotype *sbcC recBJ umuC uvrC* carried by SURE minimizes rearrangements of cloned palindromic and Z-DNA sequences (65).

[o] The *sbcA* mutation results in improved growth of *recB* strains (83).

Table 3. Restriction and modification properties of important *E. coli* strains (see also ref. 84)

Strain	EcoK[b]	McrA[c]	McrB[c]	Mrr[d]	Dam[e]	Dcm[e]
			Phenotype[a]			
BB4	$r_k^- m_k^+$				+	+
BNN93	$r_k^- m_k^+$	−	−		+	+
BNN102	$r_k^- m_k^+$	−	−		+	+
C600	$r_k^+ m_k^+$	−	+		+	+
CES200	$r_k^- m_k^+$	−	+		+	+
CES201	$r_k^- m_k^+$				+	+
DH1	$r_k^- m_k^+$	+	+		+	+
DP50supF[f]	$r_B^- m_B^-$				+	+
ED8654	$r_k^- m_k^+$	−	+		+	+
ED8767[f]	$r_B^- m_B^-$	−	−		+	+
ER1398	$r_k^- m_k^+$	+	−		+	+
ER1451	$r_k^- m_k^+$	−	−		+	+
ER1647	$r_k^- m_k^-$	−	−	−	+	+
ER1648	$r_k^- m_k^-$	−	−	−	+	+
GM2163	$r_k^- m_k^+$	−	−		−	−
HB101[f]	$r_B^- m_B^-$	+	−	−	+	+
JM83	$r_k^+ m_k^+$		+		+	+
JM101	$r_k^+ m_k^+$	+			+	+
JM103[g]	$r_k^- m_k^+$				+	+
JM105	$r_k^- m_k^+$				+	+
JM107	$r_k^- m_k^+$	−	+		+	+
JM109	$r_k^- m_k^+$	−[h]	+[h]		+	+
JM110[i]	$r_k^- m_k^+$				−	−
K802	$r_k^- m_k^+$	−	−	+	+	+
KK2186	$r_k^- m_k^+$				+	+
LE392	$r_k^- m_k^+$	−	+	+	+	+
MBM7014.5	$r_k^- m_k^+$		−		+	+
MC1061	$r_k^- m_k^+$	−	−		+	+
MM294	$r_k^- m_k^+$	+	+		+	+
NM519	$r_k^- m_k^+$	+	+		+	+
NM538	$r_k^- m_k^+$	−	+		+	+
NM539	$r_k^- m_k^+$	−[h]	+[h]		+	+
PLK strains	$r_k^- m_k^+$	−	−	+	+	+
Q358	$r_k^- m_k^+$	−	+		+	+
Q359	$r_k^- m_k^+$	−[h]	+[h]		+	+
RR1[f]	$r_B^- m_B^-$	+	−	−	+	+
SURE	$r_k^- m_k^-$	−	−	−	+	+
TAP90	$r_k^- m_k^+$				+	+
TG1	$r_k^- m_k^-$				+	+
TG2	$r_k^- m_k^-$				+	+
XL1–Blue	$r_k^- m_k^+$				+	+

Table 3. Continued

Strain	Phenotype[a]					
	EcoK[b]	McrA[c]	McrB[c]	Mrr[d]	Dam[e]	Dcm[e]
XS101	$r_k^- m_k^+$				+	+
Y1088	$r_k^- m_k^+$	−	+		+	+
Y1090	$r_k^+ m_k^+$	−[h]	+[h]		+	+
Y1090r⁻	$r_k^- m_k^+$	−	+		+	+

[a] The absence of an entry indicates that status is uncertain.

[b] The *EcoK* system (specified by the *hsdMRS* genes) recognizes the sequence 5'-AACNNNNNNGTT-3' (85). The modification component protects the host DNA by methylating the second A in each strand of the target sequence. DNA cloned in an m⁻ host will be restricted if subsequently transferred to an r⁺ host.

[c] The McrA and McrB restriction systems cleave DNA at methylated cytosines contained in the target sequences 5'-CG-3' for McrA and 5'-PuC-3' for McrB (20, 84, 86). DNA from some sources (including human DNA) may be methylated at these sites and will therefore be cloned inefficiently in strains expressing these systems.

[d] The Mrr restriction system cleaves DNA at methylated adenines, though the precise recognition sequence is not known (85, 86). DNA from some sources may be methylated at Mrr recognition sites and will therefore be cloned inefficiently in a *mrr*⁺ strain.

[e] The Dam and Dcm methylases may affect *in vitro* restriction of cloned DNA by methylating residues essential to the target sequence for certain restriction enzymes. See Chapter 4, Section 1.3.

[f] The $r_B m_B$ phenotype is given for those strains derived from *E. coli* K12 × *E. coli* B hybrids.

[g] JM103 also expresses the *EcoP*1 restriction and modification system.

[h] Inferred from the parental genotype.

[i] The *EcoK* status of JM110 is confusing. See footnote n to *Table 1*.

Table 4. *E. coli* strains classified according to application[a]

Application	Recommended strains
Cloning with plasmid vectors	DH1, DH5, DH5α, HB101, MM294, RR1, TG2
Cloning with cosmid vectors	DH1, DH5, DH5α, ED8767
Cloning with phagemid vectors	71/18, MV1184, MV1193, XS101, XS127, XL1-Blue
Cloning with M13 vectors	JM101, JM107, KK2186
Cloning with λ vectors	
General λ propagation	DP50*supF*, ED8654, ED8767, K802, LE392, NM531, NM538, Q358
Hosts for λgt10	BNN102, C600
Hosts for λgt11	Y1088, Y1089, Y1090r⁻
Hosts for Spi⁻ selection and growth	CES200, CES201, NM519, NM539, P2392, Q359, TAP90
Hosts for λZAP	BB4, XL1-Blue
Hosts for λORF8	MBM7014.5, MC1061

Table 4. Continued

Application	Recommended strains
Host for screening *lacZ* stuffer fragments	CSH18
Gene expression from the:	
T7 promoter	BL21(DE3), HMS174, JM109(DE3)
λP$_L$ promoter	M5219, MZ-1
phoA promoter	YK537
Preparation of λ packaging extracts	BHB2688, BHB2690, SMR10
High-containment hosts	χ1776 (plasmids), DP50*supF* (λ)
Preparation of uracil-DNA	CJ236
Strains allowing *lacZ* selection	71/18, BB4, ER1451, ER1647, ER1648, JM101, JM103, JM105, JM107, JM109, JM110, KK2186, MV1184, NM522, TG1, TG2, XL1-Blue, XS127
Strains designed for cloning eukaryotic DNA in *E. coli*	ER1647, ER1648, SURE
dam⁻ dcm⁻ strains	GM48, GM2163, JM110
recA⁻ strains	AG1, CES201, D1210, DH1, DH5, DH5α, DH20, DH21, ED8767, HB101, HMS174, JC9956, JM108, JM109, JM109(DE3), KLF41, NM531, NM554, PLK A and F', SK2267, XL1-Blue, XS101
recBC⁻ strains	BJ5183, CES200, CES201, JC7623, NM519, W5449
recD⁻ strain	ER1647
recBJ⁻ strain	SURE

ᵃ This table is not comprehensive. Apologies if your favorite strain is omitted.

2. BACTERIOPHAGE LAMBDA

The genetic map of wild-type phage λ is shown in *Figure 1* with details of the genes and uncharacterized ORFs in *Table 5*. For details of phage λ vectors see Chapter 6, Section 2.

Table 5. Locations and functions of phage λ genes (87, 88)

Gene	Base-pair coordinates[a]	Function
*Nu*1	191 – 733	DNA packaging and cohesive end formation
A	711 – 2633	DNA packaging and cohesive end formation
W	2633 – 2836	Modification of DNA-filled heads
B	2836 – 4434	Capsid structural component
C	4418 – 5734	Capsid structural component
*Nu*3	5132 – 5734	Transient core for capsid assembly
D	5747 – 6076	Major component of phage head
E	6135 – 7157	Major component of capsid
*F*I	7202 – 7597	DNA packaging and maturation
*F*II	7612 – 7962	Structural component of filled head
Z	7977 – 8552	Structural component of tail
U	8552 – 8944	Structural component of tail
V	8955 – 9692	Major component of tail tube
G	9711 – 10130	Assembly of tail initiator
T	10115 – 10546	Structural component of tail
H	10542 – 13100	Structural component of tail
M	13100 – 13426	Structural component of tail
L	13429 – 14124	Structural component of tail initiator
K	14276 – 14872	Tail assembly
I	14773 – 15441	Tail initiator assembly
J	15505 – 18900	Structural component of tail
lom	18965 – 19582	Membrane protein
ORF-401	19650 – 20852	No product identified
ORF-314	21029 – 21970	No product identified
ORF-194	21973 – 22554	No product identified
b*Ea*47	23918 – 22689	Unknown
b*Ea*31	25399 – 24512	Unknown
b*Ea*59	26973 – 25399	Unknown
int	28882 – 27815	Integrative and excisive recombination
xis	29078 – 28863	Excisive recombination
*Ea*8.5	29655 – 29377	Unknown
*Ea*22	30395 – 29850	Unknown
exo	32028 – 31351	General recombination: λ-exonuclease
bet	32810 – 32028	General recombination: β-protein
gam	33232 – 32819	Regulation of DNA replication
ORF-47	33330 – 33190	Probably *kil* — inhibits host division
*c*III	33463 – 33302	Establishment of lysogeny
*Ea*10	33904 – 33539	ssDNA binding protein
ral	34287 – 34090	Alleviates *Eco*K restriction system
git	34497 – 35035	Unknown
N	35360 – 35040	Positive regulation of early development
rexB	36259 – 35828	Cell growth regulation during lysogeny
rexA	37114 – 36278	Cell growth regulation during lysogeny

Table 5. Continued

Gene	Base-pair coordinates[a]	Function
cI	37940 – 37230	Maintenance of lysogeny
cro	38041 – 38238	Regulation of late stage of lytic cycle
cII	38360 – 38650	Establishment of lysogeny
O	38686 – 39582	Initiation of early DNA replication
P	39582 – 40280	Initiation of early DNA replication
ren	40280 – 40567	Unknown
ORF-146	40644 – 41081	No product identified
ORF-290	41081 – 41950	No product identified
ORF-57	41950 – 42120	No product identified
ORF-60	42090 – 42269	No product identified
ORF-56	42269 – 42436	No product identified
ORF-204	42429 – 43040	No product identified
ORF-68	43040 – 43243	No product identified
ORF-221	43224 – 43886	No product identified
Q	43886 – 44506	Positive regulation of late lysis
ORF-64	44621 – 44812	No product identified
S	45186 – 45506	Cell lysis
R	45493 – 45966	Cell lysis
Rz	45966 – 46423	Cell lysis
bor	46752 – 46462	Survival of the host in animal serum[b]

[a] The sequence coordinates are based on the complete sequence of cIind1ts857S7 (89) but with minor changes as detailed in ref. 88. The first number is the coordinate of the first nucleotide of the initiation codon and the second number is the coordinate of the third nucleotide of the last codon (not the termination triplet). The DNA strand on which the gene lies can therefore be identified: for example, gene B (2836–4434) lies on the 'plus' strand, and gene int (28882–27815) lies on the 'minus' strand.

[b] See ref. 92.

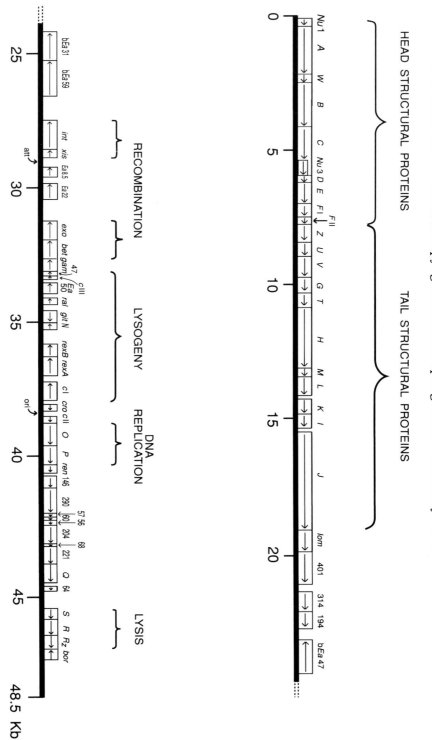

Figure 1. The genetic map of phage λ. Details of the genes and uncharacterized ORFs are given in *Table 5*. Arrows indicate the direction of transcription. Figure redrawn, with permission from *Molecular Cloning (2nd edn)* by J. Sambrook, E.F. Fritsch and T. Maniatis. Copyright 1989 Cold Spring Harbor Laboratory Press.

3. CULTURE MEDIA, SUPPLEMENTS AND PHAGE BUFFERS

A variety of different media are available for the growth of *E. coli*; recipes for the most important ones are provided in *Table 6*. The rich defined media (e.g. dYT, LB medium, N-broth) are to a certain extent interchangeable but specific media may be recommended for particular applications. For example, NZCM is often used for preparation of high-titer lambda stocks, as this medium, though rich, lacks sugars so the LamB phage receptors are not induced and phage particles are not taken out of solution by adsorption to cell debris. Formulations for soft agar overlays are given in *Table 7*; again several of these are interchangeable, though phage yields may vary according to the medium components.

Bacteria can be stored for several years as stab cultures in the agar media described in *Table 8*, or frozen in any standard medium to which glycerol has been added. Phage stocks are stored at 4°C in lambda storage buffer, SM buffer, or M13 storage medium.

Table 6. Media for *E. coli* culture

Medium	Recipe (for 1 liter)[a,b]
BBL broth	10 g BBL trypticase peptone, 5 g NaCl
dYT (2YT)	16 g bacto-tryptone, 10 g bacto-yeast extract, 5 g NaCl
λ broth	10 g bacto-tryptone, 2.5 g NaCl
LB medium	10 g bacto-tryptone, 5 g bacto-yeast extract, 10 g NaCl
N-broth	25 g bacto-nutrient broth
NZYCM[c]	10 g NZ amine (Type A casein hydrolyzate), 5 g NaCl, 5 g bacto-yeast extract, 1 g casamino acids, 2 g $MgSO_4.7H_2O$
TB medium	10 g bacto-tryptone, 5 g NaCl
YT	8 g bacto-tryptone, 5 g bacto-yeast extract, 5 g NaCl
χ1776 broth	25 g bacto-tryptone, 7.5 g bacto-yeast extract, 20 ml 1 M Tris–HCl (pH 7.5). Autoclave, cool, then add 5 ml 1 M $MgCl_2$, 10 ml 1% diaminopimelic acid, 10 ml 0.4% thymidine, 25 ml 20% glucose.
M9 minimal	6 g Na_2HPO_4, 3 g KH_2PO_4, 0.5 g NaCl, 1 g NH_4Cl. Adjust pH to 7.4, autoclave, cool, then add 2 ml 1 M $MgSO_4$, 0.1 ml 1 M $CaCl_2$, 10 ml 20% glucose.
M63 minimal	3 g KH_2PO_4, 7 g K_2HPO_4, 2 g $(NH_4)_2SO_4$, 5 mg $FeSO_4$. Before use add 0.1 ml 1 M $MgSO_4$, 0.2 ml 50 mg ml^{-1} thiamine, 1.0 ml 20% glucose.

[a] The pH should be checked and if necessary adjusted to 7.0–7.2 with NaOH. Media should be sterilized by autoclaving at 121°C, 15 lb in^{-2}, for 20 min.

[b] For agar media add 15 g bacto-agar.

[c] NZYM = NZYCM minus casamino acids. NZCM = NZYCM minus bacto-yeast extract.

Table 7. Media for overlaying agar plates ('top agar')

Medium	Recipe (for 100 ml)[a,b]
Agarose top	0.8 g agarose
BBL top	1 g BBL trypticase peptone, 0.5 g NaCl, 0.6 g bacto-agar
dYT top	1.6 g bacto-tryptone, 1 g bacto-yeast extract, 0.5 g NaCl, 0.7 g bacto-agar
F top	0.8 g NaCl, 0.7 g bacto-agar
λ top	1 g bacto-tryptone, 0.25 g NaCl, 0.7 g bacto-agar
LB top	1 g bacto-tryptone, 0.5 g bacto-yeast extract, 1 g NaCl, 0.7 g bacto-agar
TB top	1 g bacto-tryptone, 0.5 g NaCl, 0.7 g bacto-agar
Water top	0.8 g bacto-agar

[a] The pH should be checked and if necessary adjusted to 7.0 – 7.2 with NaOH. Media should be sterilized by autoclaving at 121°C, 15 lb in^{-2}, for 20 min.
[b] The amount of agar added can be varied from 0.3 to 1.2%, according to requirements.

Table 8. Media for storage of bacteria and phage

Medium[a]	Recipe (for 100 ml)[b]
λ storage	0.7 g Na$_2$HPO$_4$, 0.3 g KH$_2$PO$_4$, 0.5 g NaCl, 1 ml 1 M MgSO$_4$, 1 ml 0.01 M CaCl$_2$, 0.1 ml 1% gelatin
M13 storage	1 g bacto-tryptone, 0.5 g bacto-yeast extract, 0.63 g K$_2$HPO$_4$, 0.18 g KH$_2$PO$_4$, 45 mg Na citrate, 90 mg (NH$_4$)$_2$SO$_4$, 9 mg MgSO$_4$.7H$_2$O, 3.5 ml glycerol
N-stab agar	2.5 g bacto-nutrient broth, 0.7 g bacto-agar
SM	0.58 g NaCl, 0.2 g MgSO$_4$.7H$_2$O, 5 ml 1 M Tris–HCl (pH 7.5), 0.5 ml 2% gelatin
Stab agar	1 g bacto-tryptone, 0.7 g bacto-agar

[a] In addition to the agar media listed here, all the standard media in *Table 6* can be used to store *E. coli* cultures. Add 1.5 ml of glycerol to 0.85 ml of the bacterial culture.
[b] The pH should be checked and if necessary adjusted to 7.0 – 7.2 with NaOH. Media should be sterilized by autoclaving at 121°C, 15 lb in^{-2}, for 20 min.

Of the three indicator media listed in *Table 9*, the most important by far in molecular biology is X-gal agar. There are now a number of alternative histochemical substrates for β-galactosidase assay (see Chapter 2, Section 17), but each is used in the same way as X-gal.

A number of buffers have been devised for storage and dilution of phage λ; the most useful are shown in *Table 10*. Finally, in *Table 11* stock and working concentrations are given for antibiotic supplements used in selection of recombinant bacteria.

Table 9. Indicator media

Medium	Recipe (for 1 liter)[a]
MacConkey agar	40 g bacto-MacConkey agar base. Autoclave, cool, then add 50 ml 20% sugar.
Tetrazolium agar	25.5 g bacto-antibiotic medium 2, 50 mg 2,3,5-triphenyl-2H tetrazolium chloride. Autoclave, cool, then add 50 ml 20% sugar.
X-gal agar	X-gal and IPTG must be added to cooled top agar just before pouring. For 3 ml of top agar use: X-gal, 40 μl of a 20 mg ml^{-1} solution in dimethyl formamide (make fresh); IPTG, 4 μl of a 200 mg ml^{-1} aqueous solution (store at −20°C).

[a] The pH should be checked and if necessary adjusted to 7.0 – 7.2 with NaOH. Media should be sterilized by autoclaving at 121°C, 15 lb in^{-2}, for 20 min.

Table 10. Buffers for working with phage λ

Buffer	Recipe (for 100 ml)[a]
λ storage	0.7 g Na$_2$HPO$_4$, 0.3 g KH$_2$PO$_4$, 0.5 g NaCl, 1 ml 1 M MgSO$_4$, 1 ml 0.01 M CaCl$_2$, 0.1 ml 1% gelatin
SM	0.58 g NaCl, 0.2 g MgSO$_4$.7H$_2$O, 5 ml 1 M Tris–HCl (pH 7.5), 0.5 ml 2% gelatin
TM	5 ml 1 M Tris–HCl (pH 7.5), 0.2 g MgSO$_4$.7H$_2$O
TMG	121 mg Tris-base, 120 mg MgSO$_4$.7H$_2$O, 10 mg gelatin. Adjust pH to 7.4 with HCl.

[a] Buffers should be sterilized by autoclaving at 121°C, 15 lb in^{-2}, for 20 min.

Table 11. Antibiotics used for the selection of resistant *E. coli* cells

Antibiotic[a]	Stock solution[b] (mg ml^{-1})	Working concentration[c] (μg ml^{-1})
Ampicillin	50 in water	20 – 125
Chloramphenicol	35 in ethanol	25 – 170[d]
Kanamycin	10 in water	10 – 50
Nalidixic acid	20 in water	20
Rifampicin	50 in water	10 – 100
Streptomycin	10 in water	10 – 125
Tetracycline	5 in ethanol	10 – 50

[a] For details of modes of action see Chapter 2, Section 3.
[b] Sterilize by filtration and store stock solutions at –20°C. Add to agar media after autoclaving and cooling to 50°C.
[c] The lower concentrations are suitable for the selection of cells containing low copy number plasmids. The higher concentrations may be required for high copy number plasmids.
[d] Use 170 μg ml^{-1} chloramphenicol for plasmid amplification.

4. REFERENCES

1. Silhavy, T.J., Berman, M.L. and Enquist, L.W. (1984) *Experiments with Gene Fusions*. Cold Spring Harbor Laboratory Press, New York.

2. Perbal, B. (1988) *A Practical Guide to Molecular Cloning (2nd edn)*. Wiley, New York.

3. Yanisch-Perron, C., Vieira, J. and Messing, J. (1985) *Gene*, **33**, 103.

4. Messing, J., Gronenborn, B, Muller-Hill, B. and Hofschneider, P.H. (1977) *Proc. Natl. Acad. Sci. USA*, **74**, 3642.

5. Dente, L., Cesareni, G. and Cortese, R. (1983) *Nucleic Acids Res.*, **11**, 1645.

6. Ruther, U. and Muller-Hill, B. (1983) *EMBO J.*, **2**, 1791.

7. Clark-Curtiss, J.E. and Curtiss, R. (1983) *Methods Enzymol.*, **101**, 347.

8. Winnacker, E.-L. (1987) *From Genes to Clones: An Introduction to Gene Technology*. VCH, Weinheim.

9. Bullock, W.O., Fernandez, J.M. and Short, J.M. (1987) *BioTechniques*, **5**, 376.

10. Geider, K., Hohmeyer, C., Haas, R. and Meyer, T.F. (1985) *Gene*, **33**, 341.

11. Hohn, B. (1979) *Methods Enzymol.*, **68**, 299.

12. Hohn, B. and Murray, K. (1977) *Proc. Natl. Acad. Sci. USA*, **74**, 3259.

13. Hanahan, D. (1983) *J. Mol. Biol.*, **166**, 557.

14. Studier, F.W. and Moffat, B.A. (1986) *J. Mol. Biol.*, **189**, 113.

15. Young, R.A. and Davis, R.W. (1983) *Proc. Natl. Acad. Sci. USA*, **80**, 1194.

16. Huynh, T.V., Young, R.A. and Davis, R.W. (1985) in *DNA Cloning: A Practical Approach*. (D.M. Glover, ed.) IRL Press, Oxford, Vol. 1, p.49.

17. Bertani, G. and Weigle, J.J. (1953) *J. Bacteriol.*, **65**, 113.

18. Borck, K., Beggs, J.D., Brammar, W.J., Hopkins, A.S. and Murray, N.E. (1976) *Mol. Gen. Genet.*, **146**, 199.

19. Appleyard, R.K. (1954) *Genetics*, **39**, 440.

20. Raleigh, E. and Wilson, G. (1986) *Proc. Natl. Acad. Sci. USA*, **83**, 9070.

21. Bachmann, B.J. (1972) *Bacteriol. Rev.*, **36**, 525.

22. Jendrisak, J., Young, R.A. and Engel, J.D. (1987) *Methods Enzymol.*, **152**, 359.

23. Nader, W.F., Edlind, T.D., Huettermann, A. and Sauer, H.W. (1985) *Proc. Natl. Acad. Sci. USA,* **82**, 2698.

24. Wyman, A.R. and Wertman, K.F. (1987) *Methods Enzymol.,* **152**, 173.

25. Kunkel, T.A., Roberts, J.D. and Zakour, R.A. (1987) *Methods Enzymol.,* **154**, 367.

26. Miller, J.H. (1972) *Experiments in Molecular Genetics.* Cold Spring Harbor Laboratory Press, New York.

27. Williams, B.G. and Blattner, F.R. (1979) *J. Virol.,* **29**, 555.

28. Sadler, J.R., Tecklenburg, M. and Betz, J.L. (1980) *Gene,* **8**, 279.

29. Low, B. (1968) *Proc. Natl. Acad. Sci. USA,* **60**, 160.

30. Leder, P., Tiemeyer, D. and Enquist, L. (1977) *Science,* **196**, 175.

31. Murray, N.E., Brammar, W.J. and Murray, K. (1977) *Mol. Gen. Genet.,* **150**, 53.

32. Woodcock, D.M., Crowther, P.J., Doherty, J., Jefferson, S., DeCruz, E., Noyer-Weidner, M., Smith, S.S., Michael, M.Z. and Graham, M.W. (1989) *Nucleic Acids Res.,* **17**, 3469.

33. Marinus, M., Carraway, M., Frey, A.Z. and Arraj, J.A. (1983) *Mol. Gen. Genet.,* **192**, 288.

34. Boyer, H.W. and Roulland-Dussoix, D. (1969) *J. Mol. Biol.,* **41**, 459.

35. Bolivar, F. and Backman, K. (1979) *Methods Enzymol.,* **68**, 245.

36. Campbell, J.L., Richardson, C.C. and Studier, F.W. (1978) *Proc. Natl. Acad. Sci. USA,* **75**, 2276.

37. Kushner, S.R., Nagaishi, H., Templin, A. and Clark, A.J. (1971) *Proc. Natl. Acad. Sci. USA,* **68**, 824.

38. Maniatis, T., Fritsch, E.F. and Sambrook, J. (1982) *Molecular Cloning: A Laboratory Manual.* Cold Spring Harbor Laboratory Press, New York.

39. Vieira, J. and Messing, J. (1982) *Gene,* **19**, 259.

40. Messing, J. (1979) *Recomb. DNA Tech. Bull.,* **2(2)**, 43.

41. Messing, J., Crea, R. and Seeburg, P.H. (1981) *Nucleic Acids Res.,* **9**, 309.

42. Felton, J. (1983) *BioTechniques,* **1**, 42.

43. Anon (1989) *Promega Notes,* **20**, 2.

44. Wood, W.B. (1966) *J. Mol. Biol.,* **16**, 118.

45. Zagursky, R.J. and Berman, M.L. (1984) *Gene,* **27**, 183.

46. Guarente, L. and Ptashne, M. (1981) *Proc. Natl. Acad. Sci. USA,* **78**, 2199.

47. Remaut, E., Stanssens, P. and Fiers, W. (1981) *Gene,* **15**, 81.

48. Shimatake, H. and Rosenberg, M. (1981) *Nature,* **292**, 128.

49. Casadaban, M.J. and Cohen, S.N. (1980) *J. Mol. Biol.,* **138**, 179.

50. Meissner, P.S., Sisk, W.P. and Berman, M.L. (1987) *Proc. Natl. Acad. Sci. USA,* **84**, 4171.

51. Zoller, M.J. and Smith, M. (1987) *Methods Enzymol.,* **154**, 329.

52. Nagai, K. and Thogerson, H.C. (1984) *Nature,* **309**, 810.

53. Gottesman, M.E., Adhya, S. and Das, A. (1980) *J. Mol. Biol.,* **140**, 57.

54. Gross, J. and Gross, M. (1969) *Nature,* **224**, 1166.

55. Arber, W., Enquist, L., Hohn, B., Murray, N.E. and Murray, K. (1983) in *Lambda II.* (R.W. Hendrix, J.W. Roberts, F.W. Stahl and R.A. Weisberg, eds) Cold Spring Harbor Laboratory Press, New York. p.433.

56. Gough, J.A. and Murray, N.E. (1983) *J. Mol. Biol.,* **166**, 1.

57. Frischauf, A.-M., Lehrach, H., Poustka, A. and Murray, N. (1983) *J. Mol. Biol.,* **170**, 827.

58. Karn, J., Brenner, S., Barnett, L. and Cesareni, G. (1980) *Proc. Natl. Acad. Sci. USA,* **77**, 5172.

59. Campbell, A. (1965) *Virology,* **27**, 329.

60. Brent, R. and Ptashne, M. (1981) *Proc. Natl. Acad. Sci. USA,* **78**, 4204.

61. Bolivar, F., Rodriguez, R.L., Greene, P.J., Betlach, M.C., Heyneker, H.L., Boyer, H.W., Crosa, J.H. and Falkow, S. (1977) *Gene,* **2**, 95.

62. Peacock, S.L., McIver, C.M. and Monahan, J.J. (1981) *Biochim. Biophys. Acta,* **655**, 243.

63. Kushner, S.R. (1978) in *Genetic Engineering.* (H.W. Boyer and S. Nicosia eds) Elsevier, Amsterdam. p.17.

64. Rosenberg, S.M. (1985) *Gene,* **39**, 313.

65. Greener, A. (1990) *Stratagies,* **3**, 5.

66. Patterson, T.A. and Dean, M. (1987) *Nucleic Acids Res.,* **15**, 6298.

67. Gibson, T.J. (1984) *PhD Thesis.* Cambridge University, UK.

68. Sambrook, J., Fritsch, E.F. and Maniatis, T. (1989) *Molecular Cloning: A Laboratory Manual (2nd edn).* Cold Spring Harbor Laboratory Press, New York.

69. Levinson, A., Silver, D. and Seed, B. (1984) *J. Mol. Appl. Genet.,* **2**, 507.

70. Young, R.A. and Davis, R.W. (1983) *Science*, **222**, 778.

71. Oka, T., Sakamoto, S., Miyoshi, K., Fuwa, T., Yoda, K., Yamasaki, M., Tamura, G. and Miyake, T. (1985) *Proc. Natl. Acad. Sci. USA*, **82**, 7212.

72. Bachmann, B.J. (1983) *Microbiol. Rev.*, **47**, 180.

73. Bachmann, B.J. and Low, K.B. (1980) *Microbiol. Rev.*, **44**, 1.

74. Bachmann, B.J., Low, K.B. and Taylor, A.L. (1976) *Bacteriol. Rev.*, **40**, 116.

75. Hattman, S., Brooks, J.E. and Masurekar, M. (1978) *J. Mol. Biol.*, **126**, 367.

76. Marinus, M.G. and Morris, N.R. (1973) *J. Bacteriol.*, **114**, 1143.

77. May, M.S. and Hattman, S. (1975) *J. Bacteriol.*, **123**, 768.

78. Muller-Hill, B., Crapo, L. and Gilbert, W. (1968) *Proc. Natl. Acad. Sci. USA*, **59**, 1259.

79. Ullman, A. and Perrin, D. (1970) in *The Lactose Operon*. (J.R. Beckwith and D. Zipser, eds) Cold Spring Harbor Laboratory Press, New York. p.143.

80. Bedbrook, J.R. and Ausubel, F.M. (1976) *Cell*, **9**, 707.

81. Kushner, S.R., Nagaishi, H., Templin, A. and Clarke, A.J. (1971) *Proc. Natl. Acad. Sci. USA*, **68**, 824.

82. Leach, D.R.F. and Stahl, F.W. (1983) *Nature*, **305**, 448.

83. Kushner, S.R., Nagaishi, H. and Clark, A.J. (1974) *Proc. Natl. Acad. Sci. USA*, **71**, 3593.

84. Raleigh, E.A., Murray, N.E., Revel, H., Blumenthal, R.M., Westaway, D., Reith, A.D., Rigby, P.W.J., Elhai, J. and Hanahan, D. (1988) *Nucleic Acids Res.*, **16**, 1563.

85. Raleigh, E.R. (1987) *Methods Enzymol.*, **152**, 130.

86. Heitman, J. and Model, P. (1987) *J. Bacteriol.*, **169**, 3243.

87. Echols, H. and Murialdo, H. (1978) *Microbiol. Rev.*, **42**, 577.

88. Daniels, D.L., Schroeder, J.L., Szybalski, W., Sanger, F. and Blattner, F.R. (1987) in *Genetic Maps*. (S.J. O'Brien, ed.) Cold Spring Harbor Laboratory Press, New York, Vol. 5, p.1.3.

89. Sanger, F., Coulson, A.R, Hong, G.-F., Hill, D.F. and Petersen, G.B. (1982) *J. Mol. Biol.*, **162**, 729.

90. Anon (1989) *Product catalog*. Stratagene, La Jolla, USA.

91. Kretz, P.L. and Short, J.M. (1989) *Stratagies*, **2**, 25.

92. Barondess, J.J. and Beckwith, J. (1990) *Nature*, **346**, 871.

Note added in proof

According to New England BioLabs (1990–1991 catalog), the full genotypes for these strains are:

ER1647   *fhuA2 Δ(lacZ)rl supE44 trp31 mcrA1272::*Tn*10 his–1 recD1014 rpsL104 xyl–7 mtl–2 metB1 Δ(mcrCB-hsdSMR-mrr)2::*Tn*10*

ER1648   as ER1647 but *recD⁺*.

CHAPTER 2
CHEMICALS AND REAGENTS

This chapter provides the essential physical, chemical and biological data pertaining to the chemicals and reagents commonly used in molecular biology. To aid the reader, the data have been classified into the sections given below; the page numbers cited after each section heading can be used to locate the relevant data rapidly. In addition, cross-references are given, where appropriate, to chapters that cover specific applications.

Chemicals and reagents are classified into the following sections according to their primary use to the molecular biologist. No compound is included in more than one section and the sections themselves are organized alphabetically. If in doubt consult the main index.

For hazardous chemicals the necessary precautions are noted by numbers referring to the European Commission Risk and Safety Phrases, which are listed in Chapter 11, *Table 3*. The EC phrases are used as they are informative and have been adopted by several major suppliers. However, there is a problem in that the assignment of phrases to individual chemicals is carried out by the suppliers, rather than by an independent legislative body, so different suppliers may give different phrases to the same chemical. For instance, BDH assign R4-5-28-29-32 and S28 to sodium azide, whereas Serva consider the compound to be R26/27/28 and S1-36-45, and Fluka give R28-32 and S28. However, in most cases the distinctions are fine ones. In this chapter the Risk and Safety Phrases given for individual compounds are those assigned by BDH. For chemicals not in the BDH catalog the phrases assigned by either Serva or Fluka are used. For chemicals not available from BDH, Serva or Fluka, less comprehensive warnings (such as 'irritant') are given when appropriate. The absence of a general or specific warning should not be taken as indicating that a risk does not exist.

1. ACIDS AND BASES (p.28)

Table 1 provides data for the commonly used acids and bases.

2. AMINO ACIDS (pp. 29–30)

Physical and chemical data for the 20 L-amino acids involved in protein synthesis are given in *Table 2* and their molecular structures are shown in *Figure 1*.

3. ANTIBIOTICS AND INHIBITORS (pp. 31–37)

The chemical and biological properties of important antibiotics and inhibitors are given in *Table 3* and their molecular structures are shown in *Figure 2*. See Chapter 1, *Table 11* for the preparation of stock and working solutions of antibiotics.

4. BUFFERS

4.1 Zwitterionic (Good's) buffers (pp. 38–39)

Good's buffers are described in *Table 4* and their chemical structures are given in *Figure 3*. For more general information on Good's buffers see refs 28 and 29.

4.2 Non-zwitterionic buffers (pp. 40–41)

Table 5 gives physical and chemical data for compounds used in non-zwitterionic buffer systems.

5. CHELATING AGENTS (COMPLEXANES) (p. 42)

Only two chelating agents – EDTA and EGTA – are used to any great extent in molecular biology. Relevant data, including the stability constants for their complexes with various metal ions, are given in *Table 6*.

6. DENATURANTS (p. 43)

Physical and chemical data for common denaturants are given in *Table 7* and their structures are shown in *Figure 4*.

7. DETERGENTS AND SURFACTANTS (pp. 44–47)

Although not central to molecular biology, a number of detergents and surfactants are used in DNA and RNA extractions. *Table 8* provides relevant information for the important compounds and their structures are displayed in *Figure 5*.

8. ENZYMES (pp. 48–54)

This section does not include restriction endonucleases (see Chapter 4) or DNA and RNA modifying enzymes (see Chapter 5). Enzymes active on carbohydrates and sugars are listed in *Table 9*, proteases in *Table 10* and ribonucleases relevant to enzymatic RNA sequencing in *Table 11*. RNases with more general roles in recombinant DNA techniques (e.g. RNase A, RNase H) are covered in Chapter 5.

9. NUCLEOTIDES AND RELATED COMPOUNDS (pp. 55–60)

Physical and chemical data for nucleic acid bases and nucleosides are given in *Table 12* and for the standard unmodified nucleotides in *Table 13*. Their molecular structures are shown in *Figure 6*. The positions at which nucleotides are radiolabeled are shown in Chapter 3, *Figure 2*. In *Table 14* and *Figure 7*, data and structures are provided for important modified nucleotides.

10. ORGANIC CHEMICALS, GENERAL (p. 61)

Data for a variety of frequently used organic chemicals are given in *Table 15*. Organic solvents primarily used for scintillation counting are described in Chapter 3, *Table 10*.

11. POLYMERIC COMPOUNDS (p. 62)

Several high molecular weight polymers are used in molecular biology. Molecular weights and solubilities are given in *Table 16*, and structures are described in *Figure 8*.

12. PROTEASE INHIBITORS (pp. 63–66)

Table 17 provides data for protease inhibitors including their biological specificities. Structures are shown in *Figure 9*.

13. PROTEINS, GENERAL (p. 67)

Table 18 covers BSA, avidin, Protein A and streptavidin, none of which fit conveniently into any other category.

14. RIBONUCLEASE INHIBITORS (p. 67)

Biological and structural data for diethyl pyrocarbonate, human placental RNase inhibitor and vanadyl ribonucleoside complex are given in *Table 19*.

15. SALTS (pp. 68–69)

Inorganic salts not covered elsewhere are described in *Table 20*.

16. STAINS, INDICATORS AND TRACKING DYES (pp. 70–72)

Data for these compounds are given in *Table 21* and their structures are shown in *Figure 10*. The mobilities of tracking dyes in gel electrophoresis systems are given in Chapter 8, *Figures 2–4* and *Table 5*.

17. SUBSTRATES FOR ENZYME ASSAYS (pp. 73–75)

Alkaline phosphatase, β-galactosidase, β-glucuronidase and peroxidase assays are used in detection systems of various kinds. Their substrates are described in *Table 22* and *Figure 11*.

Note on molecular weights: Many organic chemicals exist as different salts or hydrates. The molecular weights given in the tables may not be correct for the particular form of the chemical you are using.

Table 1. Acids and bases: physical and chemical data

	Formula	Mol. wt	m.p. (°C)	b.p. (°C)	Specific gravity (20°C)	Molarity of the conc. solution	ml liter^{-1} to give a 1 M solution	Risk	Safety
Acids									
Acetic	CH_3COOH	60.05	17	118	1.05	17.4	57.5	10[a]-35	2-23-26
Formic	HCOOH	46.03	8	100	1.22	23.6 (90%) 25.9 (98%)	42.4 (90%) 38.5 (98%)	35	2-3-23-26
Hydrochloric	HCl	36.46	–	–	1.18	11.6	85.9	34-37	2-26
Nitric	HNO_3	63.01	–42	86	1.42	15.7	63.7	35	2-23-26-27
Orthophosphoric[b]	H_3PO_4	98.00	22	261	1.70	14.7	67.8	34	26
Sulfuric	H_2SO_4	98.07	10	330	1.84	18.3	54.5	35	2-26-30
Bases									
Ammonium hydroxide	NH_4OH	17.03[c]	–	–	(25%) 0.91 (35%) 0.88	13.3 18.1	75.1 55.2	34-36/37/38	2-3-7-26
Potassium hydroxide	KOH	56.11	360	1320	2.0	(solid) <	>	35	2-26-37/39
Sodium hydroxide	NaOH	40.00	318	1390	2.12	(solid) <	>	35	2-26-37/39

[a] Acetic acid: flash point 40°C, auto-ignition point 426°C.
[b] Commonly called 'phosphoric acid'.
[c] This is the molecular weight of NH_3.aq. The formula weight is 35.05.

Table 2. L-amino acids: physical and chemical data

Amino acid	Abbreviations		Mol. wt	m.p.[a] (°C)	Solubility (25°C) (g/100 ml H_2O)	pK_a (25°C) -COOH	pK_a (25°C) -NH_2	pK_a (25°C) R-group	pI (25°C)
Alanine	Ala	A	89.10	>200	16.65	2.35	9.87	–	6.11
Arginine	Arg	R	174.20	244	15.00[b]	1.82	8.99	12.48	10.76
Asparagine	Asn	N	132.12	236	2.99	2.14	8.72	–	5.43
Aspartic acid	Asp	D	133.11	270	0.5	1.99	9.90	3.90	2.98
Cysteine	Cys	C	121.16	240	v. sol.	1.92	10.70	8.37	5.15
Glutamic acid	Glu	E	147.13	248	0.86	2.10	9.47	4.07	3.08
Glutamine	Gln	Q	146.15	185	3.6[b]	2.17	9.13	–	5.65
Glycine	Gly	G	75.07	292	25.0	2.35	9.78	–	6.06
Histidine	His	H	155.16	287	4.16	1.80	9.33	6.04	7.64
Isoleucine	Ile	I	131.18	285	4.12	2.32	9.76	–	6.04
Leucine	Leu	L	131.18	294	2.43	2.33	9.74	–	6.04
Lysine	Lys	K	146.19	224	66.6[b]	2.16	9.06	10.54	9.47
Methionine	Met	M	149.21	281	5.14[b]	2.13	9.28	–	5.71
Phenylalanine	Phe	F	165.19	283	2.96	2.20	9.31	–	5.76
Proline	Pro	P	115.13	221	162.3	1.95	10.64	–	6.30
Serine	Ser	S	105.09	228	25.0[b]	2.19	9.21	–	5.70
Threonine	Thr	T	119.12	226	v. sol.	2.09	9.10	–	5.60
Tryptophan	Trp	W	204.23	289	1.14	2.46	9.41	–	5.88
Tyrosine	Tyr	Y	181.19	343	0.045	2.20	9.21	10.46	5.63
Valine	Val	V	117.15	315	8.85	2.29	9.74	–	6.02

[a] At these temperatures virtually all amino acids undergo some decomposition.
[b] Temperatures: Arg 21°C, Gln 18°C, Lys 20°C, Met 20°C, Ser 20°C.

Figure 1. Structures of L-amino acids.

GENERAL STRUCTURE:

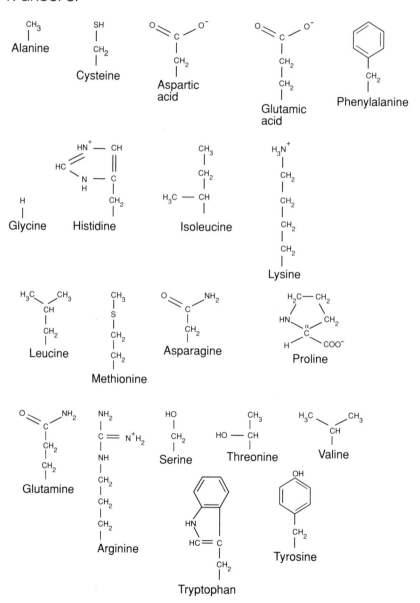

R-GROUPS:

Table 3. Antibiotics and inhibitors: physical, chemical and biological data

Compound	Mol. wt	Solubility in H$_2$O[a]	Risk	Safety	Activity	Notes	Refs
Actinomycin D	1255.5	+	25	24/25	Inhibits transcription by binding to dG(3′–5′)dC	Store dry, −20°C, in the dark	1
α-amanitin	903.0	++	26/27/28	1/36/45	Inhibits eukaryotic RNA pol II and III	–	2
Aminopterin	440.4	+	–	–	Folic acid antagonist	–	–
Ampicillin	403.5 (.3H$_2$O)	+++	< irritant	>	Disrupts bacterial cell wall synthesis	–	–
Bleomycin A2 Bleomycin B2	1415.6 1425.5	+++ +++	20/21/22–40	28–36	Inhibits DNA synthesis, causes ssDNA cleavage	= blenoxan	3
5-bromouracil	191.1	++	< mutagen	>	Thymine analog, induces GC to AT transitions	–	4
Carbenicillin	422.4	?	–	–	Inhibits bacterial cell wall synthesis	–	–
Cefotaxime	477.5	?	20/21/22	28–36	Inhibits bacterial cell wall synthesis, β-lactamase resistant	= Claforan	5, 6
Chloramphenicol	323.1	2.5 g/100 ml (room temp.)	< toxic at high concentrations	>	Blocks peptidyl transferase, at high levels inhibits eukaryotic DNA synthesis	–	7, 8
Colchicine	399.4	4.0 g/100 ml (room temp.)	26/28	1-13-45	Interferes with microtubule organization, inhibits mitosis in plants and animals	–	9
Cordycepin	251.2	++	–	–	Inhibits eukaryotic mRNA polyadenylation	= 3′-deoxy-adenosine	10
Cycloheximide	281.3	++	23-28-36/37/38	28	Inhibits eukaryotic cytoplasmic protein synthesis	Unstable at alkaline pH	11–13

Table 3. Continued

Compound	Mol. wt	Solubility in H$_2$O[a]	Risk	Safety	Activity	Notes	Refs
Cycloserine	102.1	++	–	–	Inhibits bacterial cell wall synthesis	Unstable at acid–neutral pH	–
2,4-dinitrophenol	184.1	0.56 g/100 ml (18°C)	1-23/24/25-33	28-37-44	Uncouples oxidative phosphorylation	–	14
Erythromycin	734.0(A)	2 mg/ml (slowly)	–	–	Inhibits bacterial protein synthesis	–	–
5-fluorouracil	130.1	++	23/24/25	28-44	UTP analog, causes transitions, translation errors, inhibits DNA synthesis	–	–
Hygromycin B	527.5	+++	26/27/28	1-36-45	Not clear	= hygromix	15
Kanamycin[b]	582.6	+++	–	–	Inhibits bacterial protein synthesis, binds to 30S ribosome subunit	–	–
Methotrexate	454.5	sol. in dil. acid	23/24/25-36/38-47	1-36-45	Folic acid antagonist	= amethopterin	16
Mitomycin C	334.3	++	22	24/25	Inhibits DNA synthesis	Light sensitive	–
Nalidixic acid	232.2	+	20/21/22	28-36	Inhibits bacterial DNA gyrase subunit A	–	17–20
Neomycin[b]	712.7(B)	+++	–	–	Inhibits bacterial protein synthesis, binds to 30S ribosome subunit	–	–
Novobiocin	612.6	++	–	–	Inhibits bacterial DNA gyrase	–	19, 21, 22
Penicillin G	334.4 (free acid)	+	–	–	Inhibits bacterial cell wall synthesis	–	–
Polymyxin B1	1385.8 (.5HCl)	+++	–	–	Disrupts bacterial cell membrane permeability	–	23

Table 3. Continued

Compound	Mol. wt	Solubility in H_2O[a]	Risk	Safety	Activity	Notes	Refs
Proflavin	209.3	++	irritant	>	Intercalating agent into extrachromosomal DNA	–	–
Puromycin	471.5	+	–	–	Inhibits protein synthesis by acting as an aminoacyl-tRNA analog	–	–
Rifampicin	823.0	–	–	–	Inhibits bacterial RNA polymerase	= rifampin	24
Sodium azide	65.0	++	4-5-28-29-32	28	Uncouples phosphorylase by inhibiting catalase and other Fe enzymes	–	–
Streptomycin	581.6	+++	20/21/22	28-36	Inhibits bacterial protein synthesis, binds to 30S ribosome subunit	–	–
Tetracycline	444.4	+++	toxic	>	Inhibits bacterial protein synthesis, binds to aminoacyl-tRNA	Light sensitive	25
2-thiouracil	128.2	0.06 g/100 ml (room temp.)	suspected carcinogen	>	Uracil analog, causes translation errors	–	–
Tunicamycin	816.9 ($n = 8$)[c]	++ (alkali)	26/27/28	1-36-45	Inhibits bacterial and eukaryotic N-acetyl-glucosamine transferase, involved in protein glycosylation	–	10, 26, 27

[a] +++ very soluble, ++ soluble, + slightly soluble, – insoluble.
[b] There are several different forms of these antibiotics.
[c] See *Figure 2*, where n refers to the number of side-chain groups per molecule.

CHEMICALS AND REAGENTS

Figure 2. Structures of antibiotics and inhibitors.

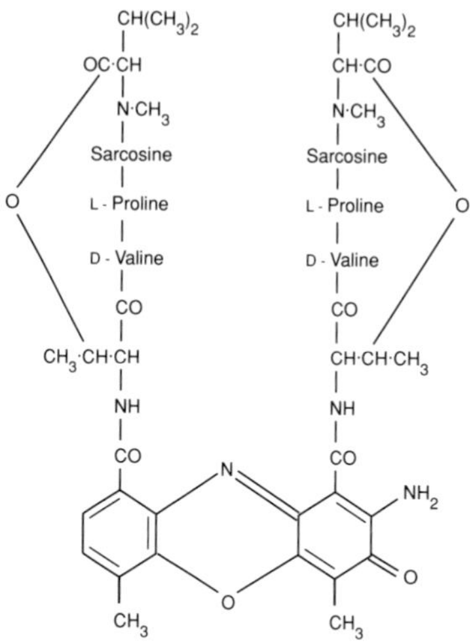

ACTINOMYCIN D

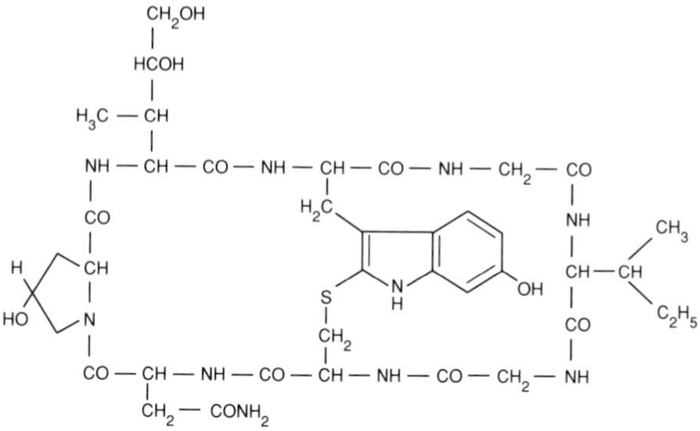

α-AMANITIN

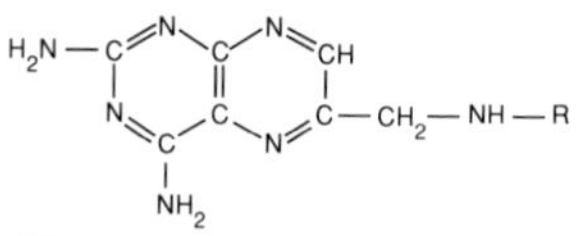

[R = benzoylglutamic acid]

AMINOPTERIN

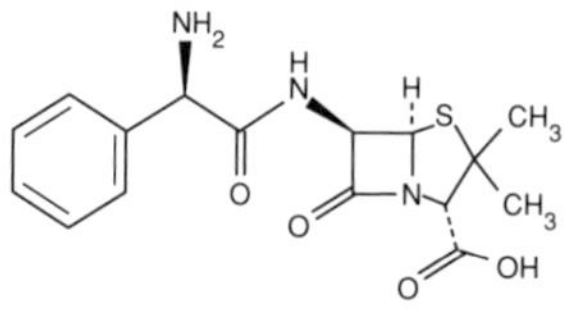

AMPICILLIN

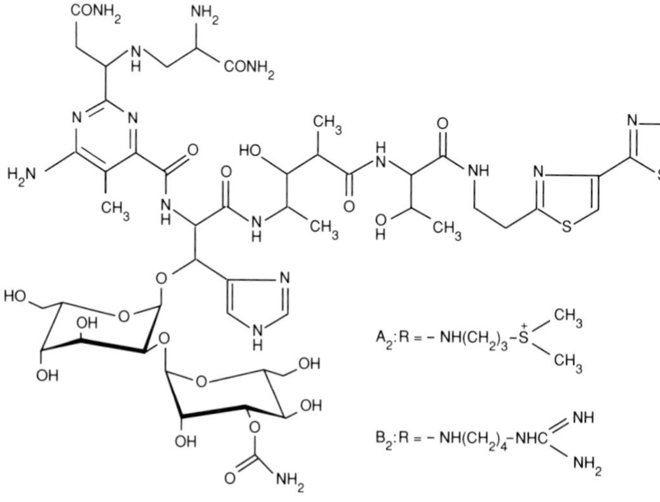

A_2:R = – NH(CH$_2$)$_3$–S$^+$(CH$_3$)CH$_3$

B_2:R = – NH(CH$_2$)$_4$–NHC(NH)NH$_2$

BLEOMYCIN

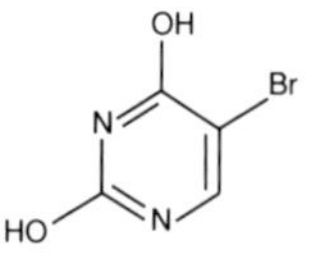

5-BROMOURACIL

Figure 2. Continued.

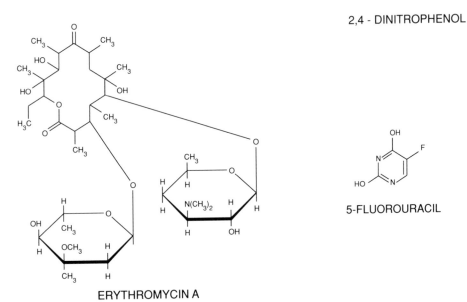

CARBENICILLIN

CEFOTAXIME

CHLORAMPHENICOL

COLCHICINE

CORDYCEPIN

CYCLOHEXIMIDE

CYCLOSERINE

2,4 - DINITROPHENOL

5-FLUOROURACIL

ERYTHROMYCIN A

Chemicals/Reagents

Figure 2. Continued.

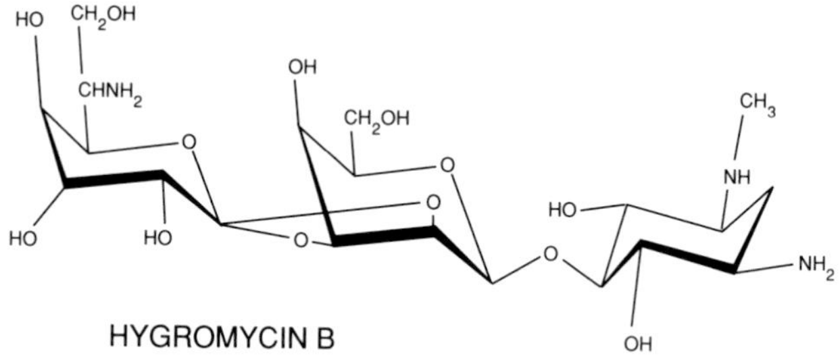

HYGROMYCIN B

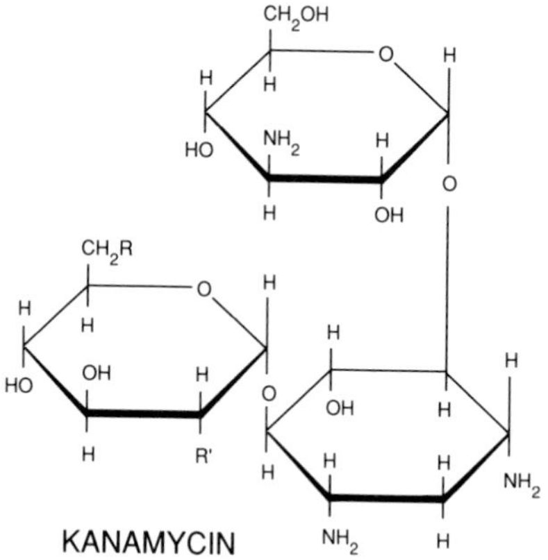

KANAMYCIN

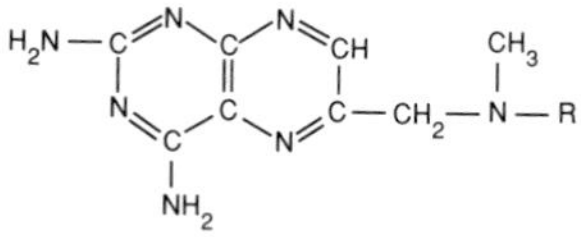

[R = benzoylglutamic acid]
METHOTREXATE

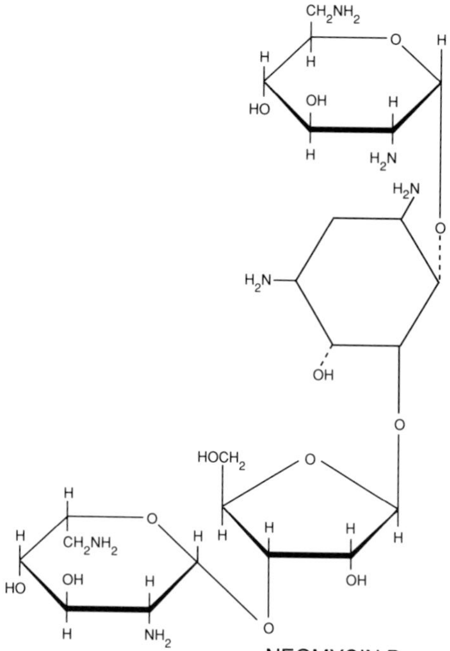

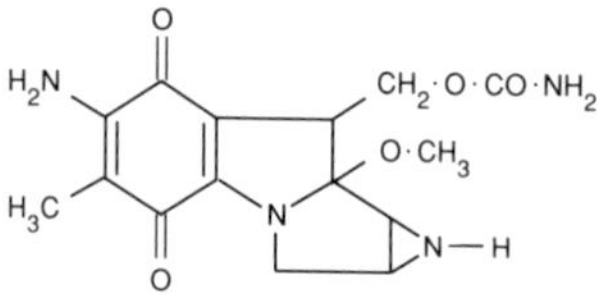

MITOMYCIN C

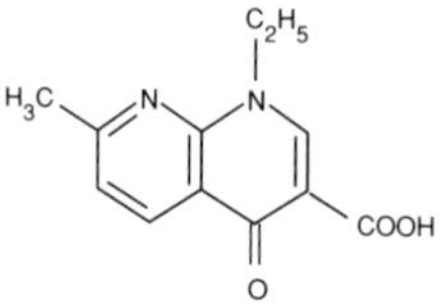

NALIDIXIC ACID

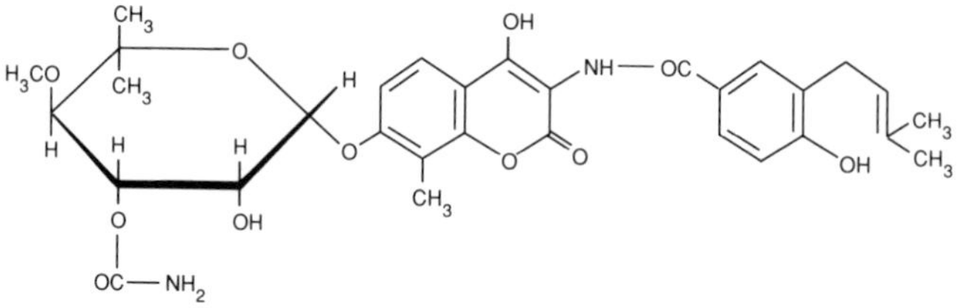

NEOMYCIN B

NOVOBIOCIN

Figure 2. Continued.

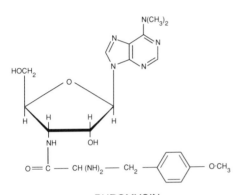

PENICILLIN G

PROFLAVIN

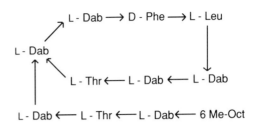

L - Dab → D - Phe → L - Leu

L - Dab

L - Thr ← L - Dab ← L - Dab

L - Dab ← L - Thr ← L - Dab ← 6 Me-Oct

Dab = 2,4-Diaminobutyric acid
6 Me-Oct = 6-Methyloctanoic acid

POLYMYXIN B1

PUROMYCIN

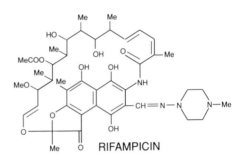

RIFAMPICIN

STREPTOMYCIN

TETRACYCLINE

TUNICAMYCIN

2-THIOURACIL

Table 4. Zwitterionic (Good's) buffers: physical and chemical data

Buffer[a]	Mol. wt	Solubility (0°C) (mol/liter H$_2$O)	pK$_a$ at 20°C	Δ/°C	Useful pH range
ACES	182.2	0.22	6.88	−0.020	6.4–7.4
ADA[b]	190.2	0.09	6.62	−0.011	6.4–7.4
BES	213.3	3.2	7.17	−0.016	6.6–7.6
Bicine	163.2	1.1	8.35	−0.018	7.8–8.8
CAPS	221.3	0.47	10.4	0.032	9.7–11.1
CHES	207.1	1.15	9.55	−0.011	9.0–10.1
Glycylglycine	132.1	1.01 (25°C)	8.25	−0.026	7.7–8.8
HEPES	238.3	2.25	7.55	−0.014	7.0–8.0
HEPPS (EPPS)	252.3	1.6	8.0	−0.011	7.6–8.6
MES	213.3	0.65	6.15	−0.011	5.8–6.5
MOPS	209.3	3.1	7.2	−0.011	6.5–7.9
PIPES	302.4	1.4	6.82	−0.009	6.4–7.2
TAPS	243.3	v. sol.	8.4 (25°C)	0.018	7.7–9.1
TES	229.3	2.6	7.5	−0.020	7.0–8.0
Tricine	179.2	0.8	8.15	−0.021	7.6–8.8
Tris	121.1	4.54 (25°C)	8.08	−0.028	7.0–9.0[c]

[a]**Chemical names**:

ACES	*N*-(2-acetamido)-2-aminoethanesulfonic acid
ADA	*N*-(2-acetamido)iminodiacetic acid
BES	*N,N*-bis(2-hydroxyethyl)-2-aminoethanesulfonic acid
Bicine	*N,N*-bis(2-hydroxyethyl)glycine
CAPS	3-(cyclohexylamino)-1-propanesulfonic acid
CHES	2-(cyclohexylamino)-ethanesulfonic acid
HEPES	4-(2-hydroxyethyl)piperazine-1-ethanesulfonic acid
HEPPS	4-(2-hydroxyethyl)piperazine-1-propanesulfonic acid
MES	2-morpholinoethanesulfonic acid monohydrate
MOPS	3-morpholinopropanesulfonic acid
PIPES	Piperazine-1,4-bis(2-ethanesulfonic acid)
TAPS	*N*-[tris(hydroxymethyl)methyl]-3-aminopropanesulfonic acid
TES	*N*-[tris(hydroxymethyl)methyl]-2-aminomethanesulfonic acid
Tricine	*N*-[tris(hydroxymethyl)methyl]glycine
Tris	Tris(hydroxymethyl)aminomethane.

[b] ADA may cause skin irritation. Avoid contact with the dust.
[c] Refers to Tris–HCl.

Figure 3. Structures of zwitterionic (Good's) buffers.

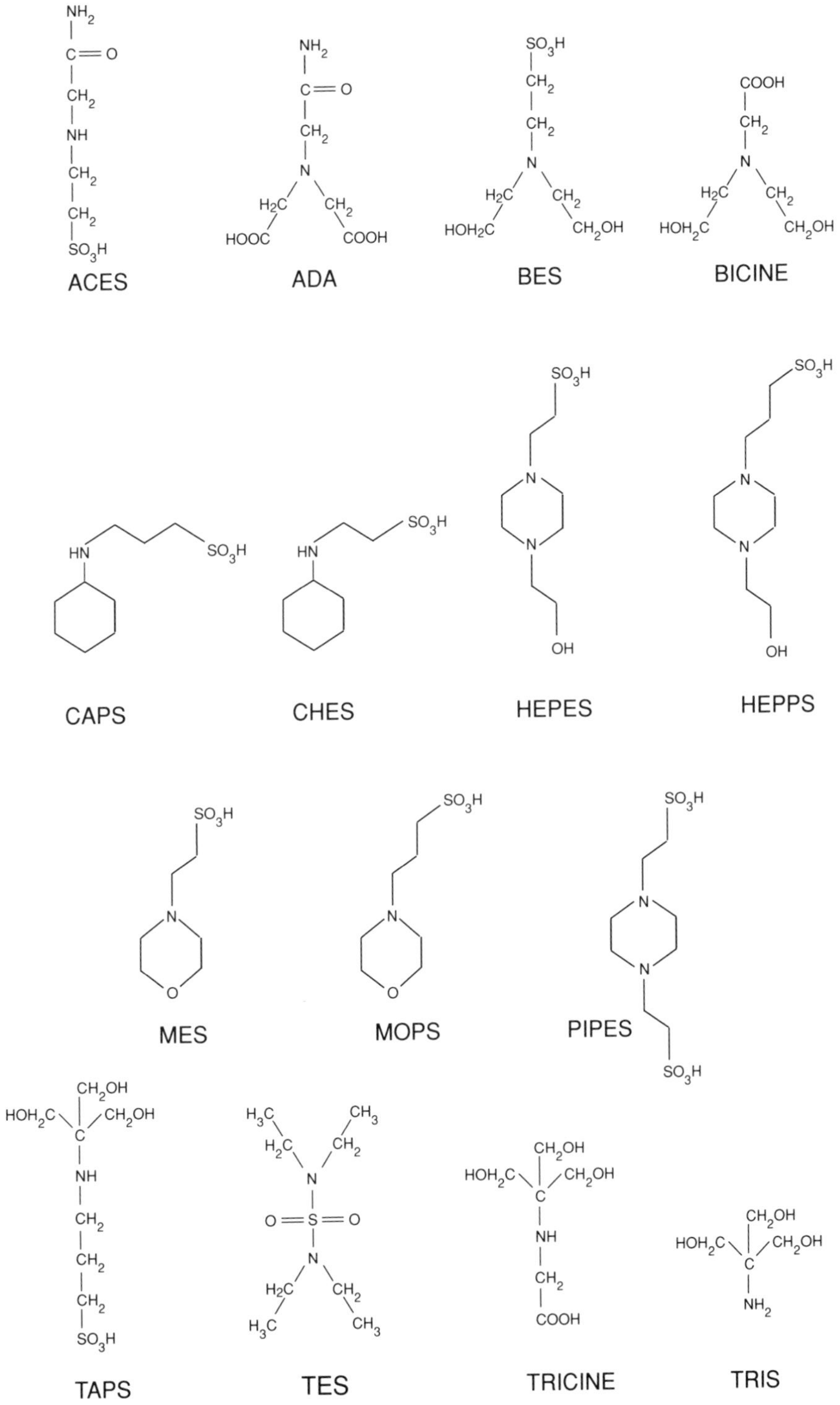

Table 5. Components of non-zwitterionic buffers: physical and chemical data

Compound	Formula	Mol. wt	Solubility (20°C) (g/100 ml H$_2$O)	pK_a (20°C) 1	2	3	Risk	Safety
Boric acid[a]	H$_3$BO$_3$	61.8	6.4	9.14	12.74	13.80	–	–
Citric acid, anhydrous	C$_6$H$_8$O$_7$[c]	192.4	146.0	3.14	4.77	6.39 (18°C)	–	–
Imidazole	C$_3$H$_4$N$_2$[c]	68.1	v. sol.	6.95 (25°C)	–	–	21/22	25
Potassium dihydrogen phosphate (potassium phosphate, monobasic)	KH$_2$PO$_4$	136.1	136.1	2.12	7.21	12.67 (25°C)	–	–
di-Potassium hydrogen phosphate (potassium phosphate, dibasic)	K$_2$HPO$_4$	174.2	174.2	2.12	7.21	12.67 (25°C)	–	–
Potassium hydrogen phthalate	C$_8$H$_5$KO$_4$[c]	204.2	204.2	2.89	5.51(25°C)	–	–	–
Sodium acetate anhydrous trihydrate	CH$_3$COONa CH$_3$COONa.3H$_2$O	82.0 136.1	119.0 (0°C) 76.2 (0°C)	4.75 (25°C)	–	–	–	–
Sodium cacodylate[b]	(CH$_3$)$_2$AsO$_2$Na.3H$_2$O	214.0	200	6.21	–	–	23/25-33	1/2-7-20/21-28-44
Sodium carbonate anhydrous decahydrate	Na$_2$CO$_3$ Na$_2$CO$_3$.10H$_2$O	106.0 286.1	7.1 (0°C) 21.5 (0°C)	6.37	10.25	–	–	–
tri-Sodium citrate	Na$_3$C$_6$H$_5$O$_7$.2H$_2$O[c]	294.1	72.0	3.09	4.75	5.41 (18°C)	–	–
Sodium dihydrogen phosphate (sodium phosphate, monobasic)	NaH$_2$PO$_4$.2H$_2$O	156.0	v. sol	2.12	4.75	7.21	–	–
Sodium hydrogen carbonate	NaHCO$_3$	84.0	9.6	6.37	10.25	12.67	–	–

Table 5. Continued

Compound	Formula	Mol. wt	Solubility (20°C) (g/100 ml H_2O)	pK_a (20°C) 1	pK_a (20°C) 2	pK_a (20°C) 3	Risk	Safety
di-Sodium hydrogen phosphate							—	—
anhydrous	Na_2HPO_4	142.0	?	2.12	7.21	12.67 (25°C)		
dihydrate	$Na_2HPO_4 \cdot 2H_2O$	178.0	60.0					
dodecahydrate	$Na_2HPO_4 \cdot 12H_2O$	358.1	4.2					
(sodium phosphate, dibasic)								
Sodium tetraborate							—	—
anhydrous	$Na_2B_4O_7$	201.2	1.1 (0°C)	9.14	12.74	13.80		
decahydrate	$Na_2B_4O_7 \cdot 10H_2O$	381.4	2.0 (0°C)					

a Also called orthoboric acid.

b Sodium cacodylate is hydroxydimethylarsine oxide, sodium salt, or dimethylarsinic acid, sodium salt.

c Structures for these compounds are:

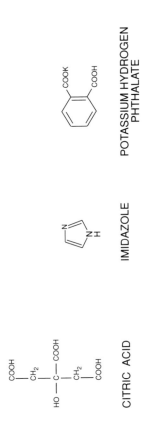

CITRIC ACID

IMIDAZOLE

POTASSIUM HYDROGEN PHTHALATE

Table 6. Chelating agents: physical and chemical data and stability constants for metal ion complexes

	EDTA[a]	EGTA[b]
Molecular weight	372.2 ($.2Na,.2H_2O$)	380.4
Solubility ($20°C$, g/100 ml H_2O)	11.1	sol.
pK_as ($25°C$)		
pK_a1	1.99	2.00
pK_a2	2.67	2.65
pK_a3	6.16	8.85
pK_a4	10.26	9.46
Absolute stability constants ($\log K_1$)		
Ag (I)	7.32	6.88
Ca (II)	10.96	11.00[c]
Cd (II)	16.46	16.70
Co (II)	16.31	12.50
Cr (III)	23.40	–
Cu (II)	18.80	17.8
Fe (II)	14.33	11.92
Fe (III)	25.1	20.5
Hg (II)	21.8	23.12
Li (I)	2.79	1.17
Mg (II)	8.69	5.21[c]
Mn (II)	14.04	12.3
Na (I)	1.66	1.38
Ni (II)	18.62	13.55
Pb (II)	18.04	14.71
Sn (II)	18.3	23.85
Tl (III)	22.5	–
Zn (II)	16.50	14.5

Structures

EDTA   EGTA

[a] EDTA = ethylenediaminetetra-acetic acid.
[b] EGTA = ethyleneglycolbis(aminoethyl)-tetra-acetic acid.
[c] EGTA displays a relatively high selectivity for Ca(II) over Mg(II) (30).

Table 7. Denaturants: physical and chemical data

Compound	Mol. wt	pK$_a$ (25°C)	Solubility in H$_2$O	Risk	Safety	Refs
Dithioerythritol	154.25	9.0, 9.9	sol.	–	–	31–34
Dithiothreitol	154.25	8.3, 9.5	sol.	–	–	31,32
Guanidine hydrochloride[a]	95.53	2.1, 4.1, 9.5	v. sol.	20/22-36/38	26-28	–
Guanidine thiocyanate[b]	118.16	–	v. sol.	20/21/22-32	13	35–40
Trichloracetic acid	163.39	–	sol.	35	24/25-26	–
Urea	60.06	–	119.3 g/100 ml (25°C)	–	–	41

[a] Also called guanidium chloride and guanidinium chloride.
[b] Also called guanidinium rhodanide and guanidinium thiocyanate.

Figure 4. Structures of denaturants.

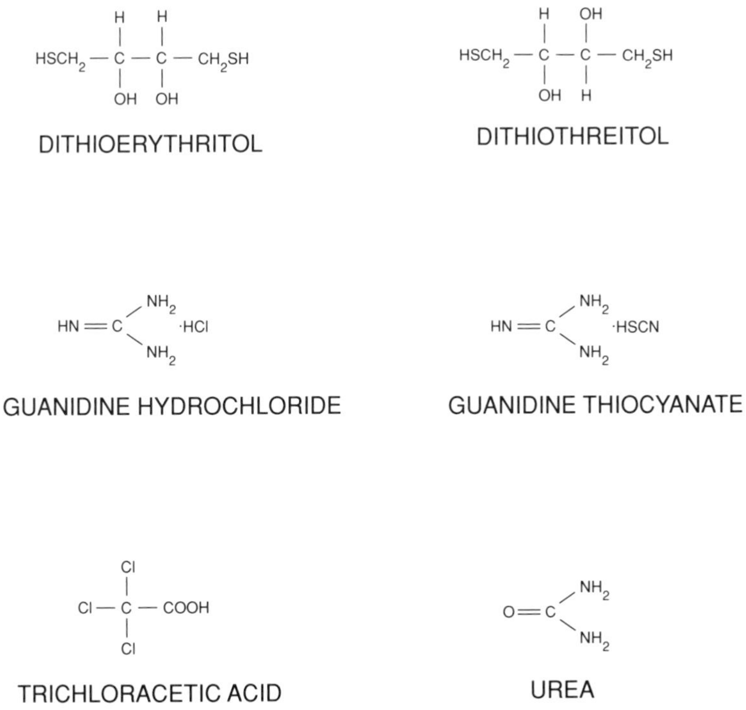

DITHIOERYTHRITOL

DITHIOTHREITOL

GUANIDINE HYDROCHLORIDE

GUANIDINE THIOCYANATE

TRICHLORACETIC ACID

UREA

Table 8. Detergents and surfactants: physical and chemical data

Compound[a]	Mol. wt	CMC[b] μM (20°C, water)	Aggregation No. (20–25°C, 0–0.1 M Na+)	Micellar weight	Risk	Safety	Refs
Anionic detergents							
Sarkosyl	293.4	–	–	–	–	–	–
SDS	288.4	8200 / 520[c]	62	18 000	22-36/38	–	42–44
Cationic detergents							
CPB	402.5	–	–	–	21/22	24/25	–
CTAB	364.5	1000	170	62 000	21/22	26	–
Zwitterionic detergents							
CHAPS	614.9	6000–10 000	10	6150	irritant	∧	45–47
CHAPSO	630.9	8000	11	9960	irritant	∧	48
Zwittergent series:							
Zwittergent 3.08	280	330 000	–	–	–	–	49
Zwittergent 3.10	308	25 000–40 000	41	–	–	–	
Zwittergent 3.12	336	2000–4000	55	–	–	–	
Zwittergent 3.14	364	100–400	83	–	–	–	
Zwittergent 3.16	392	10–60	155	–	–	–	
Non-ionic detergents							
Brij series:							
Brij 35	~1200	75	40	–	irritant	∧	50,51
Brij 56	682	2	40	–	irritant	∧	
Brij 58	1122	77	40	–	irritant	∧	

Table 8. Continued

Compound[a]	Mol. wt	CMC[b] μm (20°C, water)	Aggregation No. (20–25°C, 0–0.1 M Na$^+$)	Micellar weight	Risk	Safety	Refs
Octyl-β-D-glucopyranoside octylglucoside	292.4	25 000	84	–	–	–	47,52,53
Triton series:							50,51
Triton X-100	625	250	100–155	–	22–36	26	
Nonidet P40	625	250	100–155	–	36	24/25-26	
Tween series:							52,53
Tween 20	1228	59	–	–	–	–	
Tween 80	1310	12	58	–	–	–	

[a] Chemical names are as follows:
Sarkosyl, N-laurylsarcosine, sodium salt; SDS, sodium dodecyl (lauryl) sulfate; CPB, cetylpyridinium bromide; CTAB, cetyltrimethylammonium bromide; CHAPS, 3-[(3-cholamidopropyl)-dimethylammonio]-1-propane- sulfonate; CHAPSO, 3-[(3-cholamidopropyl)-dimethylammonio]-2-hydroxy-1-propanesulfonate; Zwittergent, N-alkyl-N,N-dimethyl-3-ammonio-1-propanesulfonate; Brij, polyoxyethylene lauryl and cetyl alcohols; Nonidet P40, polyoxyethylene(9)-p-t-octylphenol; Triton, polyoxyethylene-p-t-octylphenols; Tween, polyoxyethylene sorbitol esters.
[b] CMC, critical micelle concentration.
[c] In 0.5 M NaCl.

Figure 5. Structures of detergents and surfactants.

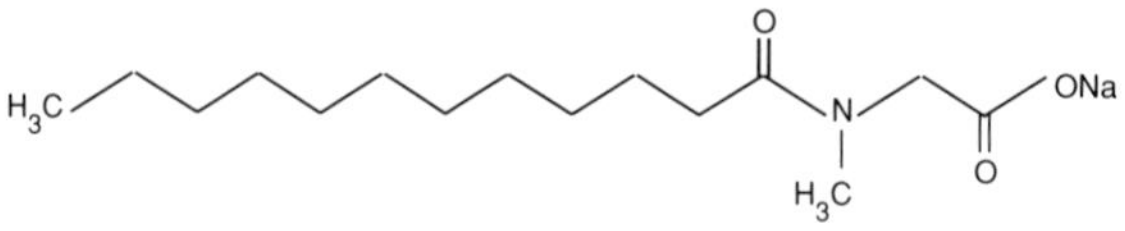

SARKOSYL

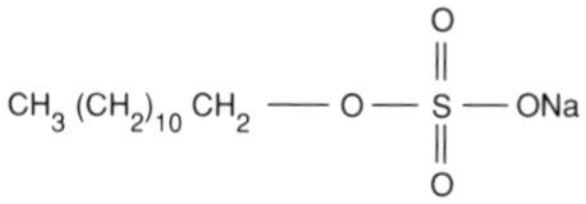

SDS

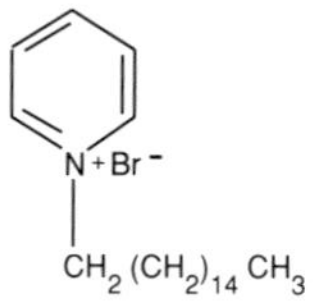

$CH_3 (CH_2)_{10} CH_2 - O - S - ONa$

CPB

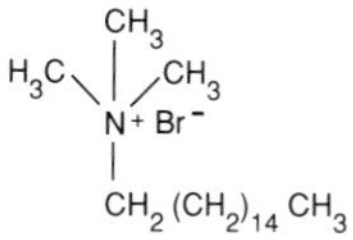

$CH_2 (CH_2)_{14} CH_3$

CTAB

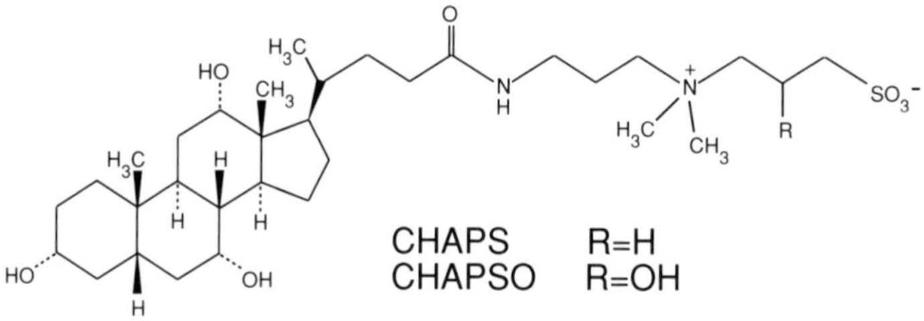

$CH_2 (CH_2)_{14} CH_3$

CHAPS   R=H
CHAPSO   R=OH

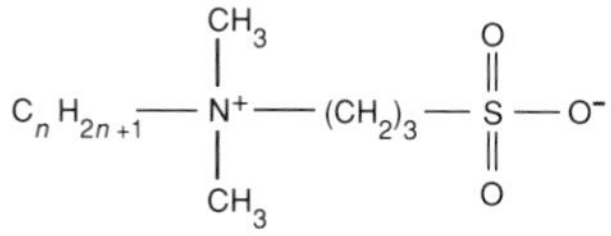

$C_n H_{2n+1}$ — ZWITTERGENT 3.n

$CH_3 - (CH_2)_n - O - [CH_2 - CH_2 - O]_m H$

BRIJ 35 $m=23$ $n=10$
      56 $m=10$ $n=15$
      58 $m=20$ $n=15$

Figure 5. Continued.

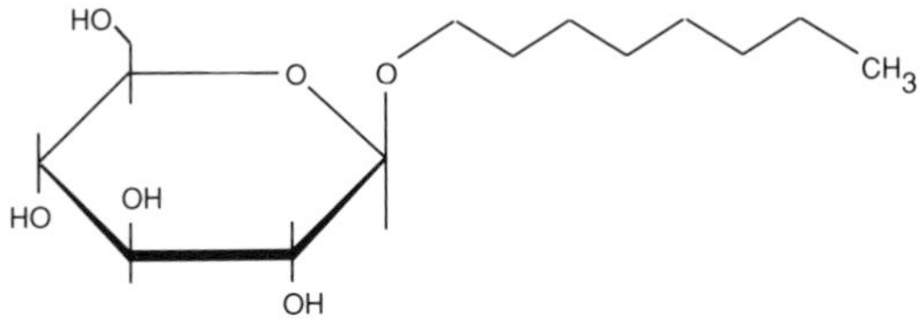

OCTYL - β - D - GLUCOPYRANOSIDE

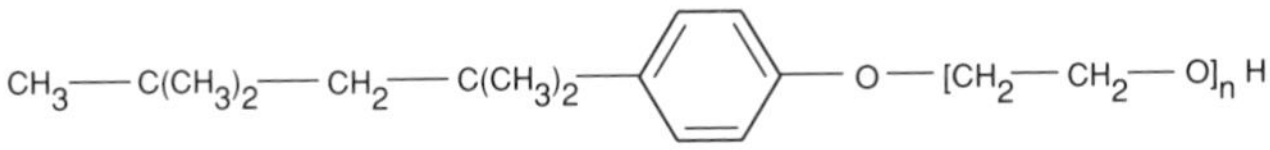

NONIDET P40 $n = 9$

TRITON X-100 $n = 9,10$

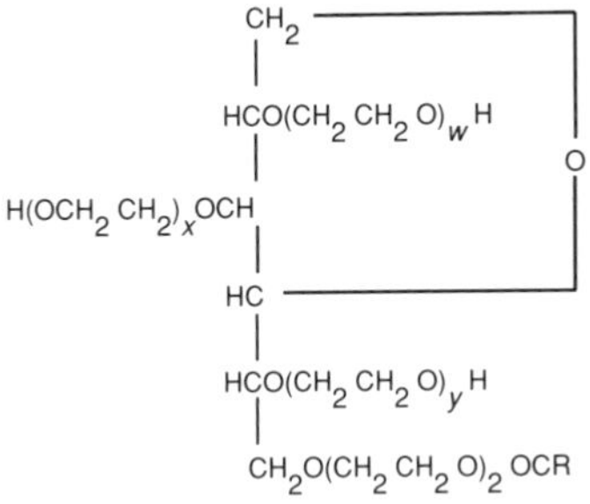

TWEEN   20 $w + x + y = 20$
R = monolaurate

TWEEN   80 $w + x + y = 20$
R = monooleate

Table 9. Enzymes active on carbohydrates and sugars

Enzyme	Source	EC number	Mol. wt	Optimum pH
β-*N*-acetyl-D-glucosaminidase	Beef kidney	3.2.1.30	155 000	4.5
N-acetylneuraminic acid aldolase	*E. coli*	4.1.3.3	–	–
α-amylase	*Bacillus subtilis*	3.2.1.1	97 000	5.7
	Pig pancreas		50 000	7.0
β-amylase	Various plants	3.2.1.2	206 000	4.8
Amyloglucosidase	*Aspergillus niger*	3.2.1.3	97 000	4.7
Cellulase	Various fungi	3.2.1.4	various	4.8
Chitinase	*Streptomyces antibioticus*	3.2.1.14		
Endo-α-*N*-acetyl galactosaminidase	*Diplococcus pneumoniae*	3.2.1.97	–	–
Endo-β-galactosidase	*Bacteroides fragilis*	3.2.1.103	32 000	5.8
Endoglycosidase D	*Diplococcus pneumoniae*	3.2.1.96	280 000	6.5
Endoglycosidase F	*Flavobacterium meningosepticum*	3.2.1.96	32 200	5.7
Endoglycosidase H	*Streptomyces* sp.	3.2.1.96	29 000	5.5
β-fructosidase	*S. cerevisiae*	3.2.1.26	270 000	4.6
α-fucosidase	Beef kidney	3.2.1.51	217 000	5.0
α-galactosidase	Various	3.2.1.22	329 000	7.2
β-galactosidase	Various	3.2.1.23	540 000	7.0
α-glucosidase[b]	*S. cerevisiae*	3.2.1.20	68 500	7.2
β-glucosidase	Almonds	3.2.1.21	67 000	5.0
β-glucuronidase	*E. coli*	3.2.1.31	–	6.2
Glycopeptidase F	*Flavobacterium meningosepticum*	3.2.2.18	35 500	7.5
Hyaluronidase	Mammalian testes	3.2.1.35	55 000	5.2
Hyaluronidase	Leech	3.2.1.35	–	–
Laminarinase	*Spisula solidissima*	3.2.1.6	–	–
Lysozyme (muramidase)	Chicken egg white	3.2.1.7	14 400	6.5

Type of enzyme[a]	Specificity	Refs
Exo	Broad specificity	–
–	Cleaves N-acetylneuraminate	54–56
Endo	α-1,4-glucosidic links	57
Endo	α-1,4-glucosidic links	–
Exo	α-1,4 and α-1,6-glucan links at non-reducing terminus	–
Endo	β-1,4-glucan links in cellulose	58
Endo	α-1,4-acetamido-2-deoxy-D-glucoside links in chitin and chitosan	–
Endo	Gal-β1,3-GalNAc links in glycoproteins	59
Endo	β-1,4-galactosidic links in GlcNAc-Gal-GlcNAc	60,61
Endo	Cleaves various high mannose glycans	62,63
Endo	Cleaves various high mannose glycans	64,65
Endo	Cleaves various high mannose glycans	66,67
Exo	β-D-fructofuranosides	–
Exo	α-L-fucopyranosides	–
Exo	Terminal α-galactosyl residues from α-D-galactopyranosides	68
Exo	*E. coli* enzyme hydrolyses β-D-galactopyranosides, beef enzyme removes terminal Gal from Gal-β1,3-GlcNAc, Gal-β1,4-GlcNAc or Gal-β1,3-GalNAc	69–71
Exo	Broad specificity for α-glucosides	–
Exo	Broad specificity for β-D-glucosides	–
Exo	Broad specificity for β-D-glucuronides	–
Endo	Glycoproteins between Asn and GlcNAc	72
Endo	1,4 links between N-acetylhexosamine and D-glucuronate in hyaluronic acid, chondroitin, etc.	–
Endo	α-glucuronidase specific for hyaluronic acid	73
Endo	β-1,3 and β-1,4 links adjacent to α-1,3-link	–
Endo	β-1,4 links between GlcNAc and N-acetyl-muramic acid in peptidoglycans	–

CHEMICALS AND REAGENTS

Table 9. Continued

Enzyme	Source	EC number	Mol. wt	Optimum pH
α-mannosidase	Jack bean	3.2.1.24	–	4.0
Neuraminidase (sialidase)	*Arthrobacter ureafaciens* *Clostridium perfringens*	3.2.1.18	90 000 60 000	5.2 5.0
Novozym	*Trichoderma harzianum*	–	–	–
Novozym	*Aspergillus aculeatus*	–	–	–
Pullulanase	*Aerobacter aerogenes*	3.2.1.41	145 000	5.0

[a] Abbreviations: Endo, endoglycosidase; Exo, exoglycosidase.
[b] = maltase.

Table 10. Proteases

Enzyme	Source	EC number	Mol. wt	Optimum pH
Alkaline protease	*Streptomyces griseus*	–	–	–
Aminopeptidase M	Pig kidney	3.4.11.2	280 000	7.2
Ancrod	*Agkistrodon*	3.4.21.28	38 000	7.5
Bromelain	Pineapple	3.4.22.3	28 000	8.2
Carboxypeptidase A	Bovine pancreas	3.4.17.1	34 500	8.0
Carboxypeptidase B	Pig pancreas	3.4.17.2	34 300	8.0
Carboxypeptidase P	*Penicillium janthinellum*	3.4.16.1	51 000	3.7
Carboxypeptidase Y	*S. cerevisiae*	3.4.16.1	61 000	6.0
Cathepsin C	Bovine spleen	3.4.14.1	210 000	5.0
Chymotrypsin A$_4$ (α-chymotrypsin)	Bovine pancreas	3.4.21.1	25 000	8.0
Clostripain (clostridiopeptidase B)	*Clostridium histolyticum*	2.3.22.8	50 000	7.6
Collagenase	*Achromobacter iophagus*	3.4.24.8	70 000	7.2
Collagenase	*Clostridium histolyticum*	3.4.24.3	105 000	7.0
Collagenase	Leech	–	–	–

Type of enzyme[a]	Specificity	Refs
Exo	Broad for α-mannosides	–
Exo	NeuAc-α2,6-Gal, NeuAc-α2,6-GlcNAc or NeuAc-α2,3-Gal	–
–	Multienzyme (cellulase, chitinase, laminarinase, proteinase, xylanase) active against yeast cell walls	74
–	Multienzyme (pectolytic, cellulolytic, hemicellulolytic, proteolytic, saccharifying) active against plant cell walls	75,76
Endo	α-1,6-glucan links	–

Specificity	Refs
Broad specificity	77
Cleaves N-terminal residue from di- and tripeptides	–
Arg-X, especially arg-gly	–
Lys-X, ala-X, tyr-X	78
Cleaves C-terminal residue, preference for aromatic and aliphatic R-groups	–
Cleaves C-terminal arg, lys or ornithine	–
Cleaves C-terminal residue	–
Cleaves C-terminal residue, preference for aromatic and aliphatic R-groups	–
Cleaves N-terminal dipeptide, blocked by lys, arg and pro	–
Tyr-X, phe-X, trp-X; lower specificity for leu-X, met-X, ala-X, asp-X and glu-X	79
Arg-X	78
Predominantly X-gly of pro-X-gly-pro	–
Predominantly X-gly of pro-X-gly-pro	79
Degrades collagen to peptides	80,81

CHEMICALS AND REAGENTS

Table 10. Continued

Enzyme	Source	EC number	Mol. wt	Optimum pH
Dispase	*Bacillus polymyxa*	3.4.24.4	35 900	8.5
Elastase	Pig pancreas	3.4.21.36	25 900	8.5
Endoproteinase Arg-C (submaxillary protease)	Mouse submaxillary gland	3.4.21.40	25 000	8.2
Endoproteinase Glu-C (V8 protease)	*Staphylococcus aureus*	3.4.21.19	12 000	4.0 & 7.8
Endoproteinase Lys-C	*Lysobacter enzymogenes*	3.4.99.30	33 000	8.6
Factor Xa	Bovine plasma	3.4.21.6	55 000	8.3
Ficin	Fig tree	3.4.22.3	23 800	6.4
Kallikrein	Pig pancreas	3.4.21.8	28 000	7.5
Leucine amino-peptidase	Pig kidney	3.4.11.1	300 000	9.2
Papain	*Carica papaya*	3.4.22.2	23 000	6.5
Pepsin	Pig stomach	3.4.23.1	34 500	1.0
Plasmin	Bovine plasma	3.4.21.7	85 000	8.9
Pronase	*Streptomyces griseus*	mixture	–	~7.0
Proteinase K	*Tritirachium album*	3.4.21.14	27 000	~8.0
Pyroglutamate aminopeptidase	Calf liver	3.4.11.8	75 000	8.0
Subtilisin	*Bacillus subtilis*	3.4.21.14	27 600	~8.0
Thermolysin	*Bacillus thermoproteo-lyticus*	3.4.24.4	35 000	8.0
Thrombin	Bovine plasma	3.4.21.5	34 000	8.6
Trypsin	Bovine pancreas	3.4.21.4	23 500	8.0

Specificity	Refs
Non-specific	–
Ala-X, gly-X; lower specificity for val-X, leu-X, ile-X and ser-X	78
Arg-X	78
Glu-X, lower specificity for asp-X	78
Lys-X	78
Arg-X of gly-arg-X	–
Bonds involving uncharged and/or aromatic amino acids	78
Arg-X, preferentially of phe-arg-X or leu-arg-X	–
Cleaves N-terminal residue	–
Bonds involving arg, lys, glu, his, gly and/or tyr	78
Broad specificity but preference for phe-X, met-X, leu-X and trp-X where X is another hydrophobic residue	78
Arg-X, lys-X	78
Broad spectrum	–
Broad spectrum	82–84
Cleavage of N-terminal pyroglutamate residue	–
Broad spectrum	–
X-leu, X-phe and other non-polar residues	78
Arg-X	78
Arg-X, lys-X	78

Table 11. Ribonucleases used in enzymatic sequencing of RNA

Enzyme	Source	Specificity	Refs
B. cereus RNase	*Bacillus cereus*	U↑pN, C↑pN	85
RNase Ch3	Chicken liver	Cp↑N	86
RNase M1	*Cucumis melo*	N↑pA, N↑pU, N↑pG	–
RNase Phy1	*Physarum polycephalum*	Gp↑N, Ap↑N, Up↑N	87
RNase PhyM	*Physarum polycephalum*	Up↑N, Ap↑N	88
RNase T1	*Aspergillus oryzae*	Gp↑N	89
RNase T2	*Aspergillus oryzae*	Ap↑N > all others	89
RNase U2	*Ustilago sphaerogema*	Ap↑N > Gp↑N	89
RNase V1	Cobra venom	Unknown	–
S. aureus RNase	*Staphylococcus aureus*	pH 7.5: A↑pN, U↑pN pH 3.5, no Ca^{2+}: C↑pN, U↑pN	–

Table 12. Nucleotide bases and nucleosides: physical and chemical data

	Mol. wt[a]	pK_a	Solubility (25°C, g/100 ml H_2O)
Nucleotide bases			
Adenine	135.1	<1.0, 4.1, 9.8	0.09
Cytosine	111.1	4.4, 12.2	0.77
Guanine	151.1	3.3, 9.2, 12	0.004 (40°C)
5-hydroxymethylcytosine	141.1	4.3, 13	sl. sol.[c]
Hypoxanthine	136.1	2.0, 8.9, 12.1	0.07 (19°C)
5-methylcytosine	125.1	4.6, 12.4	0.45
7-methylguanine	165.2	<1, 3.5, 9.9	–
Orotic acid	156.1	2.4, 9.5, 13.0	0.18 (18°C)
Thymine	126.1	~0, 9.9, >13.0	0.4
Uracil	112.1	0.5, 9.5, >13.0	0.36
Uric acid	168.1	5.4, 11.3	0.002 (20°C)
Xanthine[b]	152.1	0.8, 7.4, 11.1	0.05 (20°C)
Nucleosides			
Adenosine	267.2	3.5, 12.5	sol.
Cytidine	243.2	4.2, 12.3	v. sol.
2'-deoxyadenosine	251.2	3.8	sol.
2'-deoxycytidine	227.2	4.2, >13	–
2'-deoxyguanosine	267.2	2.5	sol.
2'-deoxythymidine	242.2	9.8, >13	–
Guanosine	283.2	1.6, 9.2, 12.4	0.08 (18°C)
5-hydroxymethyl-2'-deoxycytidine	257.2	3.5	–
Inosine	268.2	1.2, 8.9, 12.5	1.6 (20°C)
5-methylcytidine	257.2	4.3, >13	–
5-methyl-2'-deoxycytidine	241.2	4.4, >13	sol.
7-methylguanosine	297.3	7.0	–
Orotidine	288.2	–	sol.
Pseudouridine	244.2	9.0, >13	–
Uridine	244.2	9.2, 12.5	sol.

[a] Molecular weights of compounds refer to the anhydrous forms.
[b] Xanthine: R 28.
[c] sl. sol., slightly soluble.

Table 13. Unmodified nucleotides: physical and chemical data

Compound[a]	Mol. wt[b]	pK_a	Compound[a]	Mol. wt[b]	pK_a
Nucleotide-5'-phosphates			**2'-deoxynucleotide-5'-phosphates**		
AMP	347.2	3.7, 5.9	dAMP	331.2	4.4, 6.4
ADP	427.2	3.8, 6.1–6.7	dADP	411.2	–
ATP	507.2	4.1, 6.0–7.0	dATP	491.2	4.8, 6.8
CMP	323.2	4.4, 6.3	dCMP	307.2	4.6, 6.6, 13.2
CDP	403.2	4.6, 6.4	dCDP	387.2	–
CTP	483.2	4.8, 6.6	dCTP	467.2	–
GMP	363.2	2.4, 6.1, 9.4	dGMP	347.2	2.9, 6.4, 9.7
GDP	443.2	<1, 2.9, 6.3, 9.6	dGDP	427.2	–
GTP	523.2	<1, 3.3, 6.3, 9.3	dGTP	507.2	3.5, 6.5, 9.7
UMP	324.2	1, 6.4, 9.4	dTMP	322.2	1.6, 6.5, 10.0
UDP	404.2	–	dTDP	402.2	–
UTP	484.2	1, 6.6, 9.6	dTTP	482.2	–

[a] All these compounds are soluble or very soluble in aqueous buffers.
[b] Molecular weights refer to anhydrous compounds.

Figure 6. Structures of nucleotide bases, nucleosides and unmodified nucleotides.

Bases

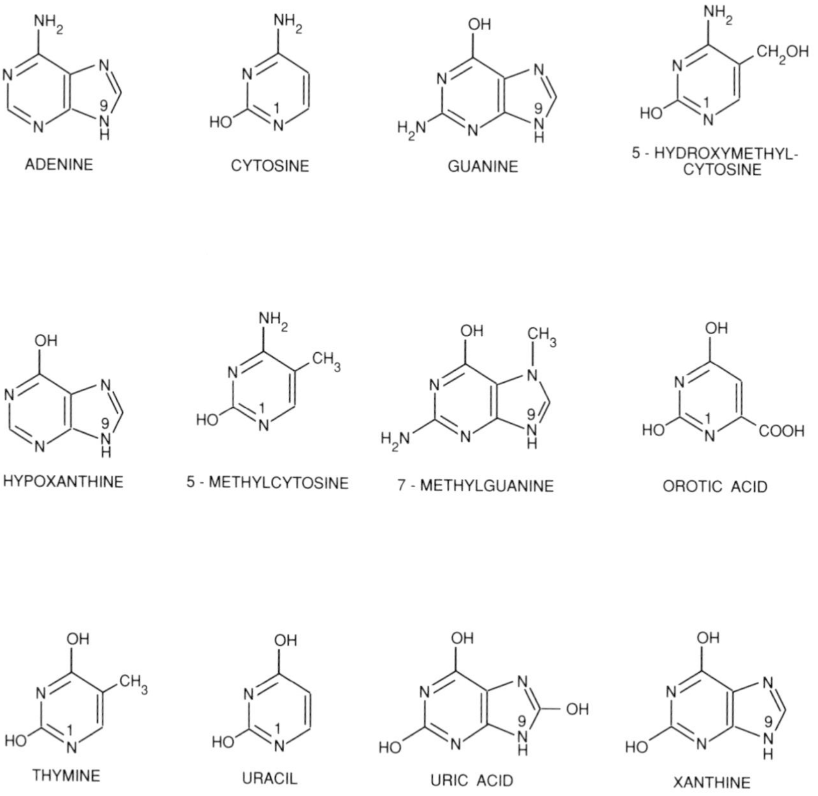

ADENINE   CYTOSINE   GUANINE   5 - HYDROXYMETHYL-CYTOSINE

HYPOXANTHINE   5 - METHYLCYTOSINE   7 - METHYLGUANINE   OROTIC ACID

THYMINE   URACIL   URIC ACID   XANTHINE

Figure 6. Continued

Nucleosides

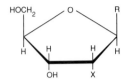

BASIC NUCLEOSIDE STRUCTURE

R = nucleotide base attached by
N^1 (pyrimidines) or N^9 (purines)

• ADENOSINE	R = adenine	X = OH
• CYTIDINE	R = cytosine	X = OH
• 2' - DEOXYNUCLEOSIDE	R = relevant base	X = H
• GUANOSINE	R = guanine	X = OH
• 5 - HYDROXYMETHYL - 2' - DEOXYCYTIDINE	R = 5-hydroxymethylcytosine	X = H
• INOSINE	R = hypoxanthine	X = OH
• 5 - METHYLCYTIDINE	R = 5-methylcytosine	X = OH
• 5 - METHYL - 2' - DEOXYCYTIDINE	R = 5-methylcytosine	X = H
• 7 - METHYLGUANOSINE	R = 7-methylguanine	X = OH
• OROTIDINE	R = orotic acid	X = OH
• PSEUDOURIDINE	R = uracil (attached to C^5)	X = OH
• URIDINE	R = uracil (attached to N^1)	X = OH

Nucleotides

BASIC NUCLEOTIDE STRUCTURE

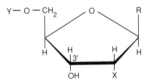

• R = nucleotide base attached by
N^1 (pyrimidines) or N^9 (purines)

• X = OH for nucleotides, H for 2'-deoxynucleotides

• For monophosphates Y =

$$^-O-\overset{\overset{\textstyle O^-}{|}}{\underset{\underset{\textstyle O}{\|}}{P}}-$$

• For diphosphates Y =

$$^-O-\overset{\overset{\textstyle O^-}{|}}{\underset{\underset{\textstyle O}{\|}}{P}}-O-\overset{\overset{\textstyle O^-}{|}}{\underset{\underset{\textstyle O}{\|}}{P}}-$$

• For triphosphates Y =

$$^-O-\overset{\overset{\textstyle O^-}{|}}{\underset{\underset{\textstyle O}{\|}}{P}}-O-\overset{\overset{\textstyle O^-}{|}}{\underset{\underset{\textstyle O}{\|}}{P}}-O-\overset{\overset{\textstyle O^-}{|}}{\underset{\underset{\textstyle O}{\|}}{P}}-$$

• For dideoxynucleotides, X=H, but the -OH attached to the 3'-C is replaced with -H.

• For 3'-deoxynucleotides, X = OH, but the -OH attached to the 3'-C is replaced with -H.

Table 14. Modified nucleotides and related compounds: physical and chemical data

Compound[a]	Mol. wt
3'-5'-cyclic AMP (cAMP)	329.2
Coenzyme A	767.6
3'-dATP (cordycepin 5'-triphosphate)	491.2
3'-dCTP	467.2
3'-dGTP	507.2
3'-dUTP	468.2
7-deaza-dGTP	521.2
7-deaza-dITP	506.2
2'-deoxyinosine-5'-triphosphate (dITP)	492.2
2'-deoxyuridine-5'-triphosphate (dUTP)	468.2
2',3'-dideoxynucleotides	
ddATP	475.2
ddCTP	451.2
ddGTP	491.2
ddTTP	466.2
Inosine-5'-triphosphate (ITP)[b]	508.2
Nicotinamide adenine dinucleotide (NAD)	663.4
Nicotinamide adenine dinucleotide, reduced (NADH)	665.4
Nicotinamide adenine dinucleotide phosphate (NADP)	743.4
Nicotinamide adenine dinucleotide phosphate, reduced (NADPH)	745.4
RNA cap structure analogs:	
G(5')ppp(5')A (Na salt)	?
G(5')ppp(5')G	?
7mG(5')ppp(5')A (diNa salt.2H$_2$O)	?
S-adenosylmethionine (SAM)	399.4

[a] All these compounds are soluble or very soluble in aqueous buffers.
[b] ITP: pK_a 6.9.

Figure 7. Structures of common modified nucleotides.

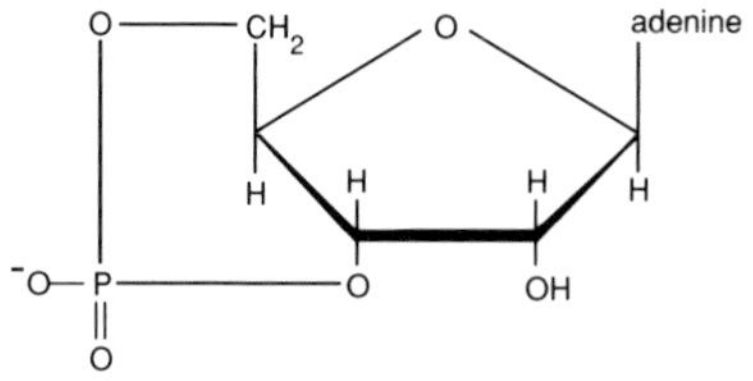

3' - 5' - CYCLIC AMP

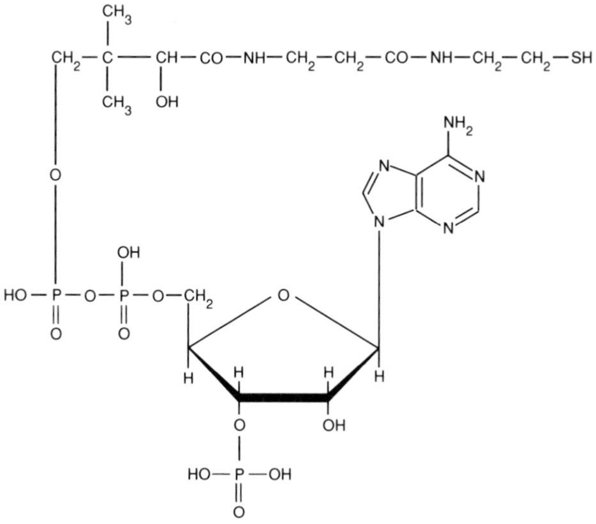

COENZYME A

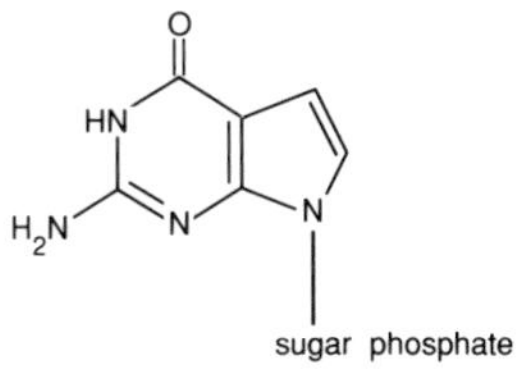

7 - DEAZA - dGTP

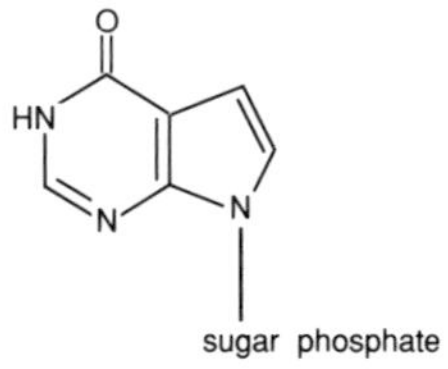

7 - DEAZA - dITP

3'-dNTPs, dITP, dUTP, ddNTPs and ITP - see Figure 6

Figure 7. Continued.

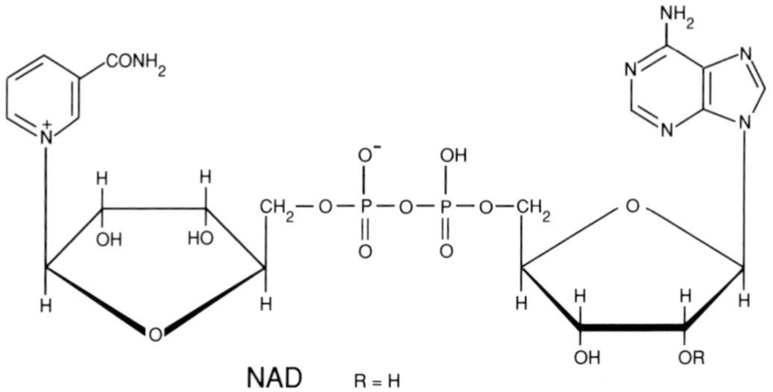

NAD   R = H

NADP R = —P=O (with OH, OH)

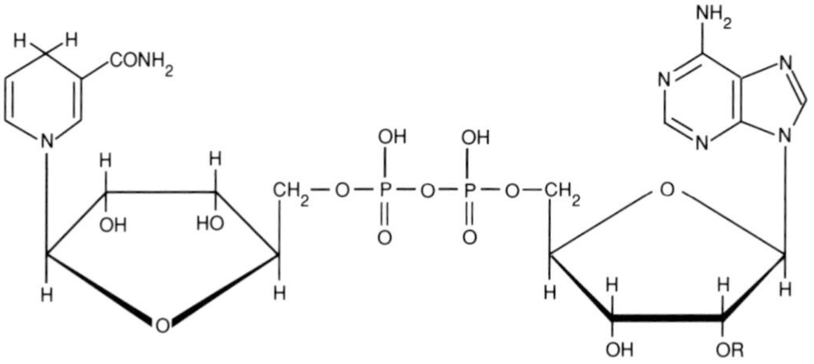

NAD   R = H

NADPH R = —P=O (with OH, OH)

5'-5' phosphodiester bond

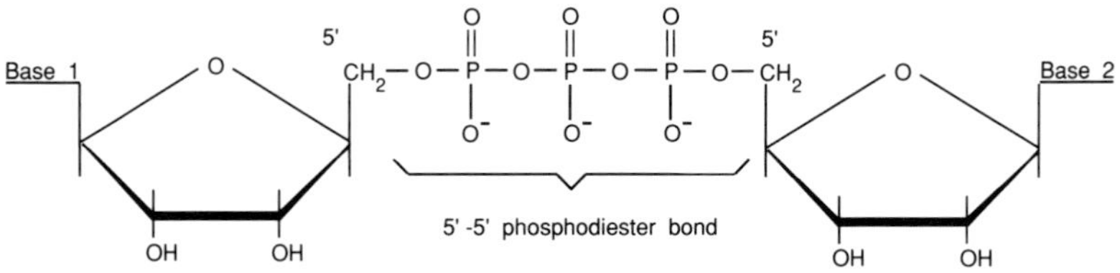

RNA CAP STRUCTURE ANALOGS

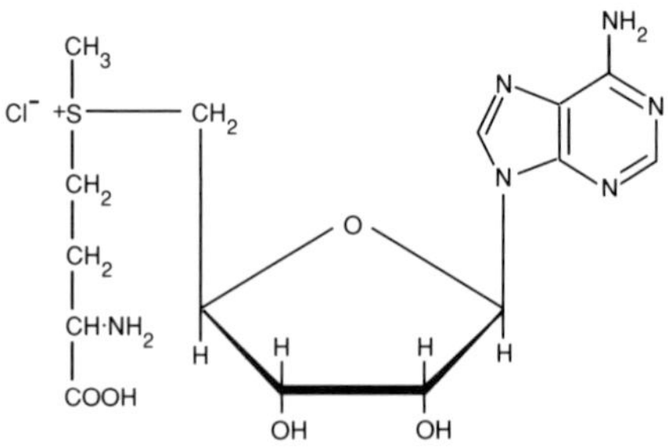

SAM   [S - adenosylmethionine]

Table 15. Organic chemicals in frequent use: physical and chemical data

Compound	Mol. wt	m.p. (°C)	b.p. (°C)	Flash point (°C)	Specific gravity	Risk	Safety
Allyl alcohol (2-propen-1-ol)	58.1	−129	97	21	0.85	11-26-36/37/38	16-39-45
n-butanol (butan-1-ol, butyl alcohol)	74.1	−89	118	29	0.81	10-18-20	16
Chloroform	119.4	−63	61	–	1.48	20	2-24/25
N,N-dimethylformamide (DMF)	73.1	−61	153	57	0.95	20/21-36	26-28-36
Dimethyl sulfate (DMS)	126.1	−27	188	83	1.33	26/27-40	7/9-24/25-26-45
Dimethyl sulfoxide (DMSO)	78.1	18	190	95	1.10	22	24/25
Ethanol (ethyl alcohol)	46.1	−117	78	13	0.8	11	7-16
Formaldehyde	30.0	–	96[a]	49[a]	1.08[a]	10-23/24/25-43	2-26-28
Formamide	45.1	2.5	210	154	1.13	23/24/25	23-24/25
Glycerol	92.1	18	290	160	1.26	–	–
Glyoxal (oxaldehyde)	58.0	15	51	–	1.27[b]	36/38	26-28
Hydrazine	50.1	−52	113	75	1.03	24/25-34-40	23-26-36/37/39
Indole-3-acetic acid (IAA)	175.2 (.H2O)	165–169	–	–	–	–	–
Isoamyl alcohol (3-methyl-1-butanol)	88.2	−117	130	45	0.81	<	irritant >
Isopropanol (propan-2-ol, isopropyl alcohol)	60.1	−89.5	82.4	22	0.79	11	7-16-33
2-mercaptoethanol (β-mercaptoethanol)	78.1	–	54	73	1.11	22-36/38	23
Phenol	94.1	41	182	80	1.07	24/25-34	2-28-44
Piperidine	85.2	−11	106	3	0.86	11-23/24/25-34	16-26-27-44
Siliconizing solution (dimethyldichlorosilane, 2% in 1,1,1-trichloroethane)	129.06	–	–	–	1.31	20/22	2-24/25
TEMED (N,N,N′,N′-tetramethyl-ethylenediamine)	116.2	−55	122	21	0.77	11-36/38	–

[a] 37.4% solution.   [b] 40% solution.

Chemicals/Reagents

Table 16. Polymers: physical and chemical data

Compound[a]		Mol. wt	Solubility in H_2O	Refs
Dextran sulfate		~500 000	–	90, 91
Ficoll 400		~400 000	50% w/v	92
PEG series (polyethylene glycol)	PEG400	340–420	v. sol.	–
	PEG4000	3500–4500	v. sol.	
Polymin P (polyethyleneimine)		30 000–40 000	v. sol.	93, 94
PVP (polyvinylpyrrolidone)		10 000–360 000	sol.	–

[a] Dextran sulfate is a highly branched glucose polymer. Ficoll 400 is a highly branched copolymer of sucrose and epichlorohydrin. For other structures see *Figure 8*.

Figure 8. Structures of polymers.

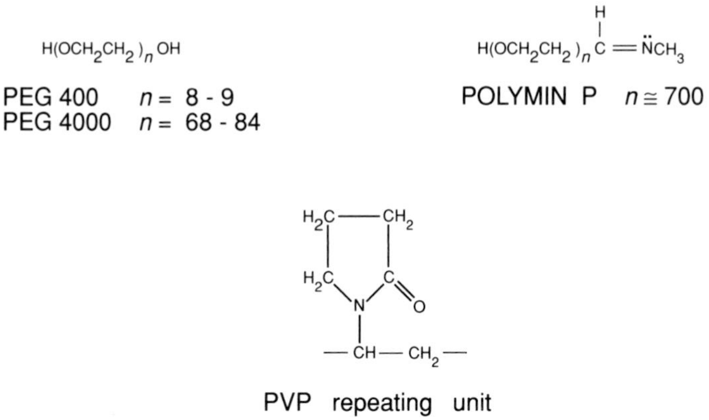

H(OCH2CH2)nOH
PEG 400 n = 8 - 9
PEG 4000 n = 68 - 84
H(OCH2CH2)nC=NCH3
POLYMIN P n ≅ 700
PVP repeating unit

Table 17. Protease inhibitors: physical, chemical and biological data

Inhibitor	Mol. wt	Solubility in H_2O	Risk	Safety	Specificity (50% inhibition level, $\mu g\ ml^{-1}$) Active against	Specificity (50% inhibition level, $\mu g\ ml^{-1}$) Inactive against	Refs
Antipain	604.7	sol.	–	–	papain (0.16) trypsin (0.26) cathepsin A (1.19) cathepsin B (0.59)	cathepsin D (125) plasmin (>93) chymotrypsin (>250) pepsin (>250)	95, 96
APMSF[a]	252.7 (.HCl)	–	–	–	serine proteases	chymotrypsin	78
Aprotinin	~6500	–	–	–	kallikrenin trypsin chymotrypsin plasmin	papain	–
Bestatin	308.2	–	–	–	aminopeptidase B leu aminopeptidase	aminopeptidase A trypsin chymotrypsin elastase papain pepsin thermolysin	97
Chymostatin	582.7	sl. sol.	–	–	chymotrypsin (0.15) papain (7.5) cathepsin B (2.6)	plasmin (>250) cathepsin A (62.5) cathepsin D (49.0) trypsin (>250)	89, 99
E-64	–	–	–	–	Thiol proteases: ficin papain bromelain		78

Table 17. Continued

Inhibitor	Mol. wt	Solubility in H$_2$O	Risk	Safety	Specificity (50% inhibition level, µg ml^{-1})		Refs
					Active against	Inactive against	
Elastatinal	512.6	sol.	–	–	elastase (0.29)	others (>250)	100, 101
Iodoacetic acid	186.0	sol.	26/27/28-35	22-36/37/39-45	< not very specific >		–
Leupeptins	426.6	sol.	–	–	plasmin (8.0) trypsin (2.0) papain (0.5) cathepsin B (0.44)	chymotrypsin (>500) pepsin (>500) cathepsin A (1680) cathepsin D (109)	102
Ovomucoid	~27 000	–	–	–	trypsin		103
Pepstatin A	685.9	sl. sol.	–	–	pepsin cathepsin D	thermolysin trypsin plasmin chymotrypsin elastase	104
Phosphoramidon	543.6	sol.	–	–	thermolysin (0.4)	collagenase (33) trypsin (>250) chymotrypsin (>250) pepsin (>250) papain (>250)	105
PMSF[a]	174.2	sl. sol.	20/21/22	26-28	chymotrypsin trypsin		–
TLCK[a]	332.9	sl. sol.	–	–	trypsin	chymotrypsin	106
TPCK[a]	351.9	insol., sol. in MeOH	–	–	chymotrypsin	trypsin	107, 108

[a] Abbreviations: APMSF, (4-amidinophenyl)methane sulfonyl fluoride; PMSF, phenylmethyl sulfonyl fluoride; TLCK, tosyllysine chloromethyl ketone, L-1-chloro-3-tosyl-amido-7-amino-2-heptanone; TPCK, tosylphenylalanine chloromethyl ketone, L-1-chloro-3-tosyl-amido-4-phenyl-2-butanone.

Figure 9. Structures of protease inhibitors.

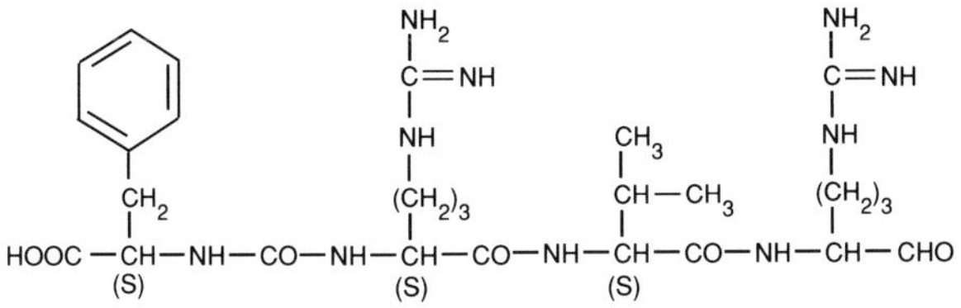

ANTIPAIN

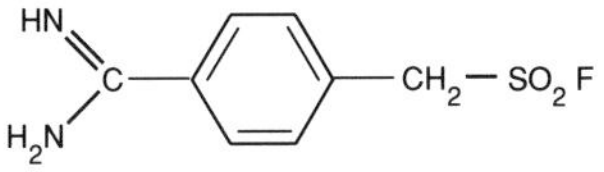

APMSF

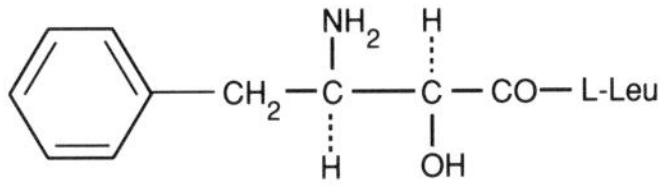

BESTATIN

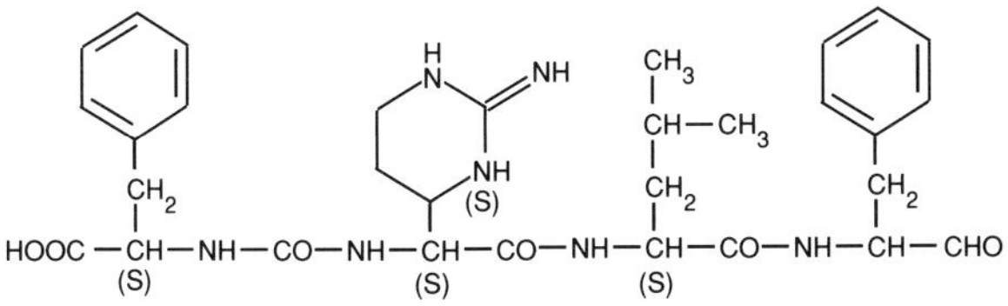

CHYMOSTATIN

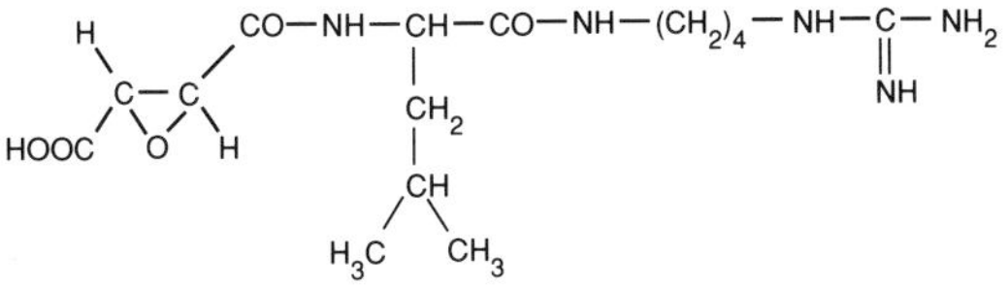

E-64

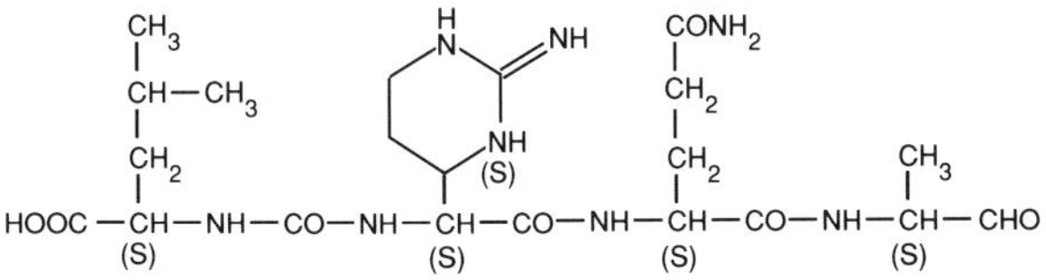

ELASTATINAL

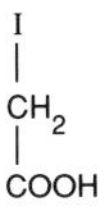

IODOACETIC ACID

Figure 9. Continued.

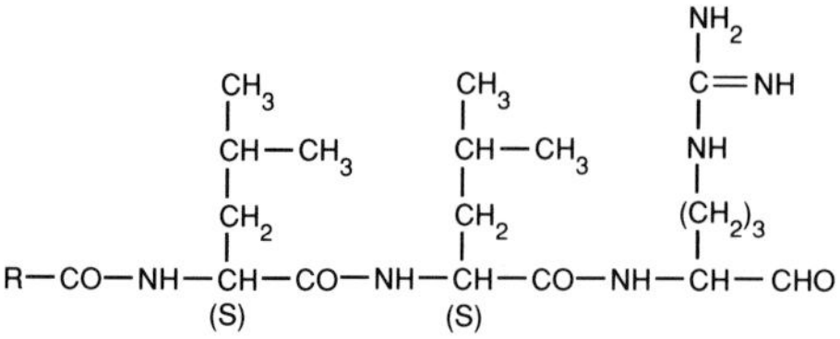

LEUPEPTIN

$R = CH_3-,\ CH_3CH_2-$

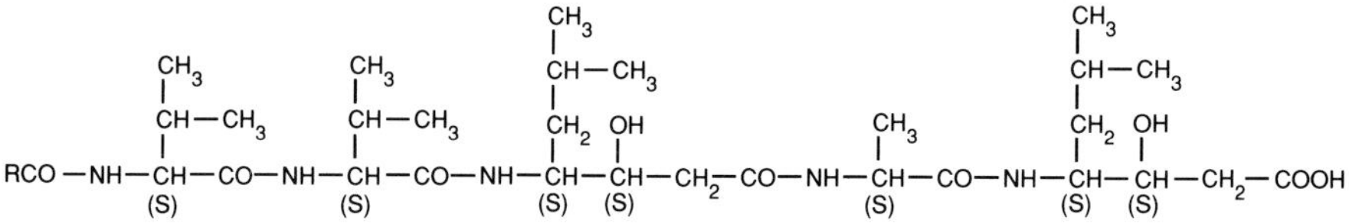

PEPSTATIN

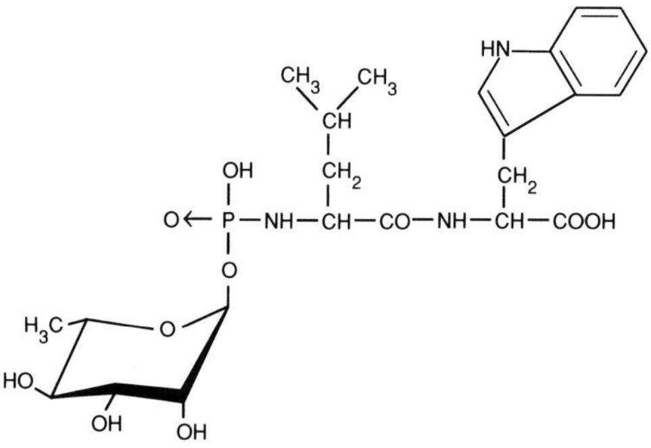

PHOSPHORAMIDON

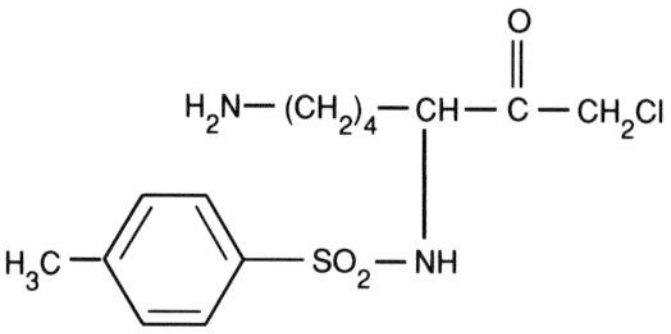

TLCK

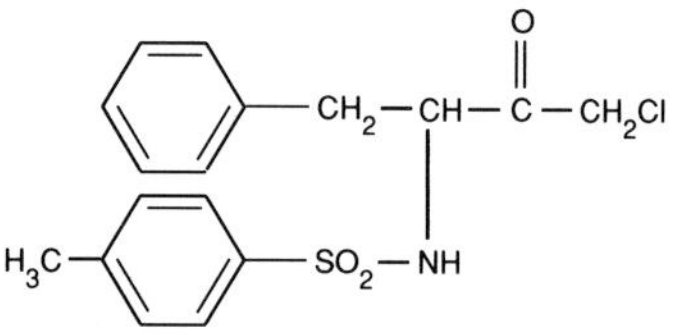

TPCK

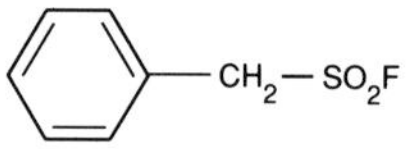

PMSF

Table 18. Proteins: physical, chemical and biological data

Protein	Source	Approx. mol. wt	Form	Use	Refs
Albumin (BSA)	bovine serum	67 000	mixed	Used to stabilize enzymes during purification and dilution	–
Avidin	egg white	66 000	tetramer	Binds biotin and biotinylated derivatives, typically 10–15 µg d-biotin per mg protein	109
Protein A	*Staphylococcus aureus*	42 000	single polypeptide	Binds the Fc fragment of IgG, typically 12–14 mg of human IgG per mg protein	110
Streptavidin	*Streptomyces avidini*	58 000	tetramer	Binds biotin and biotinylated derivatives, typically 10–15 µg d-biotin per mg protein; less non-specific binding than egg white avidin	109, 111

Table 19. Ribonuclease inhibitors: physical, chemical and biological properties

Compound	Form	Activity	Risk	Safety	Refs
Diethyl pyrocarbonate[a]	162.2 dalton organic chemical	Ethoxyformylates proteins	20	24/25	112
Human placental RNase inhibitor[b]	51 000 dalton protein	Non-competitive protein–protein inhibition	–	–	113, 114
Vanadyl ribonucleoside complex	Oxovanadium ions complexed with mixed ribonucleosides	Transition state analog		Avoid contact	115

[a] Structure:

$$H_3C-CH_2-O-\overset{\overset{O}{\|}}{C}-O-\overset{\overset{O}{\|}}{C}-O-CH_2-CH_3$$

[b] Marketed under various trade names, e.g. RNAsin (Promega), RNase block (New England Biolabs).

Chemicals/Reagents

Table 20. Salts: physical and chemical data

Compound	Formula	Mol. wt	Solubility (g/100 ml H$_2$O)		Risk	Safety
			cold (°C)	hot (°C)		
Ammonium acetate	CH$_3$COONH$_4$	77.08	148 (4)	decomposes	—	—
Ammonium chloride	NH$_4$Cl	53.49	29.7 (0)	75.8 (100)	—	—
Ammonium nitrate	NH$_4$NO$_3$	80.04	118.3 (0)	871 (100)	8-9	24/25
Ammonium persulfate (AMPS, ammonium peroxydisulfate)	(NH$_4$)$_2$S$_2$O$_8$	228.20	58.2 (0)	v. sol.	8-22	24/25
Ammonium sulfate	(NH$_4$)$_2$SO$_4$	132.13	70.6 (0)	103.8 (100)	—	—
Calcium chloride, dihydrate	CaCl$_2$.2H$_2$O	147.02	97.7 (0)	326 (60)	—	—
Calcium chloride, hexahydrate	CaCl$_2$.6H$_2$O	219.08	279 (0)	536 (20)	—	—
Calcium hypochlorite	Ca(OCl)$_2$	142.99	sol.	sol.	8-31-34	2-26-43
Lithium chloride	LiCl	42.39	63.7 (0)	130 (95)	22	24/25
Magnesium acetate, tetrahydrate	(CH$_3$COO)$_2$Mg.4H$_2$O	214.40	120 (15)	v. sol.	—	—
Magnesium chloride, hexahydrate	MgCl$_2$.6H$_2$O	203.30	167 (0)	367 (100)	—	—
Magnesium nitrate, hexahydrate	MgN$_2$O$_6$.6H$_2$O	256.41	125 (20)	v. sol.	—	—
Magnesium sulfate, heptahydrate	MgSO$_4$.7H$_2$O	246.47	71 (20)	91 (40)	—	—
Manganous chloride, anhydrous	MnCl$_2$	125.80	72.3 (25)	123.8 (100)	—	—
Manganous chloride, tetrahydrate	MnCl$_2$.4H$_2$O	197.90	151 (8)	656 (100)	—	—
Manganous sulfate, monohydrate	MnSO$_4$.H$_2$O	169.01	98.5 (48)	79.8 (100)	—	—
Manganous sulfate, heptahydrate	MnSO$_4$.7H$_2$O	223.06	172	—	—	—
Potassium acetate	CH$_3$COOK	98.14	253 (20)	492 (62)	—	—

Table 20. Continued

Compound	Formula	Mol. wt	Solubility (g/100 ml H_2O) cold (°C)	Solubility (g/100 ml H_2O) hot (°C)	Risk	Safety
Potassium chloride	KCl	74.55	34.7 (20)	56.7 (100)	—	—
Potassium iodide	KI	166.00	127.5 (0)	208.0 (100)	—	—
Potassium nitrate	KNO_3	101.10	13.3 (0)	247 (100)	8-12	—
Potassium permanganate	$KMnO_4$	158.03	6.4 (20)	25 (65)	8-22	2-24/25-27
Potassium sodium tartrate, tetrahydrate	KOCOCH(OH)CH(OH)COONa.4H_2O	282.22	26 (0)	66 (25)	—	—
Potassium sulfate	K_2SO_4	174.25	12 (25)	24.1 (100)	—	—
Sodium chloride	NaCl	58.44	35.7 (0)	39.1 (100)	—	—
Sodium metabisulfite	$Na_2S_2O_5$	190.10	54 (20)	81.7 (100)	31	—
Sodium nitrate	$NaNO_3$	84.99	92.1 (25)	180.0 (100)	8-22	—
Sodium nitrite	$NaNO_2$	69.00	81.5 (15)	163 (100)	8-25-31	44
Sodium salicylate	C_6H_4(OH).COONa	160.10	111 (15)	125 (25)	—	—
Sodium succinate, hexahydrate	$(CH_2COONa)_2$.6H_2O	270.14	21.5 (0)	86.6 (75)	—	—
Sodium sulfate, anhydrous	Na_2SO_4	142.04	4.7 (0)	42.7 (100)	—	—
Sodium sulfate, decahydrate	Na_2SO_4.10H_2O	322.19	11 (0)	92.7 (30)	—	—
Tetramethyl ammonium chloride	$(CH_3)_4$NCl	109.60	—	—	—	—
Zinc chloride	$ZnCl_2$	136.29	432 (25)	615 (100)	34	7/8-28
Zinc sulfate, heptahydrate	$ZnSO_4$.7H_2O	287.54	96.5 (20)	663.6 (100)	—	22-24/25

Table 21. Stains, indicators and tracker dyes: physical, chemical and biological data

Compound[a]	Mol. wt	Application	Risk	Safety	Refs
Acridine orange	301.8	DNA, RNA stain in gel electrophoresis	–	–	116
Bisbenzimide H33258	533.9	Fluorescent DNA stain in density gradients; chromosome stain	–	–	117–119
H33342	615.9				
Bromocresol green	698.0	Gel electrophoresis marker dye	–	–	120
Bromophenol blue	670.0	Gel electrophoresis marker dye	–	–	–
Coomassie Blue R-250[b]	826.0	Protein stain in gel electrophoresis	–	–	121–123
DAPI	350.3	Fluorescent stain for AT-rich regions of DNA in density gradients	–	–	124,125
Ethidium bromide	394.3	Fluorescent dye for DNA/RNA in density gradients and electrophoresis gels	20/22	24/25-26-28	126–128
FITC	389.4	Fluorescent label for antigen–antibody complexes	$<$ irritant	$>$	129–133
Xylene cyanol	554.6	Gel electrophoresis marker dye	–	22-24/25	134

^a Chemical names are as follows:

Acridine orange, 3,6-bis(dimethylamino)acridine hydrochloride; Bisbenzimide (Hoechst 33258), 2'-(4-hydroxyphenyl)-5-(4-methyl-1-piperazinyl)-2,5'-bi-1H-benzimidazole; Bisbenzimide (Hoeschst 33342), 2'-(4-ethoxyphenyl)-5-(4-methyl-1-piperazinyl)-2,5'-bi-1H-benzimidazole; Bromocresol green, 3',3'',5',5''-tetrabromo-m-cresolsulfonephthalein; Bromophenol blue, 3',3'',5',5''-tetrabromophenolsulfonephthalein; DAPI, 4',6-diamidino-2-phenylindole dihydrochloride; Ethidium bromide, 2,7-diamino-10-ethyl-9-phenylphenanthridium bromide; FITC, fluorescein isothiocyanate.

^b ICI Ltd, which holds the trade mark 'Coomassie' no longer produces Coomassie Blue R-250 and so other manufacturers introduced their own brand names, such as Brilliant Blue.

Figure 10. Structures of common stains, indicators and tracking dyes.

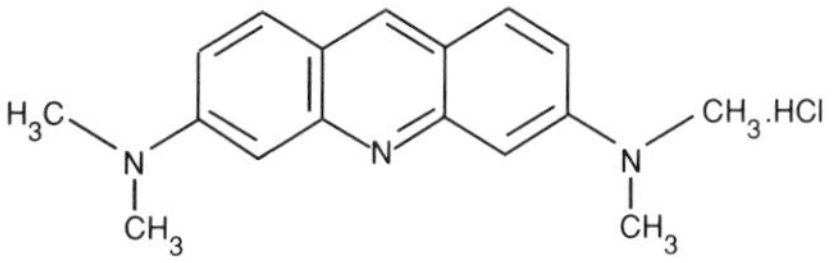

ACRIDINE ORANGE

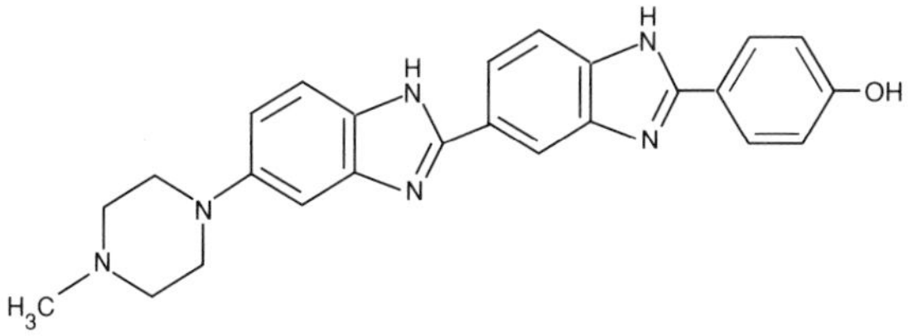

BISBENZIMIDE   H33258

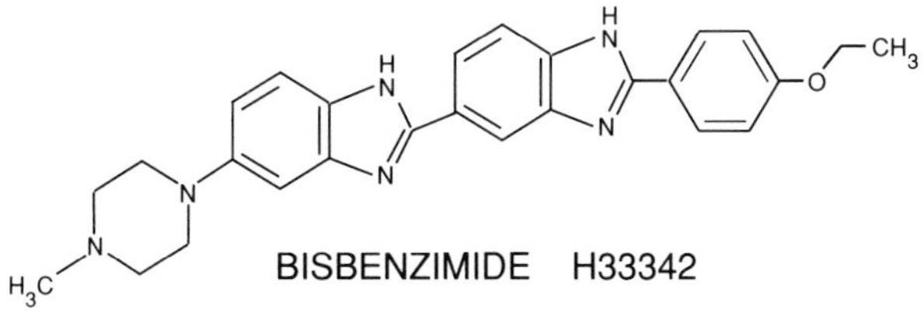

BISBENZIMIDE   H33342

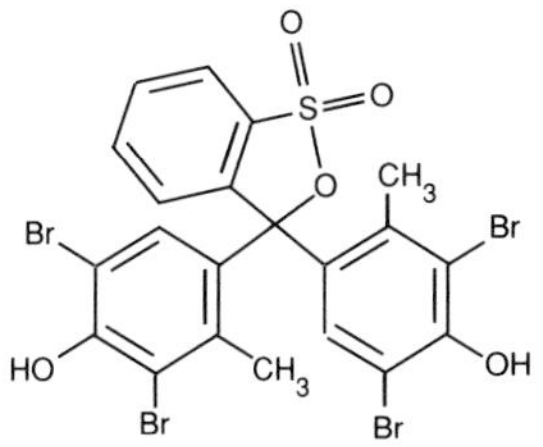

BROMOCRESOL GREEN

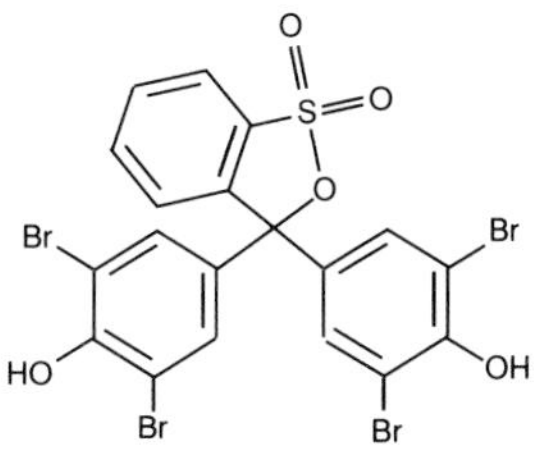

BROMOPHENOL BLUE

Figure 10. Continued.

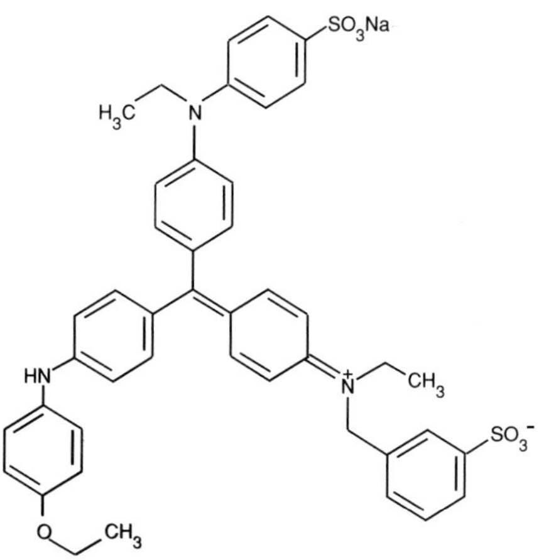

COOMASSIE BLUE
R 250

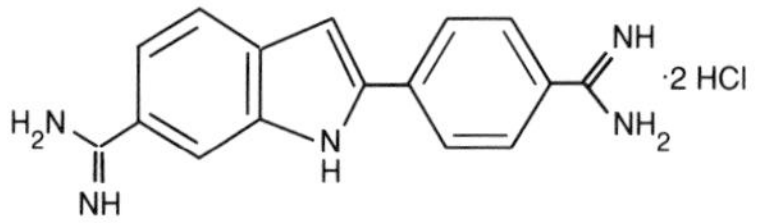

DAPI

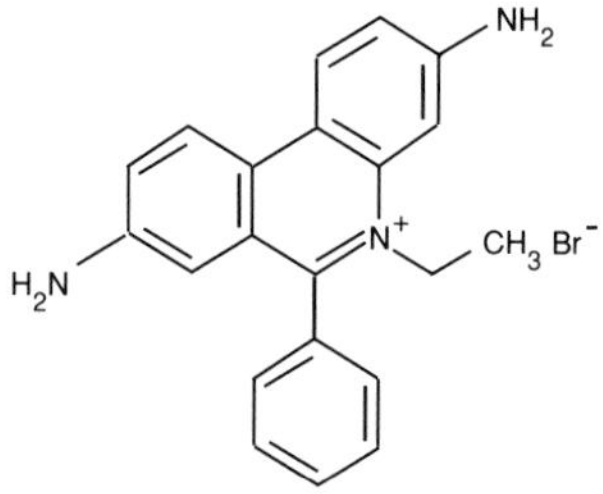

ETHIDIUM BROMIDE

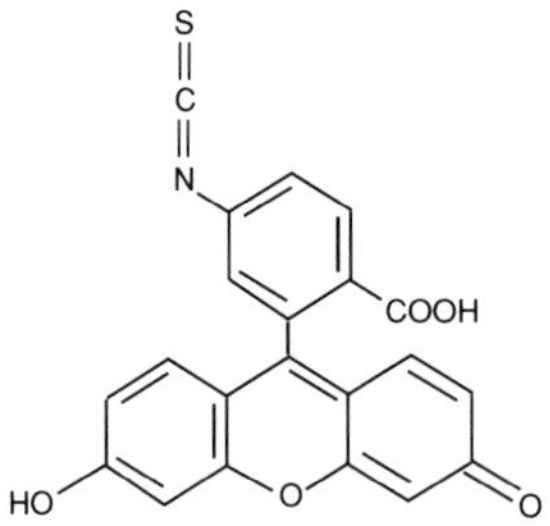

FITC

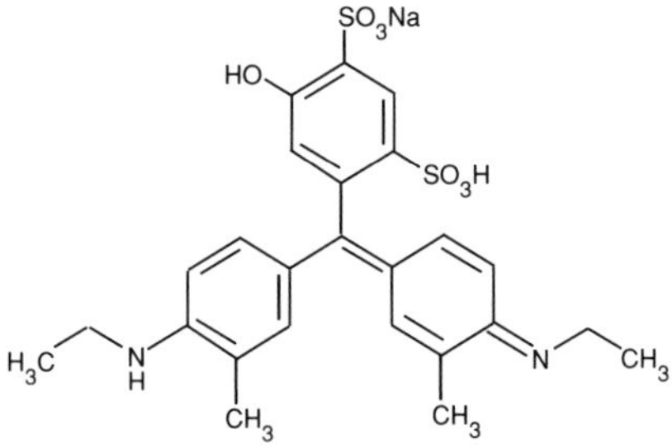

XYLENE CYANOL

Table 22. Enzyme substrates: physical, chemical and biological data

Substrate	Mol. wt	Assay[a]	Risk	Safety	Refs
Alkaline phosphatase					
5-bromo–4-chloro–3-indolyl phosphate (X-phos, BCIP)	433.6 (*p*-toluidine salt)	C	–	–	–
4-methylumbelliferyl phosphate (4-MUP)	300.1 (.2Na)	F	–	–	135
Nitroblue tetrazolium[b]	817.7	C	22	24/25	–
p-nitrophenylphosphate (PNPP)	371.1 (.2Na)	–	–	–	136
β-galactosidase					
5-bromo–4-chloro–3-indolyl-β-D-galactopyranoside (X-gal)	408.6	C	–	–	137
5-bromo–3-indolyl-β-D-galactopyranoside (Indigal)[c]	374.0	C	–	–	138, 139
Halogenated indolyl-β-D-galactopyranoside (Bluo-gal)[d]	353.3	C	–	–	–
4-methylumbelliferyl-α-D-galactopyranoside	338.3	F	–	–	140
o-nitrophenyl-β-D-galactopyranoside (ONPG)	301.2	–	–	–	141
p-nitrophenyl-β-D-galactopyranoside (PNPG)	301.2	S	–	–	141
Isopropyl-β-D-thiogalactopyranoside (IPTG)	238.3	I	–	–	142
β-glucuronidase					
5-bromo–4-chloro–3-indolyl-β-D-glucuronic acid (X-glu)	422.6	H	–	–	–
4-methylumbelliferyl-β-D-glucuronide (MUG)	352.3	F	–	–	–
p-nitrophenyl-β-D-glucuronide (PNPGlu)	315.2	S	–	–	–
resorufin glucuronide	–	H	–	–	143
trifluoromethylumbelliferyl glucuronide	–	H	–	–	143

Table 22. Continued

Substrate	Mol. wt	Assay[a]	Risk	Safety	Refs
Peroxidase					
2,2'-azino-di(3-ethylbenz-thiazoline sulfonic acid) (ABTS)	548.7	C	–	–	144
4-chloro-1-naphthol (4CN)	178.6	C	36/37/38	26	145
3',3-diaminobenzidine tetrahydrochloride (DAB)	360.1	C	20/21/22	22-23/25	146
o-phenyldiamine dihydrochloride (OPD)	181.1	C	23/24/25	28-44	–
3,3',5,5'-tetramethylbenzidine (TMB)	240.4	C	20/21/22	22-36	–

[a] C = colorimetric, F = fluorometric, S = spectrophotometric, I = inducer, H = histochemical.
[b] Nitroblue tetrazolium = 3,3'-(3,3'-dimethoxy-4,4'-biphenylylene)-bis-2-(p-nitrophenyl)-5-phenyl-2H-tetrazolium chloride.
[c] Marketed by United States Biochemical Corp.
[d] Marketed by GIBCO–BRL.

Figure 11. Structures of substrates for enzyme assays.

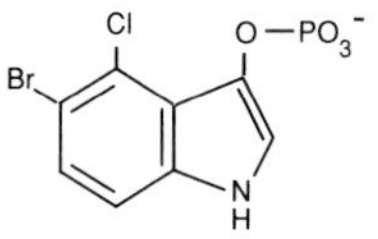

BCIP

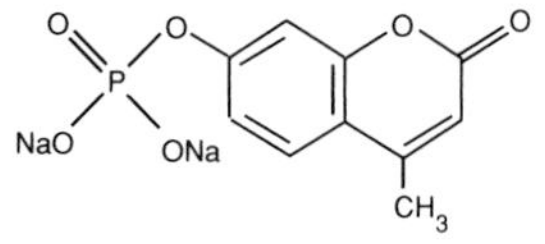

4 - MUP

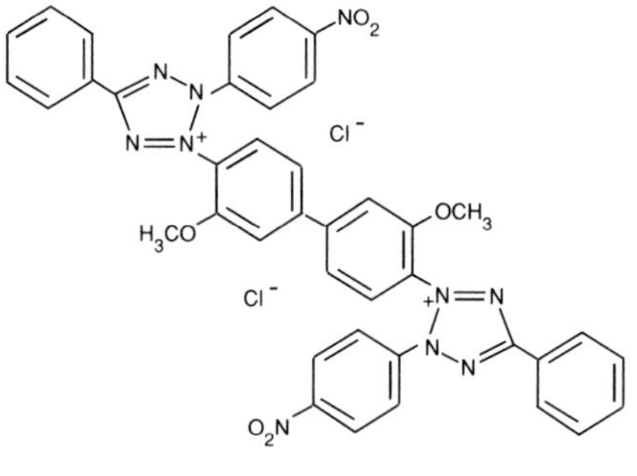

NITROBLUE TETRAZOLIUM

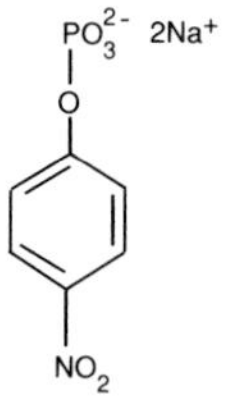

PNPP

Figure 11. Continued.

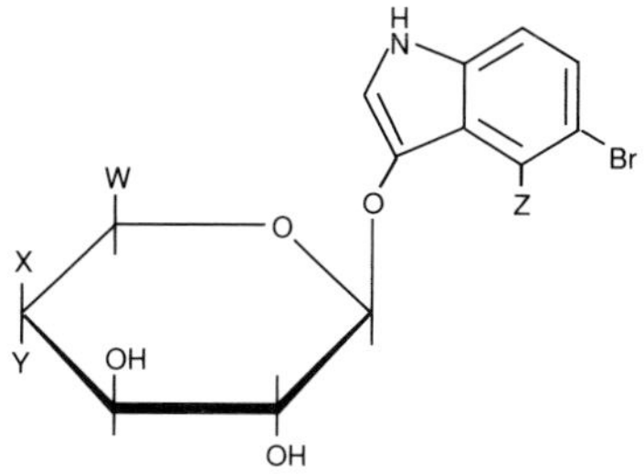

X-GAL :   W = CH₂OH,   X = OH,   Y = H,   Z = Cl
INDIGAL :   W = CH₂OH,   X = OH,   Y = H,   Z = H
BLUOGAL :   W = CH₂OH,   X = OH,   Y = H,   Z = ?, Br = ?
X-GLU :   W = COOH,   X = H,   Y = OH, Z = Cl

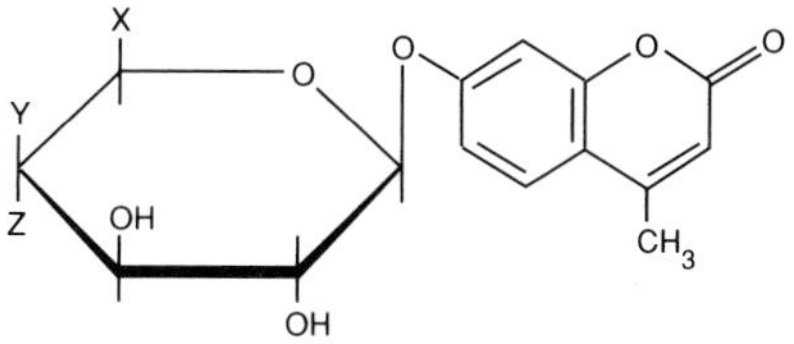

4-METHYLUMBELLIFERYL-β-D-GALACTOPYRANOSIDE
X = CH₂OH   Y = OH   Z = H

4-METHYLUMBELLIFERYL-β-D-GLUCURONIDE
X = COOH   Y = H   Z = OH

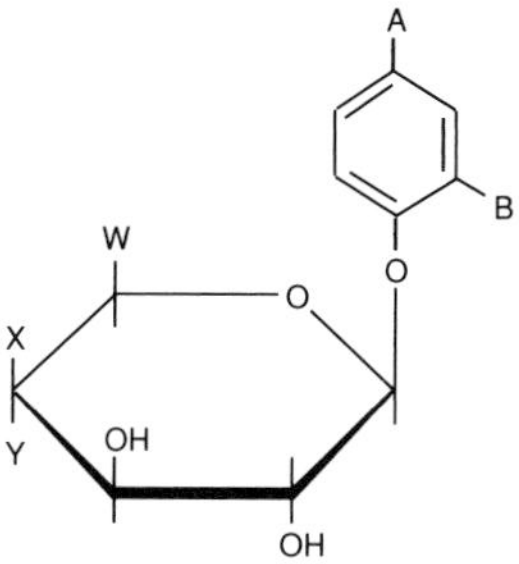

ONPG: A = H, B = NO₂, W = CH₂OH, X = OH, Y = H
PNPG: A = NO₂, B = H, W = CH₂OH, X = OH, Y = H
PNPGlu: A = NO₂, B = H, W = COOH, X = H, Y = OH

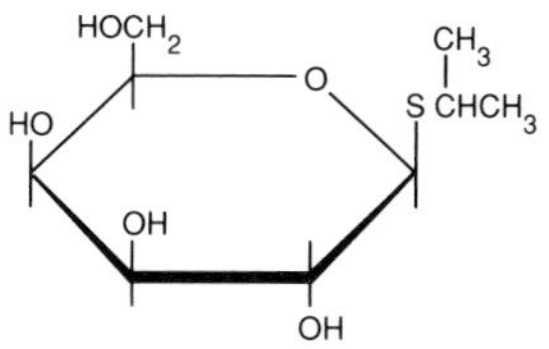

IPTG

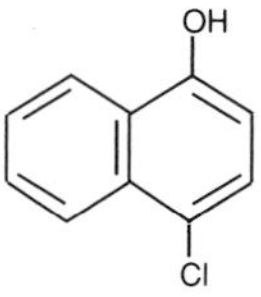

4CN

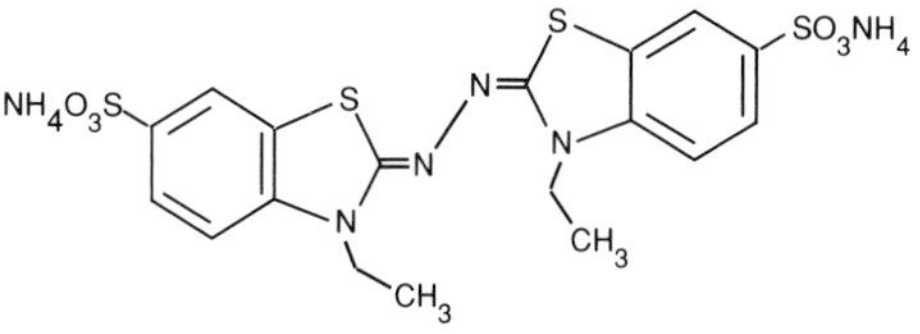

ABTS

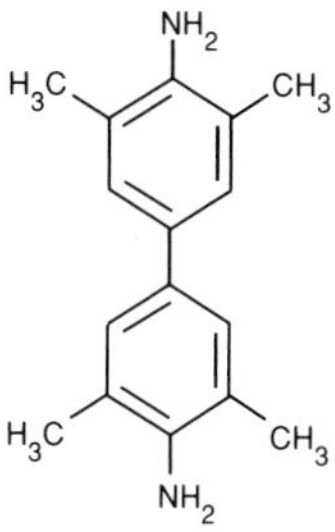

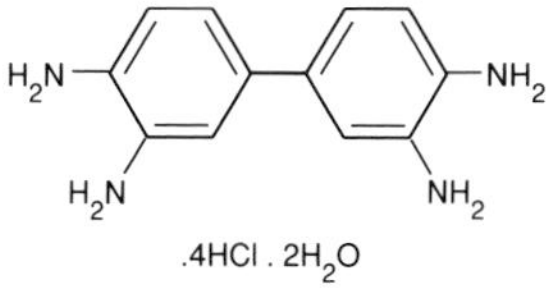

DAB

TMB

CHEMICALS AND REAGENTS

75

18. REFERENCES

1. Meienhofer, J. and Atherton, E. (1973) *Adv. Appl. Microbiol.*, **16**, 203.

2. Weiland, T. (1972) *Naturwissenschaften*, **59**, 225.

3. Lown, J.W. and Sim, S.-K. (1977) *Biochem. Biophys. Res. Commun.*, 77, 1150.

4. Litman, R.M. and Pardee, A.B. (1959) *Virology*, **8**, 125.

5. Young, P.M., Hutchins, A.S. and Canfield, M. (1984) *Plant Sci. Lett.*, **34**, 203.

6. Das Gupta, V. (1984) *J. Pharm. Sci.*, **73**, 565.

7. Monro, R.E. and Vazquez, D. (1967) *J. Mol. Biol.*, **28**, 161.

8. Freeman, K.B., Patel, H. and Haldar, D. (1977) *Mol. Pharmacol.*, **13**, 504.

9. Dustin, P. (1984) *Microtubules.* Springer, Berlin.

10. Suhadolnik, R.J. (1979) *Prog. Nucleic Acids Res. Mol. Biol.*, **22**, 193.

11. Lee, B.K. and Wilkie, D. (1965) *Nature*, **206**, 90.

12. Jackson, L.G. and Studzinski, G.P. (1968) *Exp. Cell Res.*, **52**, 408.

13. Oleinick, N.L. (1977) *Arch. Biochem. Biophys.*, **182**, 171.

14. Stockdale, M. and Selwyn, M.J. (1971) *Eur. J. Biochem.*, **21**, 565.

15. Malpartida, F., Zalacain, M., Jimenez, A. and Davies, J. (1983) *Biochem. Biophys. Res. Commun.*, **117**, 6.

16. Matthews, D., Alden, R.A., Bolin, J.T., Freer, S.T., Hamlin, R., Xuong, N., Kraut, J., Poe, M., Williams, M. and Hoogsteen, K. (1977) *Science*, **197**, 452.

17. Pedrini, A.M., Geroldi, D., Siccardi, A. and Falaschi, A. (1972) *Eur. J. Biochem.*, **25**, 359.

18. Goss, W.A. and Cook, T.M. (1975) in *Antibiotics.* (J.W. Corcorn and F.E. Hahn eds) Springer, Berlin, Vol. 3, p.174.

19. Gellert, M., Mizuuchi, K., O'Dea, M.H., Itoh, T. and Tonizawa, J.-I. (1977) *Proc. Natl. Acad. Sci. USA*, **74**, 4772.

20. Cozzarelli, N.R. (1980) *Science*, **207**, 953.

21. Staudenbauer, W.L. (1975) *J. Mol. Biol.*, **96**, 201.

22. Gellert, M., O'Dea, M.H., Itoh, T. and Tomizawa, J.-I. (1976) *Proc. Natl. Acad. Sci. USA*, **73**, 4474.

23. Storm, D.R., Rosenthal, K.S. and Swanson, P.E. (1977) *Annu. Rev. Biochem.*, **46**, 723.

24. Meilhac, M., Tysper, Z. and Chambon, P. (1972) *Eur. J. Biochem.*, **28**, 291.

25. Gordon, J. (1969) *J. Biol. Chem.*, **244**, 5680.

26. Leavith, R., Schlesinger, S. and Kornfeld, S. (1977) *J. Virol.*, **21**, 375.

27. Elbein, A.D. (1981) *Trends Biochem. Sci.*, **6**, 219.

28. Good, N.E. and Izawa, S. (1972) *Methods Enzymol.*, **24**, 53.

29. Good, N.E., Winget, G.D., Winter, W., Connolly, T.N., Izawa, S. and Singh, R.M.M. (1966) *Biochemistry*, **5**, 467.

30. Berman, M.C. (1982) *J. Biol. Chem.*, **257**, 1953.

31. Cleland, W.W. (1964) *Biochemistry*, **3**, 480.

32. Zahler, W.L. and Cleland, W.W. (1968) *J. Biol. Chem.*, **243**, 716.

33. Burstein, Y. and Patchornik, A. (1972) *Biochemistry*, **11**, 2939.

34. Lane, L.C. (1978) *Anal. Biochem.*, **86**, 655.

35. Turpin, T.H. and Griffith, O.M. (1986) *BioTechniques*, **4**, 11.

36. Chirgwin, J.M., Przybyla, A.E., MacDonald, R.J. and Rutter, W.J. (1979) *Biochemistry*, **18**, 5294.

37. Cockle, S.A., Epand, R.M. and Moscarello, M.A. (1978) *J. Biol. Chem.*, **253**, 8019.

38. Poillon, W.N. and Bertles, J.F. (1979) *J. Biol. Chem.*, **254**, 3462.

39. McCandliss, R.M., Sloma, A. and Pestka, S. (1981) *Methods Enzymol.*, **79**, 51.

40. Lizardi, P.M. (1983) *Methods Enzymol.*, **96**, 24.

41. Locker, J. (1979) *Anal. Biochem.*, **98**, 358.

42. Waehneldt, T.V. (1975) *Biosystems*, **6**, 176.

43. Weber, K. and Osborn, M. (1969) *J. Biol. Chem.*, **244**, 4406.

44. Simons, K. and Helenius, A. (1970) *FEBS Lett.*, **7**, 59.

45. Hjelmeland, L.M. (1980) *Proc. Natl. Acad. Sci. USA*, **77**, 6368.

46. Bitonti, A.J., Moss, J., Hjelmeland, L. and Vaughan, M. (1982) *Biochemistry*, **21**, 3650.

47. Gould, R.J., Ginsberg, B.H. and Spector, A.A. (1981) *Biochemistry*, **20**, 6776.

48. Hjelmeland, L.M., Nebert, D.W. and Osborne, J.C. (1983) *Anal. Biochem.*, **130**, 72.

49. Gonenne, A. and Ernst, R. (1978) *Anal. Biochem.*, **87**, 28.

50. Ashani, Y. and Catravas, G.N. (1980) *Anal. Biochem.*, **109**, 55.

51. Lever, M. (1977) *Anal. Biochem.*, **83**, 274.

52. Baron, C. and Thompson, T.E. (1975) *Biochim. Biophys. Acta*, **382**, 276.

53. Stubbs, G.W. and Litman, B.J. (1978) *Biochemistry*, 17, 215.

54. Comb, D.G. and Roseman, S. (1962) *Methods Enzymol.*, 5, 391.

55. Zeigler, D.W. and Hutchinson, H.D. (1972) *Appl. Microbiol.*, 23, 1060.

56. Uchida, Y., Tsukada, Y. and Sugimori, T. (1984) *J. Biochem.*, 96, 507.

57. Bernfeld, P. (1951) *Adv. Enzyme*, 12, 379.

58. Takebe, I., Otsuki, Y. and Aoki, S. (1968) *Plant Cell Physiol.*, 9, 115.

59. Glasgow, L.R., Paulson, J.C. and Hill, R.L. (1977) *J. Biol. Chem.*, 252, 8615.

60. Scudder, P., Uemura, K.-I., Dolby, J., Fukuda, M.N. and Feizi, T. (1983) *Biochem. J.*, 213, 485.

61. Scudder, P., Hanfland, P., Uemura, K.-I. and Feizi, T. (1984) *J. Biol. Chem.*, 259, 6586.

62. Kobata, A. (1979) *Anal. Biochem.*, 100, 1.

63. Taniguchi, T., Adler, A.J., Mizouchi, T., Kochibi, N. and Kobata, A. (1986) *J. Biol. Chem.*, 261, 1730.

64. Elder, J.H. and Alexander, S. (1982) *Proc. Natl. Acad. Sci. USA*, 79, 4540.

65. Tarentino, A.L., Gomez, C.M. and Plummer, T.H. (1985) *Biochemistry*, 24, 4665.

66. Tarentino, A.L. and Maley, F. (1974) *J. Biol. Chem.*, 249, 811.

67. Trimble, R.B. and Maley, F. (1984) *Anal. Biochem.*, 141, 515.

68. Schmid, K. and Schmitt, R. (1976) *Eur. J. Biochem.*, 67, 95.

69. Distler, J.J. and Jourdian, G.W. (1973) *J. Biol. Chem.*, 248, 6772.

70. Niemann, H., Geyer, R., Klenk, H.-D., Linder, D., Stirm, S. and Wirth, M. (1984) *EMBO J.*, 3, 665.

71. O'Sullivan, M.J. and Marks, V. (1981) *Methods Enzymol.*, 73, 147.

72. Plummer, T.H., Elder, J.H., Alexander, S., Phelan, A.W. and Tarentino, A.L. (1984) *J. Biol. Chem.*, 259, 10700.

73. Yuki, H. and Fishman, W.H. (1963) *J. Biol. Chem.*, 238, 1877.

74. Hamlyn, P.F., Bradshaw, R.E., Melson, F.M., Santiago, C.M., Wilson, J.M. and Peberdy, J.F. (1981) *Enzyme Microbiol. Tech.*, 3, 321.

75. Murao, S. (1979) *J. Ferment. Technol.*, 57, 151.

76. Murao, S. and Sakamoto, R. (1979) *Agric. Biol. Chem.*, 43, 1791.

77. Nakaniski, T. and Yamamoto, T. (1974) *Agric. Biol. Chem.*, 38, 2391.

78. Carrey, E.A. (1989) in *Protein Structure: A Practical Approach*. (T.E. Creighton ed.) IRL Press, Oxford, p.117.

79. Miller, R.L. and Udenfriend, S. (1970) *Arch. Biochem. Biophys.*, 139, 104.

80. Chavira, R., Burnett, T.J. and Hageman, J.H. (1984) *Anal. Biochem.*, 136, 446.

81. Rigbi, M., Levy, H., Iraqi, F., Teitelbaum, M., Orevi, M., Alajoutsijarvi, A., Horovitz, A. and Galun, R. (1987) *Comp. Biochem. Physiol.*, 87B, 567.

82. Ebeling, W., Hennrich, N., Klockow, M., Metz, H., Orth, H.D. and Lang, H. (1974) *Eur. J. Biochem.*, 47, 91.

83. Jany, K.D. and Mayer, B. (1985) *Biol. Chem. Hoppe-Seyler*, 366, 485.

84. Wiegers, U. and Hilz, H. (1971) *Biochem. Biophys. Res. Commun.*, 44, 513.

85. Lockard, R.E., Alzner-DeWeerd, B., Heckman, J.E., MacGee, J., Tabor, M.W. and RajBhandary, U.L. (1978) *Nucleic Acids Res.*, 5, 37.

86. Levy, C.C. and Karpetsky, T.P. (1980) *J. Biol. Chem.*, 255, 2153.

87. Pilly, D., Niemeyer, A., Schmidt, M. and Bargetzi, J.P. (1978) *J. Biol. Chem.*, 253, 437.

88. Donis-Keller, H. (1980) *Nucleic Acids Res.*, 8, 3133.

89. Donis-Keller, H., Maxam, A. and Gilbert, W. (1977) *Nucleic Acids Res.*, 4, 2527.

90. Ascione, R. and Arlinghaus, R.B. (1970) *Biochim. Biophys. Acta*, 204, 478.

91. Wahl, G.M., Stern, M. and Stark, G.R. (1979) *Proc. Natl. Acad. Sci. USA*, 76, 3683.

92. Pretlow, T.G., Boone, C.W., Shrager, R.I. and Weiss, G.H. (1969) *Anal. Biochem.*, 29, 230.

93. Zillig, W., Zechel, K. and Halbwachs, H.J. (1970) *Hoppe Seyler's Z. Physiol. Chem.*, 351, 221.

94. Hodo, H.G. and Blatti, S.P. (1977) *Biochemistry*, 16, 2334.

95. Suda, H., Aoyagi, T., Hamada, M., Takeuchi, T. and Umezawa, H. (1972) *J. Antibiot.*, 25, 263.

96. Umezawa, S., Tatsuta, K., Fujimoto, K., Tsuchiya, T., Umezawa, H. and Naganawa, H. (1972) *J. Antibiot.*, 25, 267.

97. Umezawa, H. (1982) *Annu. Rev. Microbiol.*, 36, 75.

98. Umezawa, H., Aoyagi, T., Morishima, H., Kunimoto, S., Matsuzaki, M., Hamada, M. and Takeuchi, T. (1970) *J. Antibiot.*, 23, 425.

99. Tatsuta, K., Mikami, N., Fujimoto, K., Umezawa, S., Umezawa, U. and Aoyagi, T. (1973) *J. Antibiot.*, 26, 625.

100. Umezawa, H., Aoyagi, T., Okura, A., Morishima, H., Takeuchi, T. and Okami, Y. (1973) *J. Antibiot.*, 26, 787.

101. Okura, A., Morishima, H., Takita, T., Aoyagi, T., Takeuchi, T. and Umezawa, H. (1975) *J. Antibiot.*, 28, 337.

102. Shimizu, B., Saito, A., Ito, A., Tokawa, K., Maeda, K. and Umezawa, H. (1972) *J. Antibiot.*, **25**, 515.

103. Lineweaver, H. and Murray, C.W. (1947) *J. Biol. Chem.*, **171**, 565.

104. Workman, R.J. and Burkitt, D.W. (1979) *Arch. Biochem. Biophys.*, **194**, 157.

105. Suda, H., Aoyagi, T., Takeuchi, T. and Umezawa, H. (1973) *J. Antibiot.*, **26**, 621.

106. Shaw, E., Mares-Guia, M. and Cohen, W. (1965) *Biochemistry*, **4**, 2219.

107. Kostka, V. and Carpenter, F.H. (1964) *J. Biol. Chem.*, **239**, 1799.

108. Shaw, E. (1972) *Methods Enzymol.*, **25**, 655.

109. Green, N.M. (1975) *Adv. Protein Chem.*, **29**, 85.

110. Sjodahl, J. (1977) *Eur. J. Biochem.*, **73**, 343.

111. Haeuptle, M.-T., Aubert, M.L., Djiane, J. and Kraehenbuhl, J.-P. (1983) *J. Biol. Chem.*, **258**, 305.

112. Williamson, R., Morrison, M., Lanyon, G., Enson, R. and Paul, J. (1971) *Biochemistry*, **10**, 3014.

113. Scheele, G. and Blackburn, P. (1979) *Proc. Natl. Acad. Sci. USA*, **76**, 4898.

114. de Martynoff, G., Pays, E. and Vassart, G. (1980) *Biochem. Biophys. Res. Commun.*, **93**, 645.

115. Berger, S.L. and Birkenmeier, C.S. (1979) *Biochemistry*, **18**, 5143.

116. Richards, E.G., Coll, J.A. and Gratzer, W.B. (1965) *Anal. Biochem.*, **12**, 452.

117. Hilwig, I. and Gropp, A. (1972) *Exp. Cell Res.*, **75**, 122.

118. Stokke, T. and Steen, H.B. (1985) *J. Histochem. Cytochem.*, **33**, 333.

119. Arndt-Jovin, D.J. and Jovin, T.M. (1977) *J. Histochem. Cytochem.*, **25**, 585.

120. McDonell, M.W., Simon, M.N. and Studier, F.W. (1977) *J. Mol. Biol.*, **110**, 119.

121. Holbrook, I.B. and Leaver, A.G. (1976) *Anal. Biochem.*, **75**, 634.

122. Chrambach, A., Reisfeld, R.A., Wyekoff, M. and Zaccari, J. (1967) *Anal. Biochem.*, **20**, 150.

123. Diezel, W., Kopperschlager, G. and Hofmann, E. (1972) *Anal. Biochem.*, **48**, 617.

124. Schweizer, D. (1976) *Chromosoma*, **58**, 307.

125. van der Kooy, D. and Kuypers, H.G.J.M. (1979) *Science*, **204**, 873.

126. Waring, M.J. (1966) *Biochim. Biophys. Acta*, **114**, 234.

127. LePecq, J.-B. and Paoletti, C. (1967) *J. Mol. Biol.*, **27**, 87.

128. Douthart, R.J., Burnett, J.P., Beasley, F.W. and Frank, B.H. (1973) *Biochemistry*, **12**, 214.

129. Cherry, W.B., McKinney, R.M., Emmel, V.M., Spillane, J.T., Herbert, G.A. and Pittman, B. (1969) *Stain Technol.*, **44**, 179.

130. Reisher, J.I. and Orr, H.C. (1968) *Anal. Biochem.*, **26**, 178.

131. Mann, K.G. and Fish, W.W. (1972) *Methods Enzymol.*, **26**, 28.

132. Sinsheimer, J.E., Jagodic, V. and Burckhalter, J.H. (1974) *Anal. Biochem.*, **57**, 227.

133. Weber, K., Pollack, R. and Bibring, T. (1975) *Proc. Natl. Acad. Sci. USA*, **72**, 459.

134. Maxam, A.M. and Gilbert, W. (1977) *Proc. Natl. Acad. Sci. USA*, **74**, 560.

135. Cornish, C.J., Neale, F.C. and Posen, S. (1970) *Am. J. Clin. Pathol.*, **53**, 68.

136. Snyder, S.L., Wilson, I. and Bauer, W. (1972) *Biochim. Biophys. Acta*, **258**, 178.

137. Horwitz, J.P., Chau, J., Curby, R.J., Tomson, A.J., DaRooge, M.A., Fisher, B.E., Mauricio, J. and Klundt, I. (1964) *J. Med. Chem.*, 7, 574.

138. Anderson, F.B. and Leaback, D.H. (1961) *Tetrahedron*, **12**, 236.

139. Su, H.C.F. and Tsou, K.C. (1960) *J. Am. Chem. Soc.*, **82**, 1187.

140. Leaback, D.H. (1961) *Biochem. J.*, **78**, 22P.

141. Halvorson, H. (1966) *Methods Enzymol.*, **8**, 559.

142. Donner, J., Caruthers, M.H. and Gill, S.J. (1982) *J. Biol. Chem.*, **257**, 14826.

143. Jefferson, R.A. (1989) *Nature*, **342**, 837.

144. Groome, N.P. (1980) *J. Clin. Chem. Biochem.*, **18**, 345.

145. Frickhofen, N., Bross, K.J., Heit, W. and Heimpel, H. (1985) *J. Clin. Pathol.*, **38**, 671.

146. Graham, R.C. and Karnovsky, M.J. (1966) *J. Histochem. Cytochem.*, **14**, 291.

CHAPTER 3
RADIOCHEMICALS

This chapter concentrates primarily on those aspects of radiochemistry that are directly relevant to the molecular biologist. Information for this chapter comes from a number of sources, including refs 1–7. See also Chapter 11, Section 3 (safety).

1. DEFINITIONS

Becquerel (Bq). This is the SI unit of radioactivity. 1 Bq = one disintegration per second (1 d.p.s.).

Counts per minute (c.p.m.). The number of β particles detected per minute; c.p.m. = d.p.m. $\times$ efficiency of the counting device.

Curie (Ci). The Curie was the standard unit of radioactivity before the introduction of the Becquerel. Nevertheless, it is still widely used in laboratories through widespread familiarity. 1 Ci = 2.22 $\times$ 10^{12} d.p.m.

Disintegrations per minute (d.p.m.) The number of atoms disintegrating per minute.

Electron volt (eV). The energy acquired by an electron when accelerating along a potential gradient of 1 volt. 1 eV = 1.6×10^{-12} erg. 1 MeV = 10^6 eV.

Gray. The SI unit of absorbed dose. 1 Gray = 1 joule per kg tissue = 100 rads.

Rad (radiation absorbed dose). The earlier (non-SI) unit of absorbed dose. 1 rad = 100 erg per g tissue.

RBE (relative biological effectiveness). The ratio of body damage caused by a particular radiation dose to that caused by the same dose of X-rays (average specific ionization of 100 ion pairs per μm of water).

Rem (rad equivalent man). Rem = dose in rad $\times$ RBE.

Sievert (SV). An SI unit equivalent to 100 rem.

Specific activity. The radioactivity of an element per unit mass.

2. CONVERSION BETWEEN UNITS

2.1. Conversion between Curies and d.p.m.

1 Curie (Ci)	$= 2.22 \times 10^{12}$ d.p.m.
1 milliCurie (mCi)	$= 2.22 \times 10^{9}$ d.p.m.
1 microCurie (μCi)	$= 2.22 \times 10^{6}$ d.p.m.
1 nanoCurie (nCi)	$= 2.22 \times 10^{3}$ d.p.m.
1 picoCurie (pCi)	$= 2.22$ d.p.m.

2.2. Conversion between Curies and Becquerels

1 Ci	$= 3.7 \times 10^{10}$ Bq $= 37$ GBq (gigaBq)
1 mCi	$= 3.7 \times 10^{7}$ Bq $= 37$ MBq (megaBq)
1 μCi	$= 3.7 \times 10^{4}$ Bq $= 37$ kBq (kiloBq)
1 GBq	$= 2.7 \times 10^{-4}$ Ci $= 27.027$ mCi (milliCi)
1 MBq	$= 2.7 \times 10^{-7}$ Ci $= 27.027$ μCi (microCi)
1 kBq	$= 2.7 \times 10^{-10}$ Ci $= 27.027$ nCi (nanoCi)

3. RADIONUCLIDES

A comprehensive list of biological radionuclides, with details of half-lives and emissions, is given in *Table 1*. Further details concerning the radionuclides most commonly used in molecular biology are provided in *Tables 2–7* and *Figure 1*.

Table 1. Decay characteristics of biological radionuclides

Radionuclide	Half-life		Type of emission
^{3}H	12.4	years	β^{-}
^{7}Be	53.3	days	EC[a]
^{11}C	20.5	min	β^{+}
^{14}C	5760	years	β^{-}
^{22}Na	2.6	years	$\beta^{+}\gamma$
^{24}Na	14.8	h	$\beta^{-}\gamma$
^{28}Mg	21.4	h	β^{-}
^{31}Si	170	min	β^{-}
^{32}P	14.3	days	β^{-}
^{33}P	25	days	β^{-}
^{35}S	87.4	days	β^{-}
^{36}Cl	310 000	years	$\beta^{-}\kappa$
^{38}Cl	38.5	min	$\beta^{-}\gamma$
^{42}K	12.4	h	$\beta^{-}\gamma$
^{45}Ca	165	days	β^{-}
^{47}Ca	4.54	days	β^{-}
^{51}Cr	28	days	$\gamma\kappa$
^{52}Mn	5.8	days	$\beta^{+}\kappa$
^{54}Mn	310	days	$\gamma\kappa$
^{55}Fe	2.94	years	κ
^{57}Co	270	days	γ
^{58}Co	72	days	$\beta^{+}\gamma$
^{59}Fe	46.3	days	$\beta^{-}\gamma$

Table 1. Continued

Radionuclide	Half-life	Type of emission
^{60}Co	5.3 years	$\beta^-\gamma$
^{63}Ni	100 years	β^-
^{64}Cu	12.8 h	$\beta^-\beta^+\gamma\,\kappa$
^{65}Zn	250 days	$\beta^+\kappa$
^{75}Se	121 days	$\gamma\,\kappa$
^{76}As	26.8 h	$\beta^-\gamma$
^{82}Br	36 h	$\beta^-\gamma$
^{86}Rb	18.7 days	$\beta^-\gamma$
^{89}Sr	51 days	β^-
^{90}Sr	28.5 years	β^-
^{99}Mo	68 h	$\beta^-\gamma$
^{109}Cd	462 days	EC[a]
^{110}Ag	249.8 days	β^-
^{111}Ag	7.47 days	β^-
^{113}Sn	115.1 days	EC[a]
^{115}Cd	44.6 days	β^-
^{125}I	60.0 days	γ EC[a]
^{131}I	8.04 days	$\beta^-\gamma$
^{133}Ba	10.8 years	EC[a]
^{134}Cs	2.06 years	β^-
^{135}I	9.7 h	β^-
^{137}Cs	30.17 years	β^-
^{185}W	75.1 days	β^-
^{197}Hg	64.4 h	EC[a]
^{198}Au	2.696 days	β^-
^{199}Au	3.13 days	β^-
^{203}Hg	46.6 days	β^-

[a] EC, electron capture.

Table 2. Emission energies for important radionuclides (6)

Radionuclide	Particle	Emission energy (MeV)	% of total disintegrations
^{3}H	β^-	0.018	100
^{14}C	β^-	0.155	100
^{32}P	β^-	1.71	100
^{35}S	β^-	0.167	100
^{125}I	EC[a]	–	–
	γ	0.035	–
	Te K X-rays	0.027–0.032	–

Table 2. Continued

Radionuclide	Particle	Emission energy (MeV)	% of total disintegrations
^{131}I	β^-	0.25	3
	β^-	0.33	9
	β^-	0.61	87
	β^-	0.81	1
	γ	0.08	2
	γ	0.28	5
	γ	0.36	80
	γ	0.64	9
	γ	0.72	3

[a]EC, electron capture

Table 3. Decay table for ^3H

						Months						
Years	0	1	2	3	4	5	6	7	8	9	10	11
0	1.000	0.995	0.991	0.986	0.982	0.977	0.973	0.968	0.963	0.959	0.955	0.950
1	0.946	0.941	0.937	0.933	0.928	0.924	0.920	0.916	0.911	0.907	0.903	0.899
2	0.894	0.890	0.886	0.882	0.878	0.874	0.870	0.866	0.862	0.858	0.854	0.850
3	0.846	0.842	0.838	0.834	0.831	0.827	0.823	0.819	0.815	0.811	0.808	0.804
4	0.800	0.796	0.793	0.789	0.786	0.782	0.778	0.775	0.771	0.767	0.764	0.760
5	0.757	0.753	0.750	0.746	0.743	0.739	0.736	0.733	0.729	0.726	0.722	0.719
6	0.716	0.712	0.709	0.706	0.703	0.699	0.696	0.693	0.690	0.686	0.683	0.680
7	0.677	0.674	0.671	0.668	0.665	0.661	0.658	0.655	0.652	0.649	0.646	0.643
8	0.640	0.637	0.634	0.631	0.629	0.626	0.623	0.620	0.617	0.614	0.611	0.608
9	0.606	0.603	0.600	0.597	0.595	0.592	0.589	0.586	0.583	0.581	0.578	0.575
10	0.573	0.570	0.568	0.565	0.562	0.560	0.557	0.554	0.552	0.549	0.547	0.544
11	0.542	0.539	0.537	0.534	0.532	0.529	0.527	0.524	0.522	0.520	0.517	0.515
12	0.512	0.510	0.508	0.505	0.503	0.501	0.498					

Table 4. Decay table for ^{35}S

					Days					
Days	0	1	2	3	4	5	6	7	8	9
0	1.000	0.992	0.984	0.976	0.969	0.961	0.954	0.946	0.939	0.931
10	0.924	0.916	0.909	0.902	0.895	0.888	0.881	0.874	0.867	0.860
20	0.853	0.847	0.840	0.833	0.827	0.820	0.814	0.807	0.801	0.795
30	0.788	0.782	0.776	0.770	0.764	0.758	0.752	0.746	0.740	0.734
40	0.728	0.722	0.717	0.711	0.705	0.700	0.694	0.689	0.683	0.678
50	0.673	0.667	0.662	0.657	0.652	0.646	0.641	0.636	0.631	0.626
60	0.621	0.616	0.612	0.607	0.602	0.597	0.592	0.588	0.583	0.579
70	0.574	0.569	0.565	0.560	0.556	0.552	0.547	0.543	0.539	0.534
80	0.530	0.526	0.522	0.518	0.514	0.510	0.506	0.502	0.498	

Table 5. Decay table for ^{32}P

Days	Hours							
	0	6	12	18	24	30	36	42
0	1.000	0.988	0.976	0.964	0.953	0.941	0.930	0.919
2	0.908	0.897	0.886	0.876	0.865	0.854	0.844	0.834
4	0.824	0.814	0.804	0.794	0.785	0.775	0.766	0.757
6	0.748	0.739	0.730	0.721	0.712	0.704	0.695	0.687
8	0.679	0.671	0.662	0.654	0.646	0.639	0.631	0.623
10	0.616	0.609	0.601	0.594	0.587	0.580	0.573	0.566
12	0.559	0.552	0.546	0.539	0.533	0.526	0.520	0.513
14	0.507	0.501	0.495					

Table 6. Decay table for ^{125}I

Days	Days									
	0	1	2	3	4	5	6	7	8	9
0	1.000	0.988	0.977	0.966	0.955	0.944	0.933	0.922	0.912	0.901
10	0.891	0.881	0.871	0.861	0.851	0.841	0.831	0.821	0.812	0.803
20	0.794	0.785	0.776	0.767	0.758	0.750	0.741	0.732	0.724	0.716
30	0.707	0.699	0.691	0.683	0.675	0.667	0.660	0.653	0.645	0.637
40	0.630	0.623	0.616	0.609	0.602	0.595	0.588	0.581	0.574	0.567
50	0.561	0.555	0.548	0.542	0.536	0.530	0.524	0.518	0.512	0.506
60	0.500									

Table 7. Decay table for ^{131}I

Days	Hours							
	0	3	6	9	12	15	18	21
0	1.000	0.989	0.979	0.969	0.958	0.947	0.937	0.927
1	0.917	0.908	0.898	0.888	0.879	0.870	0.860	0.851
2	0.842	0.833	0.824	0.815	0.806	0.797	0.789	0.781
3	0.772	0.764	0.756	0.748	0.740	0.732	0.724	0.716
4	0.708	0.701	0.693	0.685	0.678	0.671	0.664	0.657
5	0.650	0.643	0.636	0.629	0.622	0.616	0.609	0.602
6	0.596	0.590	0.583	0.577	0.571	0.565	0.559	0.553
7	0.547	0.541	0.535	0.529	0.524	0.519	0.513	0.507
8	0.502	0.496						

Figure 1. Energy curves for ^{3}H, ^{14}C, ^{32}P and ^{35}S.

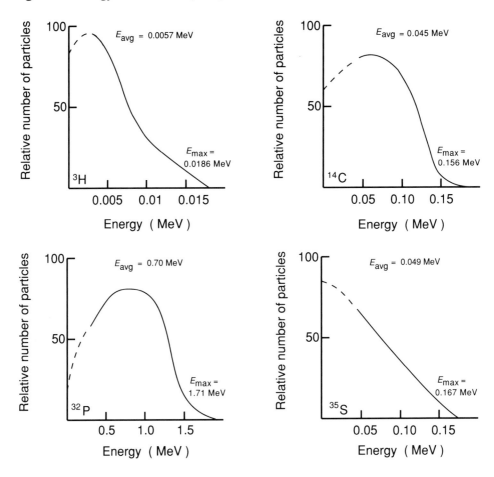

4. RADIOLABELED NUCLEOTIDES

The atomic positions at which nucleotides are usually radiochemically labeled are shown in *Figure 2. Table 8* lists the various methods for introducing radiolabeled nucleotides into nucleic acids, together with indications of the specific activities expected under good to ideal conditions, and references to the literature where practical details of these procedures can be found.

Figure 2. Nucleotides are labeled with ^{32}P at either the α, β or γ position. A ^{35}S label replaces the O⁻, usually at the α or γ position.

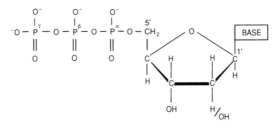

Table 8. Typical specific activities obtained by nucleic acid labeling procedures[a]

Label		Substrate	Labeled product	Procedure	Specific activity of probe (d.p.m. μg^{-1})	Refs
Nucleotide	Specific activity (Ci mmol^{-1})					
[α-^{32}P]dNTP	400–6000	dsDNA	dsDNA	nick translation	5×10^8	9
		dsDNA	dsDNA	random priming	5×10^9	10–13
		dsDNA	dsDNA	end-filling with Klenow polymerase	5×10^6	14
		dsDNA	dsDNA	replacement synthesis with T4 polymerase	$>7 \times 10^8$	15
		ssDNA	dsDNA	primer extension	10^9	16
		dsDNA	ssRNA	transcription with T3, T7 or SP6 polymerase	3×10^9	17–20
		ssRNA	cDNA	cDNA synthesis	4×10^8	14
[γ-^{32}P]ATP	3000–6000	ss/dsDNA	ss/dsDNA	5′-end labeling with T4 polynucleotide kinase	5×10^6	21–23
[α-^{32}P]ddATP	3000	ss/dsDNA	ss/dsDNA	3′-end labeling with terminal deoxynucleotidyl transferase	5×10^6	24
[α-^{35}S]dNTP	400–1500	dsDNA	dsDNA	nick translation	1×10^8	9
		dsDNA	dsDNA	random priming	7×10^8	10, 11
		dsDNA	ssRNA	transcription with T3, T7 or SP6 polymerase	2×10^9	17
[1′,2′,2,8-^{3}H]dATP, [1′,2′,5-^{3}H]dCTP, [1′,2′-^{3}H]dGTP or [methyl,1′,2′-^{3}H]dTTP	25–100	dsDNA	dsDNA	nick translation	5×10^7	9
		dsDNA	dsDNA	random priming	1.5×10^8	10, 11
		dsDNA	ssRNA	transcription with T3, T7 or SP6 polymerase	5×10^7	17
[^{125}I]dCTP	1000–2000	dsDNA	dsDNA	nick translation	5×10^7	9
		dsDNA	dsDNA	random priming	1.5×10^9	10, 11
		dsDNA	dsDNA	direct iodination	2×10^8	17

[a] These figures are taken from experiments conducted in the editor's laboratory and from a number of personal communications. They should be readily attainable with clean DNA and good quality enzymes. The table is based on a less comprehensive version from ref. 25; reproduced with permission from Oxford University Press.

5. DETECTION OF RADIOACTIVITY

5.1. Scintillation counting

Many scintillation cocktails are available commercially or can be readily prepared in the laboratory, each suited to particular types of sample (e.g. aqueous samples versus filters). Physical and chemical data on common fluors and solvents for scintillation cocktails are given in *Tables 9* and *10*, respectively. The structures of the common fluors are shown in *Figure 3*.

Table 9. Scintillants: physical and chemical data

Compound[a]	Mol. wt	Fluorescent maxima (nm)			m.p. (°C)	b.p. (°C)	Risk	Safety
		1	2	3				
Primary scintillants								
Anthracene	178.2	405	428	454	218	342	36/37/38-43	22-36
BBOT	430.6	438	–	–	200	–	–	–
Butyl-PBD	354.5	367	382	–	135	–	–	–
Naphthalene	128.2	325	336	351	80	218[b]	–	–
PBD	298.4	364	377	–	168	–	–	–
PPO	221.3	365	380	–	73	360	–	–
Secondary scintillants								
Bis-MSB	310.4	423	–	–	182	–	–	–
DMPOPOP	392.5	430	–	–	234	–	–	–
POPOP	364.4	420	441	–	245	–	–	–

[a] Abbreviations: BBOT, 2,5-bis(5-*t*-butylbenzoxazol-2-yl)thiophene;
Butyl-PBD, 2-(4-*t*-butylphenyl)-5-(4-biphenylyl)-1,3,4-oxadiazole;
PBD, 2-phenyl-5-(4-biphenylyl)-1,3,4-oxadiazole;
PPO, 2,5-diphenyloxazole; Bis-MSB, *p*-bis(*o*-methylstyryl)benzene;
DMPOPOP, 1,4-bis(4-methyl-5-phenyloxazol-2-yl)benzene;
POPOP, 1,4-bis(5-phenyloxazol-2-yl)benzene.

[b] Naphthalene flash point: 80°C.

Figure 3. Structures of common scintillation fluors.

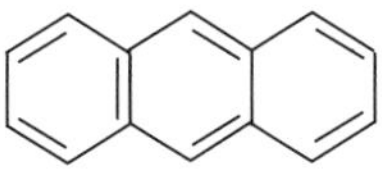

ANTHRACENE

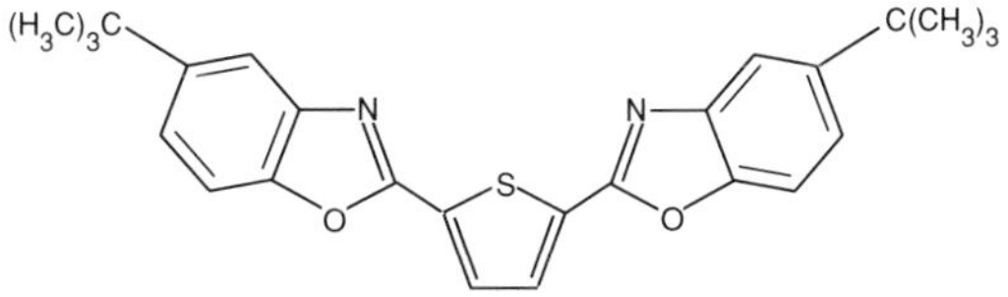

BBOT

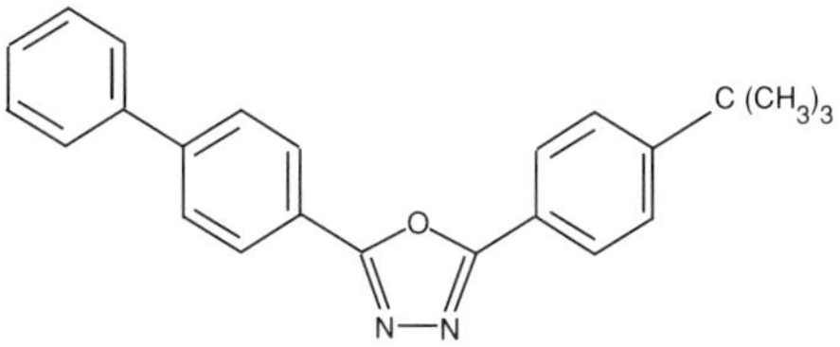

BUTYL - PBD

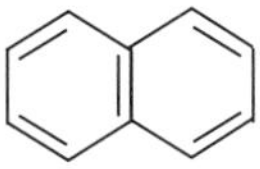

NAPHTHALENE

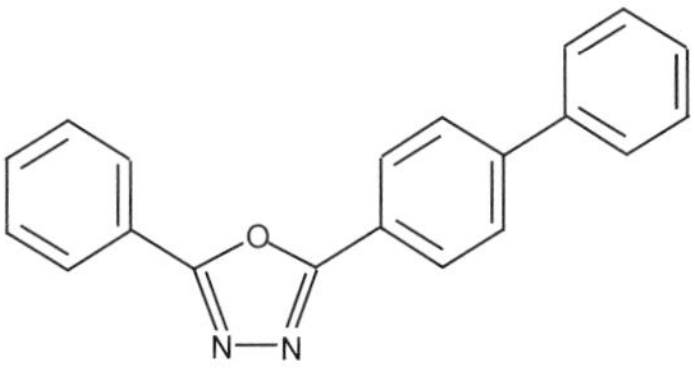

PBD

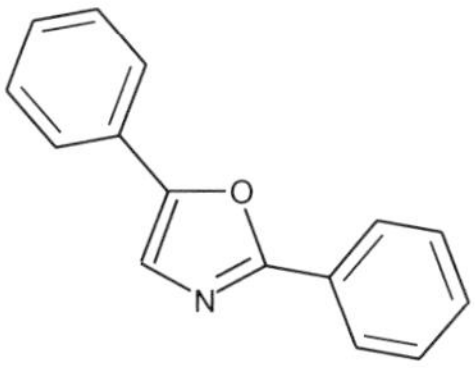

PPO

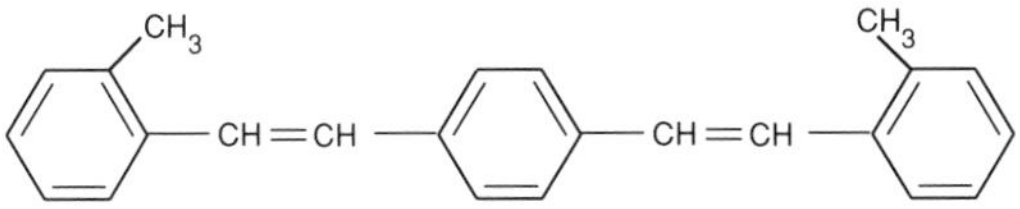

Bis - MSB

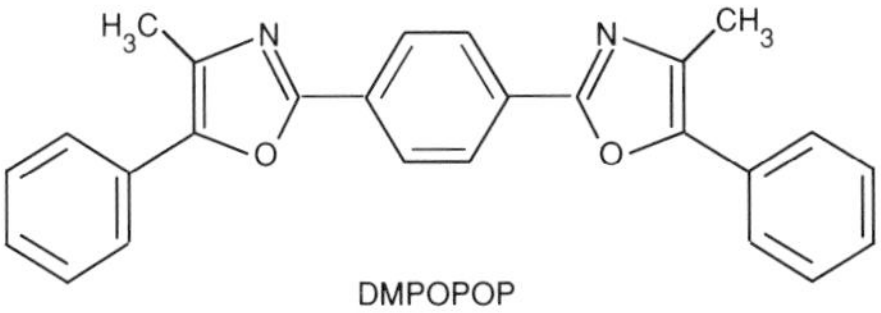

DMPOPOP

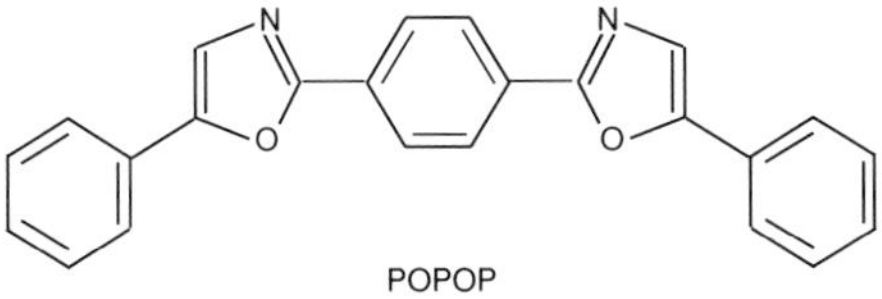

POPOP

Table 10. Solvents for scintillation counting: physical and chemical data

Compound	Mol. wt	m.p. (°C)	b.p. (°C)	f.p.[a] (°C)	Specific gravity	Risk	Safety
Acetone	58.1	−95	56	−20	0.79	11	9-16-23-33
Anisole (methyl phenyl ether)	108.1	−37	155	41	0.99	10	24/25
1,3-dimethoxybenzene	138.2	−	86	87	1.06	38	36/39
1,2-dimethoxyethane (ethylene glycol dimethyl ether)	90.1	−58	85	1	0.87	11-19-20	24/25
1,4-dioxan	88.1	11	101	12	1.03	11-19-20	9-16-33
Methanol (methyl alcohol)	32.0	−98	65	12	0.79	11-23/25	2-7-16-24
2-methoxyethanol (ethylene glycol monomethyl ether)	76.1	−86	125	46	0.96	10-20/21/22-37	24/25
Tetrahydrofuran	72.1	−108	66	−17	0.89	11-19-36/37	16-29-33
Toluene	92.1	−95	111	7	0.86	11-20	16-29-33
Xylene (mixed isomers)	106.2	−95	140	25	0.86	10-20	16-24/25-29-33

[a] f.p., flash point.

5.2. ^{32}P Cerenkov counting

High energy β-particles traveling through water cause the polarization of molecules along their trajectory which then emit photons of light (350–600 nm) as their energy returns to the ground state (the Cerenkov effect). Data relevant to counting ^{32}P in aqueous buffer in a scintillation counter without added scintillation fluid (Cerenkov counting) are given below.

Percent of ^{32}P energy spectrum above 0.5 MeV (the threshold for Cerenkov counting)	= 80%
Counting efficiency: glass vials	= ~50%
plastic vials	= ~60%

A low energy window must be used when counting under these conditions. The ^{3}H channel is sometimes used but it is better to calibrate the counter specifically.

5.3. X-ray film detection

Most experiments in molecular biology involve detection of radiolabeled nucleic acids and proteins using X-ray film. In direct autoradiography, labeled components are detected simply by placing the sample (e.g. a dried gel, filter, etc.) in contact with one of the several commercial brands of 'direct' X-ray film (e.g. Kodak direct exposure film, Amersham Hyperfilm β-max) and then exposing at room temperature. However, with ^{3}H-labeled components in gels, the low energy β-particles are unable to penetrate the gel matrix to reach the X-ray film. In this case, detection can be achieved by impregnating the gel with a scintillator, such as PPO. The β-particles then interact with the scintillator to cause the emission of light which exposes blue-sensitive X-ray film (so called 'screen type' X-ray film) placed next to the gel, forming a detectable image. Many brands of suitable screen-type film are available e.g. Fuji RX, Kodak XAR-5, Amersham Hyperfilm-MP. This procedure, fluorography (8), is also widely used for the detection of ^{14}C and ^{35}S in gels. Exposure is at $-70°$C to stabilize latent image formation during long exposures to the light generated by fluorography; this stabilization can increase the sensitivity of detection at least 4-fold. Unfortunately, using untreated X-ray film the absorbance of the fluorographic image is not proportional to the amount of radioactivity in the sample, small amounts of radioactivity producing disproportionately faint images. This is overcome by exposing the film to a flash of light (<1 msec) prior to fluorography. The procedure for preflashing X-ray film is given elsewhere (8).

The detection of ^{32}P and ^{125}I with X-ray films involves a quite different problem. Here the energy of many of the β-particles is such that the emissions pass completely through the film. However, they can be trapped by placing a calcium tungstate intensifying screen on the other side of the X-ray film. Emissions reaching this screen produce multiple flashes of light which now cause the production of a photographic image on the X-ray film, superimposed over the autoradiographic image. This procedure, indirect autoradiography, is far more sensitive for ^{32}P and ^{125}I detection than direct autoradiography (8). As with fluorography, exposure is again carried out at $-70°$C.

Table 11 gives the sensitivities of detection of direct autoradiography, indirect autoradiography and fluorography with the relevant isotopes. Note, however, that although fluorography or the use of an intensifying screen for autoradiography gives maximum sensitivity for the detection of ^{14}C/^{35}S and ^{32}P/^{125}I respectively, these methods decrease resolution by secondary scattering. Hence for maximum resolution, direct autoradiography should be chosen. *Table 12* lists the most appropriate detection method for different types of sample and *Table 13* gives suitable detection methods for typical molecular biology procedures.

Table 11. Sensitivities of autoradiography and fluorography with different radionuclides

Isotope[a]	Method	Temp. (°C)	Intensifying screen?	Preflashed film?	Sensitivity (d.p.m. cm^{-2} [b,c])
^{3}H	fluorography	-70	no	yes	8000
^{14}C/^{35}S	direct autoradiography	RT[d]	no	no	6000
	fluorography	-70	no	yes	400
^{32}P	direct autoradiography	RT[d]	no	no	500
	autoradiography with intensifying screen	-70	yes	yes	50
^{125}I	direct autoradiography	RT[d]	no	no	1600
	autoradiography with intensifying screen	-70	yes	yes	100

[a] ^{14}C and ^{35}S β-particle emissions have similar energies (see *Figure 1*) and so can be considered together here.
[b] Data from ref. 8, with permission from Oxford University Press.
[c] Defined as the production of a detectable image (0.02 A_{540}).
[d] Room temperature.

Table 12. Choice of isotope detection method[a]

Isotope	Matrix	Detection method
^3H	agarose gels, polyacrylamide gels or filters[b]	fluorography
$^{14}C/^{35}S$	agarose gels or polyacrylamide gels	fluorography (for max. sensitivity) or autoradiography (for max.resolution)
	filters[b]	direct autoradiography
$^{32}P/^{125}I$	agarose gels, polyacrylamide gels or filters[b]	autoradiography with intensifying screen (for max. sensitivity) or direct autoradiography (for max.resolution)

[a] Data from ref. 8, with permission from Oxford University Press.
[b] Any membrane or paper filter, e.g. nitrocellulose, Gene-Screen etc.

Table 13. Application of film detection methods for different molecular biology procedures (see also ref. 6)

Procedure	Isotope	Detection method
DNA/RNA dot blot hybridization	^{32}P	autoradiography, intensifying screen
DNA sequencing	^{32}P	direct autoradiography
	^{35}S	direct autoradiography
In situ hybridization	^{32}P	direct autoradiography
	^{35}S	direct autoradiography
	^3H	direct autoradiography
In vitro protein synthesis	$^{14}C/^{35}S$	fluorography
	^3H	fluorography
Plaque and colony hybridization	^{32}P	autoradiography, intensifying screen
	^{35}S	direct autoradiography
Southern and Northern hybridization	^{32}P	autoradiography, intensifying screen
	^{35}S	direct autoradiography
Western blots	^3H	fluorography
	$^{14}C/^{35}S$	direct autoradiography
	^{125}I	autoradiography, intensifying screen

6. REFERENCES

1. Robyt, J.F. and White, B.J. (1987) *Biochemical Techniques: Theory and Practice*. Brooks/Cole, Monterey.

2. Cooper, T.G. (1977) *The Tools of Biochemistry*. Wiley, New York.

3. Goulding, K.H. (1986) in *Principles and Techniques of Practical Biochemistry (3rd edn)*. (K. Wilson and K.H. Goulding, eds) Edward Arnold, London, p. 314.

4. Dawson, R.M.C., Elliott, D.C., Elliott, W.H. and Jones, K.M. (1986) *Data for Biochemical Research (3rd edn)*. Clarendon Press, Oxford.

5. Sambrook, J., Fritsch, E.F. and Maniatis, T. (1989) *Molecular Cloning: A Laboratory Manual (2nd edn)*. Cold Spring Harbor Laboratory Press, New York.

6. Anon (1989) *Life Sciences Products Catalog*. Amersham International, Amersham.

7. Bonner, W.M. (1987) *Methods Enzymol.*, **152**, 55.

8. Lasky, R.A. (1989) in *Radioisotopes in Biology: A Practical Approach* (R.J. Slater, ed.) Oxford University Press, Oxford, p. 87.

9. Rigby, P.W.J., Diecjmann, M., Rhodes, C. and Berg, P. (1977) *J. Mol. Biol.*, **113**, 237.

10. Feinberg, A.P. and Vogelstein, B. (1983) *Anal. Biochem.*, **132**, 6.

11. Feinberg, A.P. and Vogelstein, B. (1984) *Anal. Biochem.*, **137**, 266.

12. Koch, J., Kolvraa, S. and Boland, L. (1986) *Nucleic Acids Res.*, **14**, 7132.

13. Hodgson, C.P. and Fisk, R.Z. (1987) *Nucleic Acids Res.*, **15**, 6295.

14. Sambrook, J., Fritsch, E.F. and Maniatis, T. (1980) *Molecular Cloning. A Laboratory Manual*. Cold Spring Harbor Laboratory Press, New York.

15. Cunningham, M.W., Harris, D.W. and Mundy, C.R. (1990) in *Radioisotopes in Biology: A Practical Approach*. (R.J. Slater, ed.) Oxford University Press, Oxford, p. 137.

16. Ley, T.J., Anagnou, N.P., Pepe, G. and Nienhuis, A.W. (1982) *Proc. Natl. Acad. Sci. USA*, **79**, 4775.

17. Little, P.F.R. and Jackson, I.J. (1987) in *DNA Cloning: A Practical Approach*. (D.M. Glover, ed.) IRL Press, Oxford, Vol. 1, p. 1.

18. Melton, D.A., Krieg, P.A., Rebagliati, M.R., Maniatis, T., Zinn, K. and Green, M.R. (1984) *Nucleic Acids Res.*, **12**, 7035.

19. Cox, K.H., DeLeon, D.V., Angerer, L.M. and Angerer, R.C. (1984) *Dev. Biol.*, **101**, 458.

20. Butler, E.T. and Chamberlin, M.J. (1982) *J. Biol. Chem.*, **257**, 5772.

21. Arrand, J.E. (1985) in *Nucleic Acid Hybridisation: A Practical Approach*. (B.D. Hames and S.J. Higgins, eds) IRL Press, Oxford, p. 17.

22. Oommen, A., Ferrandis, I. and Wang, M.J. (1990) *BioTechniques*, **8**, 483.

23. Harrison, B. and Zimmerman, S.B. (1986) *Anal. Biochem.*, **158**, 307.

24. Yousaf, S.I., Carrol, A.R. and Clarke, B.E. (1984) *Gene*, **27**, 309.

25. Mundy, C.R., Cunningham, M.W. and Read, C.A. (1991) in *Essential Molecular Biology: A Practical Approach*. (T.A. Brown, ed.) Oxford University Press, Oxford, vol. 2. In press

Radiochemicals

CHAPTER 4
RESTRICTION AND METHYLATION

1. RESTRICTION ENDONUCLEASES

1.1. Enzymes, isoschizomers and their recognition sequences

(Data kindly provided by Richard J. Roberts, with permission from Oxford University Press.)

There are now almost 1500 known restriction endonucleases. A full alphabetical list of all Type II enzymes is given in *Table 1*. The recognition sequences for Type II enzymes, where these are known, are listed in *Table 2*. If the enzyme you are interested in is not listed in *Table 2*, use *Table 1* to identify its isoschizomer (cited in parentheses) and then refer to this enzyme in *Table 2* to determine the recognition sequence. *Tables 3* and *4* list Type I and Type III enzymes respectively.

Table 1. Complete alphabetical listing of Type II enzymes

Enzymes with known recognition sequences

*Aaa*I (*Xma*III)	*Aac*I (*Bam*HI)	*Aae*I (*Bam*HI)	*Aag*I (*Cla*I)
*Aat*I (*Stu*I)	*Aat*II	*Abr*I (*Xho*I)	*Aca*I (*Asu*II)
*Aca*II (*Bam*HI)	*Aca*III (*Mst*I)	*Aca*IV (*Hae*III)	*Acc*I
*Acc*II (*Fnu*DII)	*Acc*III (*Bsp*MII)	*Acc*38I (*Eco*RII)	*Acc*65I (*Kpn*I)
*Acc*B1I (*Hgi*CI)	*Acc*B7I (*Pfl*MI)	*Acc*EBI (*Bam*HI)	*Acr*I (*Ava*I)
*Acr*II (*Bst*EII)	*Acy*I	*Aeu*I (*Eco*RII)	*Afa*I (*Rsa*I)
*Afl*I (*Ava*II)	*Afl*II	*Afl*III	*Age*I
*Aha*I (*Cau*II)	*Aha*II (*Acy*I)	*Aha*III	*Aha*B1I (*Asu*I)
*Aha*B8I (*Kpn*I)	*Ahy*I (*Sma*I)	*Ain*I (*Pst*I)	*Ain*II (*Bam*HI)
*Ait*I (*Eco*47III)	*Ait*II (*Xho*II)	*Ait*AI (*Xho*II)	*Ali*I (*Bam*HI)
*Ali*2882I (*Pst*I)	*Ali*12257I (*Bam*HI)	*Ali*12258I (*Bam*HI)	*Ali*AJI (*Pst*I)
*Alu*I	*Alw*I (*Bin*I)	*Alw*21I (*Hgi*AI)	*Alw*26I (*Bsm*AI)
*Alw*44I (*Apa*LI)	*Alw*NI	*Alw*XI (*Bbv*I)	*Ama*I (*Nru*I)
*Ama*87I (*Ava*I)	*Ame*I (*Apa*LI)	*Ame*II (*Nae*I)	*Ani*MI (*Nae*I)
*Aoc*I (*Sau*I)	*Aoc*II (*Sdu*I)	*Aor*I (*Eco*RII)	*Aos*I (*Mst*I)
*Aos*II (*Acy*I)	*Aos*III (*Sac*II)	*Apa*I	*Apa*LI
*Ape*I (*Mlu*I)	*Ape*AI (*Nae*I)	*Apr*I (*Nae*I)	*Apu*I (*Asu*I)
*Apu*16I (*Cla*I)	*Apy*I (*Eco*RII)	*Aqu*I (*Ava*I)	*Ase*I (*Vsp*I)
*Ase*II (*Cau*II)	*Asn*I (*Vsp*I)	*Asp*I (*Tth*111I)	*Asp*1I (*Cau*II)
*Asp*16RI (*Pvu*I)	*Asp*22MI (*Pvu*I)	*Asp*36I (*Pst*I)	*Asp*47I (*Xho*I)
*Asp*52I (*Hind*III)	*Asp*78I (*Stu*I)	*Asp*697I (*Ava*II)	*Asp*700I (*Xmn*I)
*Asp*703I (*Xho*I)	*Asp*707I (*Cla*I)	*Asp*708I (*Pst*I)	*Asp*713I (*Pst*I)
*Asp*718I (*Kpn*I)	*Asp*742I (*Hae*III)	*Asp*745I (*Ava* II)	*Asp*748I (*Hpa*II)
*Asp*763I (*Sca*I)	*Asp*3065I (*Hind*III)	*Asp*AI (*Bst*EII)	*Asp*BI (*Ava*I)
*Asp*BII (*Ava*II)	*Asp*CI (*Ava*I)	*Asp*CII (*Ava*II)	*Asp*DI (*Ava*I)
*Asp*DII (*Ava*II)	*Asp*HI (*Hgi*AI)	*Asp*NI (*Nla*IV)	*Asp*TI (*Pst*I)
*Asp*TII (*Bam*HI)	*Asp*TIII (*Hae*III)	*Ast*WI (*Acy*I)	*Asu*I

Table 1. Continued

AsuII	AsuIII (AcyI)	AtuII (EcoRII)	Atu1I (EcoRII)
AtuBI (EcoRII)	AtuCI (BclI)	AvaI	AvaII
AvaIII	AviI (AsuII)	AviII (MstI)	AvrI (AvaI)
AvrII	AxyI (SauI)	BacI (SacII)	Bac36I (AsuI)
Bac465I (SacII)	BadI (XhoI)	BalI	Bal228I (AsuI)
Bal475I (HaeIII)	Bal3006 (HaeIII)	BamFI (BamHI)	BamHI
BamKI (BamHI)	BamNI (BamHI)	BamNxI (AvaII)	BanI (HgiCI)
BanII (HgiJII)	BanIII (ClaI)	BavI (PvuII)	BavAI (PvuII)
BavAII (AsuI)	BbeI (NarI)	BbeAI (NarI)	BbfI (XhoI)
Bbf7411I (BspMII)	BbiI (PstI)	BbiII (AcyI)	BbiIII (XhoI)
BbrI (HindIII)	BbrPI (PmaCI)	BbsI (BbvII)	BbuI (SphI)
BbvI	BbvII	Bbv12I (HgiAI)	Bbv16I (BbvII)
BbvAI (XmnI)	BbvAII (ClaI)	BbvAIII (BspMII)	BcaI (HhaI)
Bca1259I (BamHI)	Bce22I (AsuI)	Bce71I (HaeIII)	Bce170I (PstI)
Bce243I (MboI)	Bce751I (BamHI)	BceFI (FnuDII)	BceRI (FnuDII)
BcefI	BcgI	BclI	BcmI (ClaI)
BcnI (CauII)	BcoI (AvaI)	Bco33I (HaeIII)	Bco35I (GsuI)
Bco10278I (BamHI)	BcoAI (PmaCI)	BcrI (NlaIV)	BdiI (ClaI)
BdiSI (SfeI)	BepI (FnuDII)	Bfi458I (HaeIII)	BfrI (AflI)
BglI	BglII	Bim19I (AsuII)	Bim19II (HaeIII)
BinI	BinSI (EcoRII)	BinSII (NarI)	Bka1125I (SduI)
Bla7920I (BspMII)	BliI (HaeIII)	Bli41I (ClaI)	Bli49I (Eco31I)
Bli86I (ClaI)	BliRI (ClaI)	BluI (XhoI)	BluII (HaeIII)
BmaI (PvuI)	BmaAI (PvuI)	BmaBI (PvuI)	BmaCI (PvuI)
BmaDI (PvuI)	Bme12I (MboI)	Bme18I (AvaII)	Bme142I (HaeII)
Bme216I (AvaII)	BmyI (SduI)	BnaI (BamHI)	BpeI (HindIII)
BpuI (HgiJII)	Bpu10I	Bpu14I (AsuII)	Bpu95I (FnuDII)
BsaI (Eco31I)	BsaAI	BsaBI	BsaJI (SecI)
BsaPI (MboI)	BscI (ClaI)	BscAI (SfaNI)	BseI (HaeIII)
BseII (HpaI)	Bse21I (SauI)	BseAI (BspMII)	BsePI
BshI (HaeIII)	BshAI (HaeIII)	BshBI (HaeIII)	BshCI (HaeIII)
BshDI (HaeIII)	BshEI (HaeIII)	BshFI (HaeIII)	BshGI (EcoRII)
BshKI (AsuI)	BsiI	BsiAI (HaeIII)	BsiCI (AsuII)
BsiDI (HaeIII)	BsiHI (HaeIII)	BsiKI (BstEII)	BsiLI (EcoRII)
BsiMI (BspMII)	BsiOI (BspMII)	BsiQI (BclI)	BsiSI (HpaII)
BsmI	BsmAI	BsoPI (BsePI)	Bsp2I (ClaI)
Bsp4I (ClaI)	Bsp5I (HpaII)	Bsp6I (Fnu4HI)	Bsp6II (Eco57I)
Bsp7I (CauII)	Bsp8I (CauII)	Bsp9I (MboI)	Bsp12I (SacII)
Bsp13I (BspMII)	Bsp16I (EcoRV)	Bsp17I (PstI)	Bsp18I (MboI)
Bsp19I (NcoI)	Bsp21I (Cfr10I)	Bsp22I (GsuI)	Bsp28I (GsuI)
Bsp29I (NlaIV)	Bsp30I (BamHI)	Bsp43I (PstI)	Bsp46I (BamHI)
Bsp47I (HpaII)	Bsp48I (HpaII)	Bsp49I (MboI)	Bsp50I (FnuDII)
Bsp51I (MboI)	Bsp52I (MboI)	Bsp53I (ScrFI)	Bsp54I (MboI)
Bsp55I (CauII)	Bsp56I (EcoRII)	Bsp57I (MboI)	Bsp58I (MboI)
Bsp59I (MboI)	Bsp60I (MboI)	Bsp61I (MboI)	Bsp63I (PstI)
Bsp64I (MboI)	Bsp65I (MboI)	Bsp66I (MboI)	Bsp67I (MboI)
Bsp68I (NruI)	Bsp70I (FnuDII)	Bsp71I (HaeIII)	Bsp72I (MboI)
Bsp73I (ScrFI)	Bsp74I (MboI)	Bsp76I (MboI)	Bsp78I (PstI)
Bsp81I (PstI)	Bsp82I (AsuII)	Bsp84I (ClaI)	Bsp87I (PmaCI)

Table 1. Continued

*Bsp*91I (*Mbo*I)	*Bsp*92I (*Xho*I)	*Bsp*93I (*Pst*I)	*Bsp*98I (*Bam*HI)
*Bsp*100I (*Ava*II)	*Bsp*103I (*Eco*RII)	*Bsp*105I (*Mbo*I)	*Bsp*106I (*Cla*I)
*Bsp*107I (*Pst*I)	*Bsp*108I (*Pst*I)	*Bsp*116I (*Hpa*II)	*Bsp*117I (*Hgi*JII)
*Bsp*119I (*Asu*II)	*Bsp*120I (*Apa*I)	*Bsp*121I (*Sph*I)	*Bsp*122I (*Mbo*I)
*Bsp*211I (*Hae*III)	*Bsp*226I (*Hae*III)	*Bsp*423I (*Bbv*I)	*Bsp*519I (*Hgi*JII)
*Bsp*1286I (*Sdu*I)	*Bsp*2095I (*Mbo*I)	*Bsp*AI (*Mbo*I)	*Bsp*BI (*Pst*I)
*Bsp*BII (*Asu*I)	*Bsp*BRI (*Hae*III)	*Bsp*CI	*Bsp*DI (*Cla*I)
*Bsp*EI (*Bsp*MII)	*Bsp*GI	*Bsp*HI	*Bsp*J64I (*Mbo*I)
*Bsp*J67I (*Cau*II)	*Bsp*J74I (*Gsu*I)	*Bsp*J76I (*Fnu*DII)	*Bsp*J105I (*Ava*II)
*Bsp*J106I (*Kpn*I)	*Bsp*MI	*Bsp*MII	*Bsp*RI (*Hae*III)
*Bsp*VI (*Bbv*II)	*Bsp*XI (*Cla*I)	*Bsp*XII (*Bcl*I)	*Bsr*I
*Bsr*HI (*Bse*PI)	*Bsr*PII (*Mbo*I)	*Bss*I (*Nla*IV)	*Bss*CI (*Hae*III)
*Bss*GI (*Bst*XI)	*Bss*GII (*Mbo*I)	*Bss*HI (*Xho*I)	*Bss*HII (*Bse*PI)
*Bss*T1I (*Sty*I)	*Bst*I (*Bam*HI)	*Bst*31I (*Bst*EII)	*Bst*40I (*Hpa*II)
*Bst*1126I (*Bam*HI)	*Bst*2464I (*Bam*HI)	*Bst*2902I (*Bam*HI)	*Bst*BI (*Asu*II)
*Bst*CI (*Hae*III)	*Bst*DI (*Bst*EII)	*Bst*EII	*Bst*EIII (*Mbo*I)
*Bst*FI (*Hind*III)	*Bst*GI (*Bcl*I)	*Bst*GII (*Eco*RII)	*Bst*HI (*Xho*I)
*Bst*JI (*Hae*III)	*Bst*KI (*Bcl*I)	*Bst*LI (*Xho*I)	*Bst*MI (*Sca*I)
*Bst*NI (*Eco*RII)	*Bst*OI (*Eco*RII)	*Bst*PI (*Bst*EII)	*Bst*QI (*Bam*HI)
*Bst*RI (*Eco*RV)	*Bst*SI (*Ava*I)	*Bst*TI (*Bst*XI)	*Bst*UI (*Fnu*DII)
*Bst*VI (*Xho*I)	*Bst*WI (*Eco*NI)	*Bst*XI	*Bst*XII (*Mbo*I)
*Bst*YI (*Xho*II)	*Bst*ZI (*Xma*III)	*Bsu*15I (*Cla*I)	*Bsu*22I (*Bsp*MII)
*Bsu*36I (*Sau*I)	*Bsu*90I (*Bam*HI)	*Bsu*1076I (*Hae*III)	*Bsu*1114I (*Hae*III)
*Bsu*1192I (*Hpa*II)	*Bsu*1192II (*Fnu*DII)	*Bsu*1193I (*Fnu*DII)	*Bsu*1532I (*Fnu*DII)
*Bsu*1854I (*Hgi*JII)	*Bsu*6633I (*Fnu*DII)	*Bsu*8565I (*Bam*HI)	*Bsu*8646I (*Bam*HI)
*Bsu*BI (*Pst*I)	*Bsu*EII (*Fnu*DII)	*Bsu*FI (*Hpa*II)	*Bsu*MI (*Xho*I)
*Bsu*RI (*Hae*III)	*Btc*I (*Mbo*I)	*Bte*I (*Hae*III)	*Bth*I (*Xho*I)
*Bth*II (*Bin*I)	*Bti*I (*Ava*II)	*Btu*I (*Cla*I)	*Bvu*I (*Hgi*JII)
*Cac*I (*Mbo*I)	*Cau*I (*Ava*II)	*Cau*II	*Cau*III (*Pst*I)
*Cau*B3I (*Bsp*MII)	*Ccr*I (*Xho*I)	*Ccy*I (*Mbo*I)	*Cdi*27I (*Eco*RII)
*Cel*I (*Bam*HI)	*Cel*II (*Esp*I)	*Ceq*I (*Eco*RV)	*Cfl*I (*Pst*I)
*Cfo*I (*Hha*I)	*Cfr*I	*Cfr*4I (*Asu*I)	*Cfr*5I (*Eco*RII)
*Cfr*6I (*Pvu*II)	*Cfr*7I (*Bst*EII)	*Cfr*8I (*Asu*I)	*Cfr*9I (*Sma*I)
*Cfr*10I	*Cfr*11I (*Eco*RII)	*Cfr*13I (*Asu*I)	*Cfr*14I (*Cfr*I)
*Cfr*19I (*Bst*EII)	*Cfr*20I (*Eco*RII)	*Cfr*22I (*Eco*RII)	*Cfr*23I (*Asu*I)
*Cfr*24I (*Eco*RII)	*Cfr*25I (*Eco*RII)	*Cfr*27I (*Eco*RII)	*Cfr*28I (*Eco*RII)
*Cfr*29I (*Eco*RII)	*Cfr*30I (*Eco*RII)	*Cfr*31I (*Eco*RII)	*Cfr*32I (*Hind*III)
*Cfr*33I (*Asu*I)	*Cfr*35I (*Eco*RII)	*Cfr*37I (*Sac*II)	*Cfr*38I (*Cfr*I)
*Cfr*39I (*Cfr*I)	*Cfr*40I (*Cfr*I)	*Cfr*41I (*Sac*II)	*Cfr*42I (*Sac*II)
*Cfr*43I (*Sac*II)	*Cfr*45I (*Asu*I)	*Cfr*45II (*Sac*II)	*Cfr*46I (*Asu*I)
*Cfr*47I (*Asu*I)	*Cfr*48I (*Hgi*JII)	*Cfr*51I (*Pvu*I)	*Cfr*52I (*Asu*I)
*Cfr*54I (*Asu*I)	*Cfr*A4I (*Pst*I)	*Cfr*J4I (*Sma*I)	*Cfr*NI (*Asu*I)
*Cfr*S37I (*Eco*RII)	*Cfu*I (*Dpn*I)	*Cfu*II (*Pst*I)	*Chu*I (*Hind*III)
*Chu*II (*Hind*II)	*Chy*I (*Stu*I)	*Cin*1467I (*Mbo*I)	*Cja*I (*Xho*I)
*Cla*I	*Clc*I (*Pst*I)	*Clc*II (*Mst*I)	*Clt*I (*Ava*II)
*Cli*II (*Mst*I)	*Clm*I (*Hae*III)	*Clm*II (*Ava*II)	*Clt*I (*Hae*III)
*Cpa*I (*Mbo*I)	*Cpa*1150I (*Fnu*DII)	*Cpa*AI (*Fnu*DII)	*Cpe*I (*Bcl*I)
*Cpf*I (*Mbo*I)	*Cpo*I (*Rsr*II)	*Csc*I (*Sac*II)	*Csp*I (*Rsr*II)
*Csp*2I (*Hae*III)	*Csp*4I (*Cla*I)	*Csp*5I (*Mbo*I)	*Csp*6I (*Rsa*I)

Table 1. Continued

*Csp*45I (*Asu*II)	*Csp*1470I (*Hha*I)	*Cst*I (*Pst*I)	*Cte*1179I (*Mbo*I)
*Cte*1180 (*Mbo*I)	*Cth*I (*Bcl*I)	*Cth*II (*Eco*RII)	*Cty*I (*Mbo*I)
*Cvi*AI (*Mbo*I)	*Cvi*BI (*Hinf*I)	*Cvi*CI (*Hinf*I)	*Cvi*DI (*Hinf*I)
*Cvi*EI (*Hinf*I)	*Cvi*FI (*Hinf*I)	*Cvi*GI (*Hinf*I)	*Cvi*HI (*Mbo*I)
*Cvi*JI	*Cvi*KI (*Cvi*JI)	*Cvi*LI (*Cvi*JI)	*Cvi*MI (*Cvi*JI)
*Cvi*NI (*Cvi*JI)	*Cvi*OI (*Cvi*JI)	*Cvi*QI (*Rsa*I)	*Cvn*I (*Sau*I)
*Dde*I	*Dde*II (*Xho*I)	*Dds*I (*Bam*HI)	*Dpn*I
*Dpn*II (*Mbo*I)	*Dra*I (*Aha*III)	*Dra*II	*Dra*III
*Drd*I	*Drd*II	*Drd*III (*Pvu*I)	*Drd*AI (*Sac*II)
*Drd*BI (*Sac*II)	*Drd*CI (*Sac*II)	*Drd*DI (*Xho*I)	*Drd*EI (*Sac*II)
*Drd*FI (*Sac*II)	*Dsa*I	*Dsa*II (*Hae*III)	*Dsa*III (*Xho*II)
*Dsa*IV (*Ava*II)	*Dsa*V (*Scr*FI)	*Dsa*VI (*Acc*I)	*Dsp*1I (*Sac*II)
*Eae*I (*Cfr*I)	*Eae*46I (*Sac*II)	*Eae*PI (*Pst*I)	*Eag*I (*Xma*III)
*Eag*KI (*Eco*RII)	*Eag*MI (*Ava*II)	*Ear*I (*Ksp*632I)	*Eca*I (*Bst*EII)
*Eca*II (*Eco*RII)	*Ecc*I (*Sac*II)	*Eci*I	*Eci*AI (*Sna*BI)
*Eci*BI (*Cfr*I)	*Eci*CI (*Sau*I)	*Eci*DI (*Cau*II)	*Eci*EI (*Apa*I)
*Ecl*I (*Pvu*II)	*Ecl*II (*Eco*RII)	*Ecl*28I (*Sac*II)	*Ecl*37I (*Sac*II)
*Ecl*66I (*Eco*RII)	*Ecl*77I (*Pst*I)	*Ecl*133I (*Pst*I)	*Ecl*136I (*Eco*RII)
*Ecl*136II (*Sac*I)	*Ecl*137I (*Sac*I)	*Ecl*137II (*Eco*RII)	*Ecl*593I (*Pst*I)
*Ecl*JI (*Pvu*I)	*Ecl*RI (*Sma*I)	*Ecl*S39I (*Eco*RII)	*Ecl*XI (*Xma*III)
*Eco*VIII (*Hin*dIII)	*Eco*24I (*Hgi*JII)	*Eco*25I (*Hgi*JII)	*Eco*26I (*Hgi*JII)
*Eco*31I	*Eco*32I (*Eco*RV)	*Eco*35I (*Hgi*JII)	*Eco*38I (*Eco*RII)
*Eco*39I (*Asu*I)	*Eco*40I (*Eco*RII)	*Eco*41I (*Eco*RII)	*Eco*42I (*Eco*31I)
*Eco*43I (*Scr*FI)	*Eco*47I (*Ava*II)	*Eco*47II (*Asu*I)	*Eco*47III
*Eco*48I (*Pst*I)	*Eco*49I (*Pst*I)	*Eco*50I (*Hgi*CI)	*Eco*51I (*Eco*31I)
*Eco*51II (*Scr*FI)	*Eco*52I (*Xma*III)	*Eco*55I (*Sac*II)	*Eco*56I (*Nae*I)
*Eco*57I	*Eco*60I (*Eco*RII)	*Eco*61I (*Eco*RII)	*Eco*64I (*Hgi*CI)
*Eco*65I (*Hin*dIII)	*Eco*67I (*Eco*RII)	*Eco*68I (*Hgi*JII)	*Eco*70I (*Eco*RII)
*Eco*71I (*Eco*RII)	*Eco*72I (*Pma*CI)	*Eco*76I (*Sau*I)	*Eco*78I (*Nar*I)
*Eco*80I (*Scr*FI)	*Eco*81I (*Sau*I)	*Eco*82I (*Eco*RI)	*Eco*83I (*Pst*I)
*Eco*85I (*Scr*FI)	*Eco*88I (*Ava*I)	*Eco*90I (*Cfr*I)	*Eco*91I (*Bst*EII)
*Eco*92I (*Sac*II)	*Eco*93I (*Scr*FI)	*Eco*95I (*Eco*31I)	*Eco*96I (*Sac*II)
*Eco*97I (*Eco*31I)	*Eco*98I (*Hin*dIII)	*Eco*99I (*Sac*II)	*Eco*100I (*Sac*II)
*Eco*101I (*Eco*31I)	*Eco*104I (*Sac*II)	*Eco*105I (*Sna*BI)	*Eco*113I (*Hgi*JII)
*Eco*115I (*Sau*I)	*Eco*118I (*Sau*I)	*Eco*120I (*Eco*31I)	*Eco*121I (*Cau*II)
*Eco*125I (*Eco*57I)	*Eco*127I (*Eco*31I)	*Eco*128I (*Eco*RII)	*Eco*129I (*Eco*31I)
*Eco*130I (*Sty*I)	*Eco*134I (*Sac*II)	*Eco*135I (*Sac*II)	*Eco*143I (*Bse*PI)
*Eco*147I (*Stu*I)	*Eco*149I (*Kpn*I)	*Eco*153I (*Scr*FI)	*Eco*155I (*Eco*31I)
*Eco*156I (*Eco*31I)	*Eco*157I (*Eco*31I)	*Eco*158I (*Sac*II)	*Eco*158II (*Sna*BI)
*Eco*159I (*Eco*RI)	*Eco*161I (*Pst*I)	*Eco*162I (*Eco*31I)	*Eco*164I (*Cfr*I)
*Eco*167I (*Pst*I)	*Eco*168I (*Hgi*CI)	*Eco*169I (*Hgi*CI)	*Eco*170I (*Eco*RII)
*Eco*171I (*Hgi*CI)	*Eco*173I (*Hgi*CI)	*Eco*178I (*Eco*RV)	*Eco*179I (*Cau*II)
*Eco*180I (*Hgi*JII)	*Eco*182I (*Sac*II)	*Eco*185I (*Eco*31I)	*Eco*188I (*Hin*dIII)
*Eco*190I (*Cau*II)	*Eco*191I (*Eco*31I)	*Eco*193I (*Eco*RII)	*Eco*195I (*Hgi*CI)
*Eco*196I (*Sac*II)	*Eco*196II (*Asu*I)	*Eco*200I (*Scr*FI)	*Eco*201I (*Asu*I)
*Eco*203I (*Eco*31I)	*Eco*204I (*Eco*31I)	*Eco*205I (*Eco*31I)	*Eco*206I (*Eco*RII)
*Eco*207I (*Eco*RII)	*Eco*208I (*Sac*II)	*Eco*208II (*Sty*I)	*Eco*211I (*Hgi*JII)
*Eco*215I (*Hgi*JII)	*Eco*216I (*Hgi*JII)	*Eco*217I (*Eco*31I)	*Eco*225I (*Eco*31I)
*Eco*228I (*Eco*RI)	*Eco*231I (*Hin*dIII)	*Eco*232I (*Hgi*JII)	*Eco*233I (*Eco*31I)

Table 1. Continued

*Eco*237I (*Eco*RI)	*Eco*239I (*Eco*31I)	*Eco*240I (*Eco*31I)	*Eco*241I (*Eco*31I)
*Eco*246I (*Eco*31I)	*Eco*247I (*Eco*31I)	*Eco*252I (*Eco*RI)	*Eco*A41 (*Eco*31I)
*Eco*HI (*Cfr*I)	*Eco*ICRI (*Sac*I)	*Eco*NI	*Eco*O65I (*Bst*EII)
*Eco*O109I (*Dra*II)	*Eco*RI	*Eco*RII	*Eco*RV
*Eco*T14I (*Sty*I)	*Eco*T22I (*Ava*III)	*Eco*T38I (*Hgi*JII)	*Eco*T88I (*Hgi*JII)
*Eco*T93I (*Hgi*JII)	*Eco*T95I (*Hgi*JII)	*Eco*T104I (*Sty*I)	*Ehe*I (*Nar*I)
*Erh*B9I (*Pvu*I)	*Erh*B9II (*Sty*I)	*Erp*I (*Ava*II)	*Esp*I
*Esp*1I (*Hgi*CI)	*Esp*2I (*Eco*RII)	*Esp*3I	*Esp*4I (*Afl*II)
*Esp*5I (*Nae*I)	*Esp*5II (*Pst*I)	*Esp*6I (*Hgi*CI)	*Esp*7I (*Bse*PI)
*Esp*8I (*Bse*PI)	*Esp*9I (*Hgi*CI)	*Esp*10I (*Hgi*CI)	*Esp*11I (*Hgi*CI)
*Esp*12I (*Hgi*CI)	*Esp*13I (*Hgi*CI)	*Esp*14I (*Hgi*CI)	*Esp*15I (*Hgi*CI)
*Esp*19I (*Kpn*I)	*Esp*22I (*Hgi*CI)	*Esp*141I (*Pst*I)	*Fau*I
*Fba*I (*Bcl*I)	*Fbl*I (*Acc*I)	*Fbr*I (*Fnu*4HI)	*Fdi*I (*Ava*II)
*Fdi*II (*Mst*I)	*Fin*I	*Fin*II (*Hpa*II)	*Fin*SI (*Hae*III)
*Fnu*4HI	*Fnu*AI (*Hinf*I)	*Fnu*AII (*Mbo*I)	*Fnu*CI (*Mbo*I)
*Fnu*DI (*Hae*III)	*Fnu*DII	*Fnu*DIII (*Hha*I)	*Fnu*EI (*Mbo*I)
*Fok*I	*Fsc*I (*Sac*II)	*Fse*I	*Fsf*I (*Eco*57I)
*Fsp*I (*Mst*I)	*Fsp*II (*Asu*II)	*Fsp*1604I (*Eco*RII)	*Fsp*MI (*Fnu*DII)
*Fsp*MSI (*Ava*II)	*Fsu*I (*Tth*111I)	*Gal*I (*Sac*II)	*Gce*I (*Sac*II)
*Gce*GLI (*Sac*II)	*Gdi*I (*Stu*I)	*Gdi*II	*Gdo*I (*Bam*HI)
*Gin*I (*Bam*HI)	*Gox*I (*Bam*HI)	*Gse*I (*Asu*I)	*Gse*II (*Pst*I)
*Gse*III (*Bam*HI)	*Gsp*I (*Pvu*II)	*Gsp*AI (*Ava*II)	*Gsp*AII (*Mst*I)
*Gsu*I	*Hac*I (*Mbo*I)	*Hae*I	*Hae*II
*Hae*III	*Hal*B6I (*Eco*RI)	*Hal*B6II (*Pst*I)	*Hap*II (*Hpa*II)
*Hga*I	*Hgi*I (*Acy*I)	*Hgi*AI	*Hgi*BI (*Ava*II)
*Hgi*CI	*Hgi*CII (*Ava*II)	*Hgi*CIII (*Sal*I)	*Hgi*DI (*Acy*I)
*Hgi*DII (*Sal*I)	*Hgi*EI (*Ava*II)	*Hgi*EII	*Hgi*GI (*Acy*I)
*Hgi*HI (*Hgi*CI)	*Hgi*HII (*Acy*I)	*Hgi*HIII (*Ava*II)	*Hgi*JI (*Ava*II)
*Hgi*JII	*Hgi*S21I (*Cau*II)	*Hgi*S22I (*Cau*II)	*Hha*I
*Hha*II (*Hinf*I)	*Hhg*I (*Hae*III)	*Hin*1I (*Acy*I)	*Hin*1II (*Nla*II)
*Hin*2I (*Hpa*II)	*Hin*3I (*Cau*II)	*Hin*5I (*Hpa*II)	*Hin*5II (*Asu*I)
*Hin*5III (*Hind*III)	*Hin*6I (*Hha*I)	*Hin*7I (*Hha*I)	*Hin*8I (*Acy*I)
*Hin*8II (*Nla*III)	*Hin*173I (*Hind*III)	*Hin*1056I (*Fnu*DII)	*Hin*1076III (*Hind*III)
*Hin*1160II (*Hind*II)	*Hin*1161II (*Hind*II)	*Hin*GUI (*Hha*I)	*Hin*GUII (*Fok*I)
*Hin*HI (*Hae*II)	*Hin*JCI (*Hind*II)	*Hin*JCII (*Hind*III)	*Hin*P1I (*Hha*I)
*Hin*S1I (*Hha*I)	*Hin*S2I (*Hha*I)	*Hin*bIII (*Hind*III)	*Hinc*II (*Hind*II)
*Hind*II	*Hind*III	*Hinf*I	*Hinf*II (*Hind*III)
*Hja*I (*Eco*RV)	*Hpa*I	*Hpa*II	*Hph*I
*Hsp*2I (*Ava*II)	*Hsu*I (*Hind*III)	*Isp*I (*Fnu*4HI)	*Kox*I (*Bst*EII)
*Kox*II (*Hgi*JII)	*Kox*165I (*Eco*RII)	*Koy*I (*Sal*I)	*Kpn*I
*Kpn*2I (*Bsp*MII)	*Kpn*10I (*Eco*RII)	*Kpn*12I (*Pst*I)	*Kpn*13I (*Eco*RII)
*Kpn*14I (*Eco*RII)	*Kpn*16I (*Eco*RII)	*Kpn*30I (*Bse*PI)	*Kpn*K14I (*Kpn*I)
*Ksp*I (*Sac*II)	*Ksp*22I (*Bcl*I)	*Ksp*632I	*Kzo*9I (*Mbo*I)
*Kzo*49I (*Ava*II)	*Lmu*60I (*Sau*I)	*Lpl*I (*Cla*I)	*Lsp*I (*Asu*II)
*Mae*I	*Mae*II	*Mae*III	*Mam*I (*Bsa*BI)
*Mau*I (*Pst*I)	*Mav*I (*Xho*I)	*Mbo*I	*Mbo*II
*Mca*I (*Xho*I)	*Mch*I (*Nar*I)	*Mcr*I	*Mec*I (*Xho*I)
*Meu*I (*Mbo*I)	*Mfe*I	*Mfl*I (*Xho*II)	*Mfo*I (*Ava*II)
*Mis*I (*Nae*I)	*Mja*I (*Mae*I)	*Mja*II (*Asu*I)	*Mki*I (*Hind*III)

Restriction/Methylation

Table 1. Continued

*Mkr*I (*Pst*I)	*Mkr*AI (*Mbo*I)	*Mla*I (*Asu*II)	*Mla*AI (*Xho*I)
*Mle*I (*Bam*HI)	*Mli*I (*Ava*II)	*Mlt*I (*Alu*I)	*Mlu*I
*Mlu*23I (*Bam*HI)	*Mlu*2300I (*Eco*RII)	*Mlu*B2I (*Nru*I)	*Mly*I
*Mly*113I (*Nar*I)	*Mme*I	*Mme*II (*Mbo*I)	*Mni*I (*Hae*III)
*Mni*II (*Hpa*II)	*Mnl*I	*Mnn*I (*Hin*dII)	*Mnn*II (*Hae*III)
*Mnn*IV (*Hha*I)	*Mno*I (*Hpa*II)	*Mno*III (*Mbo*I)	*Mos*I (*Mbo*I)
*Mph*I (*Eco*RII)	*Mpu*I (*Xho*I)	*Mra*I (*Sac*II)	*Mrh*I (*Xho*I)
*Mro*I (*Bsp*MII)	*Msc*I (*Bal*I)	*Mse*I	*Msi*I (*Xho*I)
*Msp*I (*Hpa*II)	*Msp*20I (*Bal*I)	*Msp*24I (*Asu*I)	*Msp*67I (*Scr*FI)
*Msp*67II (*Mbo*I)	*Msp*AI (*Ava*II)	*Msp*A1I (*Nsp*BII)	*Msp*BI (*Mbo*I)
*Msp*B4I (*Hgi*CI)	*Msp*YI (*Bsa*AI)	*Mst*I	*Mst*II (*Sau*I)
*Mth*I (*Mbo*I)	*Mth*1047I (*Mbo*I)	*Mth*AI (*Mbo*I)	*Mth*TI (*Asu*I)
*Mva*I (*Eco*RII)	*Mva*AI (*Fnu*DII)	*Mvn*I (*Fnu*DII)	*Mwo*I
*Mzi*I (*Pvu*II)	*Nae*I	*Nam*I (*Nar*I)	*Nan*I (*Eco*RV)
*Nan*II (*Dpn*I)	*Nar*I	*Nas*I (*Pst*I)	*Nas*BI (*Bam*HI)
*Nas*SI (*Sac*I)	*Nas*WI (*Nae*I)	*Nba*I (*Nae*I)	*Nbl*I (*Pvu*I)
*Nbr*I (*Nae*I)	*Nca*I (*Hin*fI)	*Nci*I (*Cau*II)	*Nco*I
*Ncu*I (*Mbo*II)	*Nda*I (*Nar*I)	*Nde*I	*Nde*II (*Mbo*I)
*Nfl*I (*Mbo*I)	*Nfl*AI (*Eco*RV)	*Nfl*AII (*Mbo*I)	*Nfl*BI (*Mbo*I)
*Ngb*I (*Pst*I)	*Ngo*I (*Hae*II)	*Ngo*II (*Hae*III)	*Ngo*III (*Sac*II)
*Ngo*AIII (*Sac*II)	*Ngo*AIV (*Nae*I)	*Ngo*BI (*Hph*I)	*Ngo*DI (*Sac*II)
*Ngo*DIII (*Dpn*I)	*Ngo*MI (*Nae*I)	*Ngo*PII (*Hae*III)	*Ngo*PIII (*Sac*II)
*Ngo*SI (*Hae*III)	*Nhe*I	*Nla*I (*Hae*III)	*Nla*II (*Mbo*I)
*Nla*III	*Nla*IV	*Nla*DI (*Mbo*I)	*Nla*DII (*Asu*I)
*Nla*DIII (*Sac*II)	*Nla*SI (*Sac*II)	*Nla*SII (*Acy*I)	*Nli*I (*Ava*I)
*Nli*II (*Ava*II)	*Nme*CI (*Mbo*I)	*Nme*RI (*Pvu*II)	*Nmi*I (*Kpn*I)
*Nmu*I (*Nae*I)	*Nmu*AI (*Ava*I)	*Nmu*AII (*Ava*II)	*Nmu*DI (*Dpn*I)
*Nmu*EI (*Dpn*I)	*Nmu*EII (*Asu*I)	*Nmu*FI (*Nae*I)	*Nmu*SI (*Asu*I)
*Noc*I (*Pst*I)	*Nop*I (*Sal*I)	*Not*I	*Nov*II (*Hin*fI)
*Nph*I (*Mbo*I)	*Nru*I	*Nsi*I (*Ava*III)	*Nsi*AI (*Mbo*I)
*Nsi*CI (*Eco*RV)	*Nsi*HI (*Hin*fI)	*Nsp*I	*Nsp*II (*Sdu*I)
*Nsp*III (*Ava*I)	*Nsp*IV (*Asu*I)	*Nsp*V (*Asu*II)	*Nsp*AI (*Mbo*I)
*Nsp*BI (*Asu*II)	*Nsp*BII	*Nsp*DI (*Ava*I)	*Nsp*DII (*Ava*II)
*Nsp*EI (*Ava*I)	*Nsp*FI (*Asu*II)	*Nsp*GI (*Ava*II)	*Nsp*HI (*Nsp*I)
*Nsp*HII (*Ava*II)	*Nsp*HIII (*Mst*I)	*Nsp*JI (*Asu*I)	*Nsp*KI (*Ava*II)
*Nsp*LI (*Mst*I)	*Nsp*LII (*Asu*I)	*Nsp*MI (*Mst*I)	*Nsp*MACI (*Bgl*II)
*Nsp*SAI (*Ava*I)	*Nsp*SAII (*Bst*EII)	*Nsp*SAIII (*Nco*I)	*Nsp*SAIV (*Bam*HI)
*Nsp*WI (*Nae*I)	*Nsu*I (*Mbo*I)	*Nsu*DI (*Dpn*I)	*Nta*I (*Tth*111I)
*Nta*SI (*Stu*I)	*Nta*SII (*Nae*I)	*Nun*II (*Nar*I)	*Oco*I (*Xho*I)
*Otu*I (*Alu*I)	*Otu*NI (*Alu*I)	*Oxa*I (*Alu*I)	*Oxa*NI (*Sau*I)
*Pae*I (*Sph*I)	*Pae*177I (*Bam*HI)	*Pae*181I (*Cau*II)	*Pae*AI (*Sac*II)
*Pae*BI (*Sma*I)	*Pae*R7I (*Xho*I)	*Pai*I (*Hae*III)	*Pal*I (*Hae*III)
*Pan*I (*Xho*I)	*Pde*12I (*Asu*I)	*Pde*133I (*Hae*III)	*Pde*137I (*Hpa*II)
*Pei*9403I (*Mbo*I)	*Pfa*I (*Mbo*I)	*Pfl*AI (*Fnu*DII)	*Pfl*MI
*Pfl*NI (*Xho*I)	*Pfl*WI (*Xho*I)	*Pfu*I (*Spl*I)	*Pgl*I (*Nae*I)
*Pgl*B4I (*Cla*I)	*Ple*I	*Ple*19I (*Pvu*I)	*Pma*I (*Pst*I)
*Pma*44I (*Pst*I)	*Pma*CI	*Pme*55I (*Stu*I)	*Pml*I (*Pma*CI)
*Pmy*I (*Pst*I)	*Pov*I (*Bcl*I)	*Ppa*I (*Eco*31I)	*Pph*3215I (*Hgi*AI)
*Ppu*I (*Hae*III)	*Ppu*MI	*Pse*I (*Asu*I)	*Psh*AI
*Psp*I (*Asu*I)	*Psp*61I (*Nae*I)	*Pss*I (*Dra*II)	*Pst*I

Table 1. Continued

*Psu*161I (*Pvu*I)	*Pvu*I	*Pvu*II	*Pvu*HKUI (*Pvu*II)
*Rfl*FI (*Sal*I)	*Rhe*I (*Sal*I)	*Rhp*I (*Sal*I)	*Rhs*I (*Bam*HI)
*Rle*AI	*Rlu*I (*Nae*I)	*Rlu*1I (*Mbo*I)	*Rlu*3I (*Nla*IV)
*Rlu*4I (*Bam*HI)	*Rrh*I (*Sal*I)	*Rro*I (*Sal*I)	*Rsa*I
*Rsh*I (*Pvu*I)	*Rsh*II (*Cau*II)	*Rsp*I (*Pvu*I)	*Rsp*XI (*Bsp*HI)
*Rsr*I (*Eco*RI)	*Rsr*II	*Saa*I (*Sac*II)	*Sab*I (*Sac*II)
*Sac*I	*Sac*II	*Sac*AI (*Nae*I)	*Sak*I (*Sac*II)
*Sal*I	*Sal*1974I (*Xho*I)	*Sal*AI (*Mbo*I)	*Sal*CI (*Nae*I)
*Sal*DI (*Nru*I)	*Sal*HI (*Mbo*I)	*Sal*PI (*Pst*I)	*Sao*I (*Nae*I)
*Sar*I (*Stu*I)	*Sau*I	*Sau*10I (*Kpn*I)	*Sau*12I (*Eco*31I)
*Sau*96I (*Asu*I)	*Sau*3239I (*Xho*I)	*Sau*6782I (*Mbo*I)	*Sau*AI (*Nae*I)
*Sau*3AI (*Mbo*I)	*Sau*BI (*Asu*I)	*Sau*BMKI (*Nae*I)	*Sau*CI (*Mbo*I)
*Sau*DI (*Mbo*I)	*Sau*EI (*Mbo*I)	*Sau*FI (*Mbo*I)	*Sau*GI (*Mbo*I)
*Sau*MI (*Mbo*I)	*Sba*I (*Pvu*II)	*Sbl*AI (*Sty*I)	*Sbl*BI (*Sty*I)
*Sbl*CI (*Sty*I)	*Sbo*I (*Sac*II)	*Sbo*13I (*Nru*I)	*Sca*I
*Sca*1827I (*Xho*I)	*Sce*I (*Fnu*DII)	*Scg*2I (*Eco*RII)	*Sci*I (*Xho*I)
*Sci*1831I (*Xho*I)	*Sci*AI (*Bst*EII)	*Sci*AII (*Pvu*II)	*Sci*NI (*Hha*I)
*Sco*I (*Sac*I)	*Scr*FI	*Scu*I (*Xho*I)	*Sdu*I
*Sdy*I (*Asu*I)	*Sec*I	*Sec*II (*Hpa*II)	*Sec*III (*Sau*I)
*Sex*I (*Xho*I)	*Sfa*I (*Hae*III)	*Sfa*GUI (*Hpa*II)	*Sfa*NI
*Sfe*I	*Sfi*I	*Sfl*I (*Pst*I)	*Sfl*2aI (*Eco*RII)
*Sfl*2bI (*Eco*RII)	*Sfn*I (*Ava*II)	*Sfo*I (*Nar*I)	*Sfr*I (*Sac*II)
*Sfr*274I (*Xho*I)	*Sfr*303I (*Sac*II)	*Sfr*382I (*Sac*II)	*Sfu*I (*Asu*II)
*Sfu*1762I (*Xho*I)	*Sga*I (*Xho*I)	*Sgh*1835I (*Ava*II)	*Sgo*I (*Xho*I)
*Sgr*20I (*Eco*RII)	*Sgr*1839I (*Asu*II)	*Sgr*1841I (*Xho*I)	*Sgr*AI
*Shy*I (*Sac*II)	*Shy*1766I (*Xho*I)	*Sin*I (*Ava*II)	*Sin*AI (*Ava*II)
*Sin*BI (*Ava*II)	*Sin*CI (*Ava*II)	*Sin*DI (*Ava*II)	*Sin*EI (*Ava*II)
*Sin*FI (*Ava*II)	*Sin*GI (*Ava*II)	*Sin*HI (*Ava*II)	*Sin*JI (*Ava*II)
*Sin*MI (*Mbo*I)	*Ska*I (*Nae*I)	*Ska*II (*Pst*I)	*Sla*I (*Xho*I)
*Sle*I (*Eco*RII)	*Slu*I (*Xho*I)	*Slu*1777I (*Nae*I)	*Sma*I
*Sma*AI (*Spl*I)	*Sma*AII (*Tth*111I)	*Sma*AIII (*Pvu*I)	*Sma*AIV (*Pvu*II)
*Sna*I	*Sna*3286I (*Nru*I)	*Sna*BI	*Sno*I (*Apa*LI)
*Sol*3335I (*Pvu*II)	*Spa*I (*Xho*I)	*Spa*XI (*Sph*I)	*Spe*I
*Sph*I	*Sph*1719I (*Xho*I)	*Spi*I	*Spl*I
*Spl*II (*Tth*111I)	*Spl*III (*Hae*III)	*Spl*AI (*Spl*I)	*Spl*AII (*Tth*111I)
*Spl*AIII (*Pvu*I)	*Spl*AIV (*Pvu*II)	*Spo*I (*Nru*I)	*Sse*I (*Bcl*I)
*Sse*II (*Sac*II)	*Ssh*AI (*Sau*I)	*Sso*I (*Eco*RI)	*Sso*II (*Scr*FI)
*Ssp*I	*Ssp*1I (*Asu*II)	*Ssp*2I (*Cau*II)	*Ssp*4I (*Xho*I)
*Ssp*152I (*Asu*II)	*Ssp*1725I (*Sac*II)	*Ssp*AI (*Eco*RII)	*Ssp*JI (*Sna*BI)
*Ssp*JII (*Acy*I)	*Ssp*KI (*Spl*I)	*Ssp*M1I (*Sna*BI)	*Ssp*M1II (*Acy*I)
*Ssp*M1III (*Hgi*CI)	*Ssp*M2I (*Sna*BI)	*Ssp*M2II (*Acy*I)	*Ssr*I (*Hpa*I)
*Ssr*B6I (*Hpa*I)	*Sst*I (*Sac*I)	*Sst*II (*Sac*II)	*Sst*IV (*Bcl*I)
*Ssv*I (*Stu*I)	*Ste*I (*Stu*I)	*Sth*I (*Kpn*I)	*Sth*AI (*Kpn*I)
*Sth*BI (*Kpn*I)	*Sth*CI (*Kpn*I)	*Sth*DI (*Kpn*I)	*Sth*EI (*Kpn*I)
*Sth*FI (*Kpn*I)	*Sth*GI (*Kpn*I)	*Sth*HI (*Kpn*I)	*Sth*JI (*Kpn*I)
*Sth*KI (*Kpn*I)	*Sth*LI (*Kpn*I)	*Sth*MI (*Kpn*I)	*Sth*NI (*Kpn*I)
*Stu*I	*Sty*I	*Sua*I (*Hae*III)	*Sul*I (*Hae*III)
*Sur*2I (*Bam*HI)	*Sve*194I (*Xho*I)	*Taq*I	*Taq*II
*Taq*XI (*Eco*RII)	*Tce*I (*Mbo*II)	*Tfi*I	*Tfl*I (*Taq*I)
*Tgl*I (*Sac*II)	*Tha*I (*Fnu*DII)	*Tma*I (*Fnu*DII)	*Tmu*1I (*Cau*II)

Restriction/Methylation

Table 1. Continued

*Tru*I (*Ava*II)	*Tru*II (*Mbo*I)	*Tru*9I (*Mse*I)	*Tru*201I (*Xho*II)
*Tsp*I (*Tth*111I)	*Tsp*45I	*Tsp*EI	*Tsp*ZNI (*Hae*III)
*Tte*I (*Tth*111I)	*Tte*AI (*Hae*III)	*Tth*111I	*Tth*111II
*Tth*HB8I (*Taq*I)	*Ttn*I (*Hae*III)	*Ttr*I (*Tth*111I)	*Uba*1I (*Ppu*MI)
*Uba*6I (*Mlu*I)	*Uba*1102I (*Esp*I)	*Uba*1103I (*Ava*III)	*Uba*1103II (*Dpn*I)
*Uba*1103II (*Dpn*I)	*Uba*1104I (*Ksp*632I)	*Uba*1105I	*Uba*1106I (*Ppu*MI)
*Uba*1107I (*Sna*I)	*Uba*1108I	*Uba*1109I (*Bbv*I)	*Uba*1110I (*Bam*HI)
*Uur*960I (*Fnu*4HI)	*Van*I (*Bgl*I)	*Van*91I (*Pfl*MI)	*Vha*I (*Hae*III)
*Vha*464I (*Afl*II)	*Vne*I (*Apa*LI)	*Vne*AI (*Dra*II)	*Vni*I (*Hae*III)
*Vsp*I	*Xam*I (*Sal*I)	*Xba*I	*Xca*I (*Sna*I)
*Xci*I (*Sal*I)	*Xcm*I	*Xcy*I (*Sma*I)	*Xgl*3216I (*Pvu*I)
*Xgl*3217I (*Pvu*I)	*Xgl*3218I (*Pvu*I)	*Xgl*3219I (*Pvu*I)	*Xgl*3220I (*Pvu*I)
*Xho*I	*Xho*II	*Xma*I (*Sma*I)	*Xma*II (*Pst*I)
*Xma*III	*Xml*I (*Pvu*I)	*Xml*AI (*Pvu*I)	*Xmn*I
*Xni*I (*Pvu*I)	*Xor*I (*Pst*I)	*Xor*II (*Pvu*I)	*Xpa*I (*Xho*I)
*Xph*I (*Pst*I)	*Yen*I (*Pst*I)	*Yen*AI (*Pst*I)	*Yen*BI (*Pst*I)
*Yen*CI (*Pst*I)	*Yen*DI (*Pst*I)	*Yen*EI (*Pst*I)	*Zan*I (*Eco*RII)
*Zsp*2I (*Ava*III)			

Enzymes with unknown recognition sequences

*Aam*I	*Acy*II	*Aim*I	*Ani*I
*Atu*AI	*Atu*BVI	*Atu*IAMI	*Bbe*II
*Bbe*AII	*Bbe*SI	*Bbi*IV	*Bce*1229I
*Bce*14579I	*Blo*I	*Bme*I	*Bme*205I
*Bme*899I	*Bpr*I	*Bsi*BI	*Bsi*EI
*Bsi*FI	*Bsi*GI	*Bsi*JI	*Bsi*NI
*Bsi*PI	*Bsi*RI	*Bsi*TI	*Bsi*UI
*Bsi*VI	*Bsi*YI	*Bsp*12II	*Bsr*PI
*Bss*PI	*Bst*AI	*Bst*EI	*Bsu*1145I
*Bsu*1259I	*Cal*I	*Chi*I	*Cli*III
*Clu*I	*Csu*I	*Cve*I	*Cvi*I
*Cvi*PI	*Cvi*QII	*Dmo*I	*Eco*CKI
*Eco*O34I	*Eco*O44I	*Esp*II	*Fnu*48I
*Fsa*I	*Ggl*I	*Gsp*AIII	*Hag*I
*Hap*I	*Hcu*I	*Hgi*FI	*Hgi*KI
*Hhl*I	*Hin*1056II	*Hsa*I	*Lpn*I
*Lpn*II	*Mbv*I	*Mgl*I	*Mgl*II
*Mnn*III	*Mno*II (*Mnn*III)	*Msi*II	*Mvi*I
*Mvi*II	*Nfl*II	*Nfl*III	*Ngo*DII
*Nme*I	*Nme*II	*Nme*III	*Nme*IV
*Nop*II	*Nov*I	*Nsp*EII	*Nsp*LIII
*Nsp*LIV	*Nun*I	*Oxa*II	*Pfl*I
*Pgl*II	*Pmi*I	*Pss*II	*Rhp*II
*Rle*I	*Rme*I	*Rrb*I	*Rrh*II
*Rru*AI	*Sac*III	*Sal*II	*San*I
*Sbr*I	*Sex*II	*Sgr*I	*Shy*TI
*Sin*MII	*Sis*I	*Sod*I	*Sod*II
*Ssa*I	*Ssc*I	*Ssp*XI	*Sst*III (*Sac*III)
*Stm*I	*Sty*D4I	*Tmi*I	*Tru*III

Table 2. Type II restriction enzymes[a]

Enzyme	Isoschizomers	Recognition sequence[b]	Enzyme	Isoschizomers	Recognition sequence[b]
*Aat*II		GACGT↑C		*Cfr*8I	GGNCC
*Acc*I		GT↑MKAC		*Cfr*13I	G↑GNCC
	*Dsa*VI	GTMKAC		*Cfr*23I	GGNCC
	*Fbl*I	GT↑MKAC		*Cfr*33I	GGNCC
*Acy*I		GR↑CGYC		*Cfr*45I	GGNCC
	*Aha*II	GR↑CGYC		*Cfr*46I	GGNCC
	*Aos*II	GR↑CGYC		*Cfr*47I	GGNCC
	*Ast*WI	GR↑CGYC		*Cfr*52I	GGNCC
	*Asu*III	GR↑CGYC		*Cfr*54I	GGNCC
	*Bbi*II	GR↑CGYC		*Cfr*NI	GGNCC
	*Hgi*I	GR↑CGYC		*Eco*39I	GGNCC
	*Hgi*DI	GR↑CGYC		*Eco*47II	GGNCC
	*Hgi*GI	GR↑CGYC		*Eco*196II	GGNCC
	*Hgi*HII	GR↑CGYC		*Eco*201I	GGNCC
	*Hin*1I	GR↑CGYC		*Gse*I	GGNCC
	*Hin*8I	GRCGYC		*Hin*5II	GGNCC
	*Nla*SII	GRCGYC		*Mja*II	GGNCC
	*Ssp*JII	GRCGYC		*Msp*24I	GGNCC
	*Ssp*M1II	GRCGYC		*Mth*TI	GGNCC
	*Ssp*M2II	GRCGYC		*Nla*DII	GGNCC
*Afl*II		C↑TTAAG		*Nmu*EII	GGNCC
	*Esp*4I	C↑TTAAG		*Nmu*SI	GGNCC
	*Vha*464I	C↑TTAAG		*Nsp*IV	G↑GNCC
*Afl*III		A↑CRYGT		*Nsp*LII	GGNCC
*Age*I		A↑CCGGT		*Pde*12I	G↑GNCC
*Aha*III		TTT↑AAA		*Pse*I	GGNCC
	*Dra*I	TTT↑AAA		*Psp*I	GGNCC
*Alu*I		AG↑CT		*Sau*96I	G↑GNCC
	*Mlt*I	AG↑CT		*Sau*BI	GGNCC
	*Otu*I	AGCT		*Sdy*I	GGNCC
	*Otu*NI	AGCT	*Asu*II		TT↑CGAA
	*Oxa*I	AGCT		*Aca*I	TTCGAA
*Alw*NI		CAGNNN↑CTG		*Avi*I	TTCGAA
*Apa*I		GGGCC↑C		*Bim*19I	TT↑CGAA
	*Bsp*120I	G↑GGCCC		*Bpu*14I	TT↑CGAA
	*Eci*EI	GGGCCC		*Bsi*CI	TTCGAA
*Apa*LI		G↑TGCAC		*Bsp*82I	TTCGAA
	*Alw*44I	G↑TGCAC		*Bsp*119I	TTCGAA
	*Ame*I	GTGCAC		*Bst*BI	TT↑CGAA
	*Sno*I	G↑TGCAC		*Csp*45I	TT↑CGAA
	*Vne*I	G↑TGCAC		*Fsp*II	TT↑CGAA
*Asu*I		G↑GNCC		*Lsp*I	TT↑CGAA
	*Aha*B1I	G↑GNCC		*Mla*I	TT↑CGAA
	*Apu*I	GGNCC		*Nsp*V	TTCGAA
	*Bac*36I	G↑GNCC		*Nsp*BI	TTCGAA
	*Bal*228I	G↑GNCC		*Nsp*FI	TTCGAA
	*Bav*AII	G↑GNCC		*Nsp*JI	TTCGAA
	*Bce*22I	G↑GNCC		*Sfu*I	TT↑CGAA
	*Bsh*KI	G↑GNCC		*Sgr*1839I	TTCGAA
	*Bsp*BII	G↑GNCC		*Ssp*1I	TT↑CGAA
	*Cfr*4I	GGNCC		*Ssp*152I	TTCGAA

Restriction/Methylation

Table 2. Continued

Enzyme	Isoschizomers	Recognition sequence[b]	Enzyme	Isoschizomers	Recognition sequence[b]
*Ava*I		C↑YCGRG		*Nmu*AII	GGWCC
	*Acr*I	CYCGRG		*Nsp*DII	GGWCC
	*Ama*87I	C↑YCGRG		*Nsp*GI	GGWCC
	*Aqu*I	C↑YCGRG		*Nsp*HII	GGWCC
	*Asp*BI	CYCGRG		*Nsp*KI	GGWCC
	*Asp*CI	CYCGRG		*Sfn*I	GGWCC
	*Asp*DI	CYCGRG		*Sgh*1835I	GGWCC
	*Avr*I	CYCGRG		*Sin*I	G↑GWCC
	*Bco*I	C↑YCGRG		*Sin*AI	GGWCC
	*Bst*SI	C↑YCGRG		*Sin*BI	GGWCC
	*Eco*88I	CYCGRG		*Sin*CI	GGWCC
	*Nli*I	CYCGRG		*Sin*DI	GGWCC
	*Nmu*AI	CYCGRG		*Sin*EI	GGWCC
	*Nsp*III	C↑YCGRG		*Sin*FI	GGWCC
	*Nsp*DI	CYCGRG		*Sin*GI	GGWCC
	*Nsp*EI	CYCGRG		*Sin*HI	GGWCC
	*Nsp*SAI	C↑YCGRG		*Sin*JI	GGWCC
*Ava*II		G↑GWCC		*Tru*I	GGWCC
	*Afl*I	G↑GWCC	*Ava*III		ATGCAT
	*Asp*697I	GGWCC		*Eco*T22I	ATGCA↑T
	*Asp*745I	G↑GWCC		*Nsi*I	ATGCA↑T
	*Asp*BII	GGWCC		*Uba*103I	ATGCA↑T
	*Asp*CII	GGWCC		*Zsp*2I	ATGCA↑T
	*Asp*DII	GGWCC	*Avr*II		C↑CTAGG
	*Bam*NxI	G↑GWCC	*Bal*I		TGG↑CCA
	*Bme*18I	G↑GWCC		*Msc*I	TGG↑CCA
	*Bme*216I	G↑GWCC		*Msp*20I	TGG↑CCA
	*Bsp*100I	GGWCC	*Bam*HI		G↑GATCC
	*Bsp*J105I	GGWCC		*Aac*I	GGATCC
	*Bti*I	GGWCC		*Aae*I	GGATCC
	*Cau*I	G↑GWCC		*Aca*II	GGATCC
	*Cli*I	GGWCC		*Acc*EBI	G↑GATCC
	*Clm*II	GGWCC		*Ain*II	GGATCC
	*Dsa*IV	G↑GWCC		*Ali*I	G↑GATCC
	*Eag*MI	G↑GWCC		*Ali*12257I	GGATCC
	*Eco*47I	G↑GWCC		*Ali*12258I	GGATCC
	*Erp*I	G↑GWCC		*Asp*TII	GGATCC
	*Fdi*I	G↑GWCC		*Bam*FI	GGATCC
	*Fsp*MSI	G↑GWCC		*Bam*KI	GGATCC
	*Gsp*AI	GGWCC		*Bam*NI	GGATCC
	*Hgi*BI	G↑GWCC		*Bca*1259I	GGATCC
	*Hgi*CII	G↑GWCC		*Bce*751I	G↑GATCC
	*Hgi*EI	G↑GWCC		*Bco*10278I	GGATCC
	*Hgi*HIII	G↑GWCC		*Bna*I	G↑GATCC
	*Hgi*JI	G↑GWCC		*Bsp*30I	GGATCC
	*Hsp*2I	GGWCC		*Bsp*46I	GGATCC
	*Kzo*49I	G↑GWCC		*Bsp*98I	GGATCC
	*Mfo*I	GGWCC		*Bst*I	G↑GATCC
	*Mli*I	GGWCC		*Bst*1126I	GGATCC
	*Msp*AI	GGWCC		*Bst*2464I	GGATCC
	*Nli*II	GGWCC		*Bst*2902I	GGATCC

Table 2. Continued

Enzyme	Isoschizomers	Recognition sequence[b]	Enzyme	Isoschizomers	Recognition sequence[b]
	*Bst*QI	GGATCC		*Msp*YI	YAC↑GTR
	*Bsu*90I	GG↑ATCC	*Bsa*BI		GATNN↑NNATC
	*Bsu*8565I	GGATCC		*Mam*I	GATNN↑NNATC
	*Bsu*8646I	GGATCC	*Bse*PI		GCGCGC
	*Ce*lI	GGATCC		*Bso*PI	GCGCGC
	*Dds*I	GGATCC		*Bsr*HI	GCGCGC
	*Gdo*I	GGATCC		*Bss*HII	G↑CGCGC
	*Gin*I	GGATCC		*Eco*143I	GCGCGC
	*Gox*I	GGATCC		*Esp*7I	GCGCGC
	*Gse*III	GGATCC		*Esp*8I	GCGCGC
	*Mle*I	GGATCC		*Kpn*30I	GCGCGC
	*Mlu*23I	G↑GATCC	*Bsi*I		CTCGTG(−5/−1)
	*Nas*BI	GGATCC	*Bsm*I		GAATGC(1/−1)
	*Nsp*SAIV	G↑GATCC	*Bsm*AI[d]		GTCTC(1/5)
	*Pae*177I	GGATCC		*Alw*26I	GTCTC(1/5)
	*Rhs*I	GGATCC	*Bsp*CI		GCNN↑NNGC
	*Rlu*4I	GGATCC	*Bsp*GI		CTGGAC
	*Sur*2I	G↑GATCC	*Bsp*HI		T↑CATGA
	*Uba*1110I	GGATCC		*Rsp*XI	T↑CATGA
*Bbv*I		GCAGC(8/12)	*Bsp*MI		ACCTGC(4/8)
	*Alw*XI	GCAGC(8/12)	*Bsp*MII		T↑CCGGA
	*Bsp*423I	GCAGC		*Acc*III	T↑CCGGA
	*Uba*1109I	GCAGC		*Bbf*7411I	TCCGGA
*Bbv*II		GAAGAC(2/6)		*Bbv*AIII	T↑CCGGA
	*Bbs*I	GAAGAC		*Bla*7920I	TCCGGA
	*Bbv*16I	GAAGAC(2/6)		*Bse*AI	T↑CCGGA
	*Bsp*VI	GAAGAC		*Bsi*MI	TCCGGA
*Bce*fI		ACGGC(12/13)		*Bsi*OI	TCCGGA
*Bcg*I[c]		GCANNNNNNTCG		*Bsp*13I	T↑CCGGA
*Bcl*I		T↑GATCA		*Bsp*EI	T↑CCGGA
	*Atu*CI	TGATCA		*Bsu*22I	TCCGGA
	*Bsi*QI	TGATCA		*Cau*B3I	T↑CCGGA
	*Bsp*XII	T↑GATCA		*Kpn*2I	T↑CCGGA
	*Bsi*GI	TGATCA		*Mro*I	T↑CCGGA
	*Bst*KI	TGATCA	*Bsr*I		ACTGG(1/−1)
	*Cpe*I	TGATCA	*Bst*EII		G↑GTNACC
	*Cth*I	TGATCA		*Acr*II	G↑GTNACC
	*Fba*I	TGATCA		*Asp*AI	G↑GTNACC
	*Ksp*22I	T↑GATCA		*Bsi*K1	GGTNACC
	*Pov*I	TGATCA		*Bst*31I	GGTNACC
	*Sse*I	TGATCA		*Bst*DI	GGTNACC
	*Sst*IV	TGATCA		*Bst*PI	G↑GTNACC
*Bgl*I		GCCNNNN↑NGGC		*Cfr*7I	GGTNACC
	*Van*I	GCCNNNNNGGC		*Cfr*19I	GGTNACC
*Bgl*II		A↑GATCT		*Eca*I	G↑GTNACC
	*Nsp*MACI	A↑GATCT		*Eco*91I	G↑GTNACC
*Bin*I		GGATC(4/5)		*Eco*O65I	G↑GTNACC
	*Alw*I	GGATC(4/5)		*Kox*I	GGTNACC
	*Bth*II	GGATC		*Nsp*SAII	G↑GTNACC
*Bpu*10I		CCTNAGC(−5/−2)		*Sci*AI	GGTNACC
*Bsa*AI		YAC↑GTR			

Restriction/Methylation

Table 2. Continued

Enzyme	Isoschizomers	Recognition sequence[b]	Enzyme	Isoschizomers	Recognition sequence[b]
*Bst*XI		CCANNNNN↑NTGG		*Bsp*106I	AT↑CGAT
	*Bss*GI	CCANNNNNNTGG		*Bsp*DI	AT↑CGAT
	*Bst*TI	CCANNNNNNTGG		*Bsp*XI	AT↑CGAT
*Cau*II		CC↑SGG		*Bsu*15I	AT↑CGAT
	*Aha*I	CC↑SGG		*Btu*I	ATCGAT
	*Ase*II	CC↑SGG		*Csp*4I	ATCGAT
	*Asp*1I	CCSGG		*Lpl*I	AT↑CGAT
	*Bcn*I	CC↑SGG		*Pgl*B4I	AT↑CGAT
	*Bsp*7I	CCSGG	*Cvi*JI		RG↑CY
	*Bsp*8I	CCSGG		*Cvi*KI	RGCY
	*Bsp*55I	CCSGG		*Cvi*LI	RGCY
	*Bsp*J67I	CCSGG		*Cvi*MI	RGCY
	*Eci*DI	CCSGG		*Cvi*NI	RGCY
	*Eco*121I	CCSGG		*Cvi*OI	RGCY
	*Eco*179I	CCSGG	*Dde*I		C↑TNAG
	*Eco*190I	CCSGG	*Dpn*I		GA↑TC
	*Hgi*S21I	CCSGG		*Cfu*I	GA↑TC
	*Hgi*S22I	CC↑SGG		*Nan*II	GATC
	*Hin*3I	CCSGG		*Ngo*DIII	GATC
	*Nci*I	CC↑SGG		*Nmu*DI	GATC
	*Pae*181I	CCSGG		*Nmu*EI	GATC
	*Rsh*II	CCSGG		*Nsu*DI	GATC
	*Ssp*2I	CCSGG		*Uba*1103II	GATC
	*Tmu*1I	CCSGG	*Dra*II		RG↑GNCCY
*Cfr*I		Y↑GGCCR		*Eco*O109I	RG↑GNCCY
	*Cfr*14I	YGGCCR		*Pss*I	RGGNC↑CY
	*Cfr*38I	YGGCCR		*Vne*AI	RGGNCCY
	*Cfr*39I	YGGCCR	*Dra*III		CACNNN↑GTG
	*Cfr*40I	YGGCCR	*Drd*I		GACNNNN↑NNGTC
	*Eae*I	Y↑GGCCR	*Drd*II		GAACCA
	*Eci*BI	YGGCCR	*Dsa*I		C↑CRYGG
	*Eco*90I	YGGCCR	*Eci*I		TCCGCC
	*Eco*164I	YGGCCR	*Eco*31I		GGTCTC(1/5)
	*Eco*HI	YGGCCR		*Bli*49I	GGTCTC
*Cfr*10I		R↑CCGGY		*Bsa*I	GGTCTC(1/5)
	*Bsp*21I	RCCGGY		*Eco*42I	GGTCTC
*Cla*I		AT↑CGAT		*Eco*51I	GGTCTC
	*Aag*I	AT↑CGAT		*Eco*95I	GGTCTC
	*Apu*16I	ATCGAT		*Eco*97I	GGTCTC
	*Asp*707I	ATCGAT		*Eco*101I	GGTCTC
	*Ban*III	ATCGAT		*Eco*120I	GGTCTC
	*Bbv*AII	AT↑CGAT		*Eco*127I	GGTCTC
	*Bcm*I	AT↑CGAT		*Eco*129I	GGTCTC
	*Bdi*I	AT↑CGAT		*Eco*155I	GGTCTC
	*Bli*41I	AT↑CGAT		*Eco*156I	GGTCTC
	*Bli*86I	AT↑CGAT		*Eco*157I	GGTCTC
	*Bli*RI	ATCGAT		*Eco*162I	GGTCTC
	*Bsc*I	AT↑CGAT		*Eco*185I	GGTCTC
	*Bsp*2I	ATCGAT		*Eco*191I	GGTCTC
	*Bsp*4I	ATCGAT		*Eco*203I	GGTCTC
	*Bsp*84I	ATCGAT		*Eco*204I	GGTCTC

Table 2. Continued

Enzyme	Isoschizomers	Recognition sequence[b]	Enzyme	Isoschizomers	Recognition sequence[b]
	Eco205I	GGTCTC		Cfr25I	CCWGG
	Eco217I	GGTCTC		Cfr27I	CCWGG
	Eco225I	GGTCTC		Cfr28I	CCWGG
	Eco233I	GGTCTC		Cfr29I	CCWGG
	Eco239I	GGTCTC		Cfr30I	CCWGG
	Eco240I	GGTCTC		Cfr31I	CCWGG
	Eco241I	GGTCTC		Cfr35I	CCWGG
	Eco246I	GGTCTC		CfrS37I	CCWGG
	Eco247I	GGTCTC		CthII	CC↑WGG
	EcoA4I	GGTCTC(1/5)		EagKI	CCWGG
	PpaI	GGTCTC		EcaII	CCWGG
	Sau12I	GGTCTC		EclII	CCWGG
Eco47III		AGC↑GCT		Ecl66I	CCWGG
	AitI	AGC↑GCT		Ecl136I	CCWGG
Eco57I		CTGAAG(16/14)		Ecl137II	CCWGG
	Bsp6II	CTGAAG		EclS39I	CCWGG
	Eco125I	CTGAAG		Eco38I	CCWGG
	FsfI	CTGAAG		Eco40I	CCWGG
EcoNI		CCTNN↑NNNAGG		Eco41I	CCWGG
	BstWI	CCTNNNNNAGG		Eco60I	CCWGG
EcoRI		G↑AATTC		Eco61I	CCWGG
	Eco82I	GAATTC		Eco67I	CCWGG
	Eco159I	GAATTC		Eco70I	CCWGG
	Eco228I	GAATTC		Eco71I	CCWGG
	Eco237I	GAATTC		Eco128I	CCWGG
	Eco252I	GAATTC		Eco170I	CCWGG
	HalB6I	G↑AATTC		Eco193I	CCWGG
	RsrI	G↑AATTC		Eco206I	CCWGG
	SsoI	G↑AATTC		Eco207I	CCWGG
EcoRII		↑CCWGG		Esp2I	CCWGG
	Acc38I	CCWGG		Fsp1604I	CC↑WGG
	AeuI	CC↑WGG		Kox165I	CCWGG
	AorI	CC↑WGG		Kpn10I	CCWGG
	ApyI	CC↑WGG		Kpn13I	CCWGG
	AtuII	CCWGG		Kpn14I	CCWGG
	Atu1I	CCWGG		Kpn16I	CCWGG
	AtuBI	CCWGG		Mlu2300I	CCWGG
	BinSI	CCWGG		MphI	CCWGG
	BshGI	CC↑WGG		MvaI	CC↑WGG
	BsiLI	CCWGG		Scg2I	CCWGG
	Bsp56I	CCWGG		Sfl2aI	CCWGG
	Bsp103I	CCWGG		Sfl2bI	CCWGG
	BstGII	CCWGG		Sgr20I	CCWGG
	BstNI	CC↑WGG		SleI	CCWGG
	BstOI	CCWGG		SspAI	CCWGG
	Cdi27I	CCWGG		TaqXI	CC↑WGG
	Cfr5I	CCWGG		ZanI	CC↑WGG
	Cfr11I	CCWGG	EcoRV		GAT↑ATC
	Cfr20I	CCWGG		Bsp16I	GATATC
	Cfr22I	CCWGG		BstRI	GATATC
	Cfr24I	CCWGG		CeqI	GAT↑ATC

Restriction/Methylation

Table 2. Continued

Enzyme	Isoschizomers	Recognition sequence[b]	Enzyme	Isoschizomers	Recognition sequence[b]
	Eco32I	GAT↑ATC	HaeI		WGG↑CCW
	Eco178I	GATATC	HaeII		RGCGC↑Y
	HjaI	GAT↑ATC		Bme142I	RGC↑GCY
	NanI	GATATC		HinHI	RGCGCY
	NflAI	GATATC		NgoI	RGCGCY
	NsiCI	GAT↑ATC	HaeIII		GG↑CC
EspI		GC↑TNAGC		AcaIV	GGCC
	CelII	GCTNAGC		Asp742I	GGCC
	Uba1102I	GC↑TNAGC		AspTIII	GGCC
Esp3I		GAGACG(1/5)		Bal475I	GGCC
FauI		CCCGC(4/6)		Bal3006	GGCC
FinI		GTCCC		Bce71I	GGCC
Fnu4HI		GC↑NGC		Bco33I	GGCC
	Bsp6I	GC↑NGC		Bfi458I	GGCC
	FbrI	GC↑NGC		Bim19II	GG↑CC
	IspI	GC↑NGC		BliI	GGCC
	Uur960I	GC↑NGC		BluII	GGCC
FnuDII		CG↑CG		BseI	GGCC
	AccII	CG↑CG		BshI	GGCC
	BceFI	CGCG		BshAI	GGCC
	BceRI	CGCG		BshBI	GGCC
	BepI	CG↑CG		BshCI	GGCC
	Bpu95I	CG↑CG		BshDI	GGCC
	Bsp50I	CG↑CG		BshEI	GGCC
	Bsp70I	CGCG		BshFI	GG↑CC
	BspJ76I	CGCG		BsiAI	GGCC
	BstUI	CG↑CG		BsiDI	GGCC
	Bsu1192II	CGCG		BsiHI	GGCC
	Bsu1193I	CGCG		Bsp71I	GGCC
	Bsu1532I	CG↑CG		Bsp211I	GG↑CC
	Bsu6633I	CGCG		Bsp226I	GGCC
	BsuEII	CGCG		BspBRI	GG↑CC
	Cpa1150I	CGCG		BspRI	GG↑CC
	CpaAI	CGCG		BssCI	GGCC
	FspMI	CGCG		BstCI	GGCC
	Hin1056I	CGCG		BstJI	GGCC
	MvaAI	CGCG		Bsu1076I	GGCC
	MvnI	CG↑CG		Bsu1114I	GGCC
	PflAI	CGCG		BsuRI	GG↑CC
	SceI	CGCG		BteI	GGCC
	ThaI	CG↑CG		ClmI	GGCC
	TmaI	CGCG		CltI	GG↑CC
FokI		GGATG(9/13)		Csp2I	GGCC
	HinGUII	GGATG		DsaII	GG↑CC
FseI		GGCCGG↑CC		FinSI	GGCC
GdiII		YGGCCG(−5/−1)		FnuDI	GG↑CC
GsuI		CTGGAG(16/14)		HhgI	GGCC
	Bco35I	CTGGAG		MniI	GGCC
	Bsp22I	CTGGAG		MnnII	GGCC
	Bsp28I	CTGGAG		NgoII	GGCC
	BspJ74I	CTGGAG		NgoPII	GG↑CC

Table 2. Continued

Enzyme	Isoschizomers	Recognition sequence[b]	Enzyme	Isoschizomers	Recognition sequence[b]
	*Ngo*SI	GGCC		*Bvu*I	GRGCY↑C
	*Nla*I	GGCC		*Cfr*48I	GRGCYC
	*Pai*I	GGCC		*Eco*24I	GRGCY↑C
	*Pal*I	GG↑CC		*Eco*25I	GRGCYC
	*Pde*133I	GG↑CC		*Eco*26I	GRGCYC
	*Ppu*I	GGCC		*Eco*35I	GRGCYC
	*Sfa*I	GG↑CC		*Eco*68I	GRGCYC
	*Spl*III	GGCC		*Eco*113I	GRGCYC
	*Sua*I	GG↑CC		*Eco*180I	GRGCYC
	*Sul*I	GGCC		*Eco*211I	GRGCYC
	*Tsp*ZNI	GGCC		*Eco*215I	GRGCYC
	*Tte*AI	GGCC		*Eco*216I	GRGCYC
	*Ttn*I	GGCC		*Eco*232I	GRGCYC
	*Vha*I	GGCC		*Eco*T38I	GRGCYC
	*Vni*I	GGCC		*Eco*T88I	GRGCYC
*Hga*I		GACGC(5/10)		*Eco*T93I	GRGCYC
*Hgi*AI		GWGCW↑C		*Eco*T95I	GRGCYC
	*Alw*21I	GWGCW↑C		*Kox*II	GRGCY↑C
	*Asp*HI	GWGCW↑C	*Hha*I		GCG↑C
	*Bbv*12I	GWGCW↑C		*Bca*I	GCGC
	*Pph*3215I	GWGCWC		*Cfo*I	GCG↑C
*Hgi*CI		G↑GYRCC		*Csp*1470I	GCGC
	*Acc*B1I	G↑GYRCC		*Fnu*DIII	GCG↑C
	*Ban*I	G↑GYRCC		*Hin*6I	G↑CGC
	*Eco*50I	GGYRCC		*Hin*7I	GCGC
	*Eco*64I	G↑GYRCC		*Hin*GUI	GCGC
	*Eco*168I	GGYRCC		*Hin*P1I	G↑CGC
	*Eco*169I	GGYRCC		*Hin*S1I	GCGC
	*Eco*171I	GGYRCC		*Hin*S2I	GCGC
	*Eco*173I	GGYRCC		*Mnn*IV	GCGC
	*Eco*195I	GGYRCC		*Sci*NI	G↑CGC
	*Esp*1I	GGYRCC	*Hin*dII		GTY↑RAC
	*Esp*6I	GGYRCC		*Chu*II	GTYRAC
	*Esp*9I	GGYRCC		*Hin*1160II	GTYRAC
	*Esp*10I	GGYRCC		*Hin*1161II	GTYRAC
	*Esp*11I	GGYRCC		*Hin*JCI	GTY↑RAC
	*Esp*12I	GGYRCC		*Hin*cII	GTY↑RAC
	*Esp*13I	GGYRCC		*Mnn*I	GTYRAC
	*Esp*14I	GGYRCC	*Hin*dIII		A↑AGCTT
	*Esp*15I	GGYRCC		*Asp*52I	AAGCTT
	*Esp*22I	GGYRCC		*Asp*3065I	AAGCTT
	*Hgi*HI	G↑GYRCC		*Bbr*I	AAGCTT
	*Msp*B4I	G↑GYRCC		*Bpe*I	AAGCTT
	*Ssp*M1III	GGYRCC		*Bst*FI	A↑AGCTT
*Hgi*EII		ACCNNNNNGGT		*Cfr*32I	AAGCTT
*Hgi*JII		GRGCY↑C		*Chu*I	AAGCTT
	*Ban*II	GRGCY↑C		*Eco*VIII	A↑AGCTT
	*Bpu*I	GRGCYC		*Eco*65I	AAGCTT
	*Bsp*117I	GRGCYC		*Eco*98I	AAGCTT
	*Bsp*519I	GRGCY↑C		*Eco*188I	AAGCTT
	*Bsu*1854I	GRGCY↑C		*Eco*231I	AAGCTT

Restriction/Methylation

Table 2. Continued

Enzyme	Isoschizomers	Recognition sequence[b]
	Hin5III	AAGCTT
	Hin173I	AAGCTT
	Hin1076III	AAGCTT
	HinJCII	AAGCTT
	HinbIII	AAGCTT
	HinfII	AAGCTT
	HsuI	A↑AGCTT
	MkiI	AAGCTT
HinfI		G↑ANTC
	CviBI	G↑ANTC
	CviCI	GANTC
	CviDI	GANTC
	CviEI	GANTC
	CviFI	GANTC
	CviGI	GANTC
	FnuAI	G↑ANTC
	HhaII	G↑ANTC
	NcaI	GANTC
	NovII	GANTC
	NsiHI	GANTC
HpaI		GTT↑AAC
	BseII	GTTAAC
	SsrI	GTT↑AAC
	SsrB6I	GTT↑AAC
HpaII		C↑CGG
	Asp748I	CCGG
	BsiSI	CCGG
	Bsp5I	CCGG
	Bsp47I	CCGG
	Bsp48I	CCGG
	Bsp116I	CCGG
	Bst40I	C↑CGG
	Bsu1192I	CCGG
	BsuFI	CCGG
	FinII	CCGG
	HapII	C↑CGG
	Hin2I	CCGG
	Hin5I	CCGG
	MniII	CCGG
	MnoI	C↑CGG
	MspI	C↑CGG
	Pde137I	C↑CGG
	SecII	CCGG
	SfaGUI	CCGG
HphI		GGTGA(8/7)
	NgoBI	GGTGA
KpnI		GGTAC↑C
	Acc65I	G↑GTACC
	AhaB8I	G↑GTACC
	Asp718I	G↑GTACC
	BspJ106I	GGTACC
	Eco149I	GGTACC
	Esp19I	GGTACC
	KpnK14I	GGTACC
	NmiI	GGTACC
	Sau10I	GGTACC
	SthI	G↑GTACC
	SthAI	GGTACC
	SthBI	GGTACC
	SthCI	GGTACC
	SthDI	GGTACC
	SthEI	GGTACC
	SthFI	GGTACC
	SthGI	GGTACC
	SthHI	GGTACC
	SthJI	GGTACC
	SthKI	GGTACC
	SthLI	GGTACC
	SthMI	GGTACC
	SthNI	GGTACC
Ksp632I		CTCTTC(1/4)
	EarI	CTCTTC(1/4)
	Uba1104I	CTCTTC(1/5)
MaeI		C↑TAG
	MjaI	CTAG
MaeII		A↑CGT
MaeIII		↑GTNAC
MboI		↑GATC
	Bce243I	↑GATC
	Bme12I	GATC
	BsaPI	GATC
	Bsp9I	GATC
	Bsp18I	GATC
	Bsp49I	GATC
	Bsp51I	GATC
	Bsp52I	GATC
	Bsp54I	GATC
	Bsp57I	GATC
	Bsp58I	GATC
	Bsp59I	GATC
	Bsp60I	GATC
	Bsp61I	GATC
	Bsp64I	GATC
	Bsp65I	GATC
	Bsp66I	GATC
	Bsp67I	↑GATC
	Bsp72I	GATC
	Bsp74I	GATC
	Bsp76I	GATC
	Bsp91I	GATC
	Bsp105I	↑GATC
	Bsp122I	GATC

Table 2. Continued

Enzyme	Isoschizomers	Recognition sequence[b]	Enzyme	Isoschizomers	Recognition sequence[b]
	Bsp2095I	↑GATC		Sau3AI	↑GATC
	BspAI	↑GATC		Sau6782I	GATC
	BspJ64I	GATC		SauCI	GATC
	BsrPII	GATC		SauDI	GATC
	BssGII	GATC		SauEI	GATC
	BstEIII	GATC		SauFI	GATC
	BstXII	GATC		SauGI	GATC
	BtcI	GATC		SauMI	GATC
	CacI	↑GATC		SinMI	GATC
	CcyI	↑GATC		TruII	GATC
	Cin1467I	GATC	MboII		GAAGA(8/7)
	CpaI	GATC		NcuI	GAAGA
	CpfI	↑GATC		TceI	GAAGA
	Csp5I	GATC	McrI		C↑GRYCG
	Cte1179I	GATC	MfeI		CAATTG
	Cte1180	GATC	MluI		A↑CGCGT
	CtyI	GATC		ApeI	ACGCGT
	CviAI	↑GATC		Uba6I	ACGCGT
	CviHI	GATC	MlyI		GASTC
	DpnII	GATC	MmeI		TCCRAC(20/18)
	FnuAII	GATC	MnlI		CCTC(7/7)
	FnuCI	↑GATC	MseI		T↑TAA
	FnuEI	↑GATC		Tru9I	T↑TAA
	HacI	↑GATC	MstI		TGC↑GCA
	Kzo9I	↑GATC		AcaIII	TGCGCA
	MeuI	GATC		AosI	TGC↑GCA
	MkrAI	GATC		AviII	TGC↑GCA
	MmeII	GATC		ClcII	TGCGCA
	MnoIII	GATC		CltII	TGCGCA
	MosI	GATC		FdiII	TGC↑GCA
	Msp67II	GATC		FspI	TGC↑GCA
	MspBI	GATC		GspAII	TGCGCA
	MthI	GATC		NspHIII	TGCGCA
	Mth1047I	GATC		NspLI	TGCGCA
	MthAI	GATC		NspMI	TGCGCA
	NdeII	↑GATC	MwoI		GCNNNNN↑NNGC
	NflI	GATC	NaeI		GCC↑GGC
	NflAII	GATC		AmeII	GCCGGC
	NflBI	GATC		AniMI	GCCGGC
	NlaII	↑GATC		ApeAI	GCCGGC
	NlaDI	GATC		AprI	GCCGGC
	NmeCI	↑GATC		Eco56I	G↑CCGGC
	NphI	↑GATC		Esp5I	GCCGGC
	NsiAI	GATC		MisI	GCCGGC
	NspAI	GATC		NasWI	GCCGGC
	NsuI	GATC		NbaI	GCCGGC
	Pei9403I	GATC		NbrI	GCCGGC
	PfaI	GATC		NgoAIV	G↑CCGGC
	RluII	GATC		NgoMI	GCCGGC
	SalAI	GATC		NmuI	GCCGGC
	SalHI	GATC		NmuFI	GCCGGC

Restriction/Methylation

Table 2. Continued

Enzyme	Isoschizomers	Recognition sequence[b]	Enzyme	Isoschizomers	Recognition sequence[b]
	NspWI	GCCGGC	PflMI		CCANNNN↑NTGG
	NtaSII	GCCGGC		AccB7I	CCANNNN↑NTGG
	PglI	GCCGGC		Van91I	CCANNNN↑NTGG
	Psp61I	GCCGGC	PleI		GAGTC(4/5)
	RluI	GCCGGC	PmaCI		CAC↑GTG
	SacAI	GCCGGC		BbrPI	CAC↑GTG
	SalCI	GCCGGC		BcoAI	CAC↑GTG
	SaoI	GCCGGC		Bsp87I	CACGTG
	SauAI	GCCGGC		Eco72I	CAC↑GTG
	SauBMKI	GCC↑GGC		PmlI	CAC↑GTG
	SkaI	GCCGGC	PpuMI		RG↑GWCCY
	Slu1777I	GCC↑GGC		Uba1I	RG↑GWCCY
NarI		GG↑CGCC		Uba1106I	RGGWCCY
	BbeI	GGCGC↑C	PshAI		GACNN↑NNGTC
	BbeAI	GGCGCC	PstI		CTGCA↑G
	BinSII	GGCGCC		AinI	CTGCAG
	Eco78I	GGC↑GCC		Ali2882I	CTGCAG
	EheI	GGC↑GCC		AliAJI	CTGCA↑G
	MchI	GG↑CGCC		Asp36I	CTGCAG
	Mly113I	GG↑CGCC		Asp708I	CTGCAG
	NamI	GGCGCC		Asp713I	CTGCA↑G
	NdaI	GG↑CGCC		AspTI	CTGCAG
	NunII	GG↑CGCC		BbiI	CTGCAG
	SfoI	GGCGCC		Bce170I	CTGCAG
NcoI		C↑CATGG		Bsp17I	CTGCAG
	Bsp19I	C↑CATGG		Bsp43I	CTGCAG
	NspSAIII	CCATGG		Bsp63I	CTGCA↑G
NdeI		CA↑TATG		Bsp78I	CTGCAG
NheI		G↑CTAGC		Bsp81I	CTGCAG
NlaIII		CATG↑		Bsp93I	CTGCAG
	Hin1II	CATG↑		Bsp107I	CTGCAG
	Hin8II	CATG		Bsp108I	CTGCAG
NlaIV		GGN↑NCC		BspBI	CTGCA↑G
	AspNI	GGN↑NCC		BsuBI	CTGCAG
	BcrI	GGNNCC		CauIII	CTGCAG
	Bsp29I	GGNNCC		CflI	CTGCA↑G
	BssI	GGNNCC		CfrA4I	CTGCA↑G
	Rlu3I	GGNNCC		CfuII	CTGCA↑G
NotI		GC↑GGCCGC		ClcI	CTGCAG
NruI		TCG↑CGA		CstI	CTGCA↑G
	AmaI	TCGCGA		EaePI	CTGCAG
	Bsp68I	TCGCGA		Ecl77I	CTGCAG
	MluB2I	TCG↑CGA		Ecl133I	CTGCAG
	SalDI	TCGCGA		Ecl593I	CTGCAG
	Sbo13I	TCG↑CGA		Eco48I	CTGCAG
	Sna3286I	TCGCGA		Eco49I	CTGCAG
	SpoI	TCG↑CGA		Eco83I	CTGCAG
NspI		RCATG↑Y		Eco161I	CTGCAG
	NspHI	RCATG↑Y		Eco167I	CTGCAG
NspBII		CMG↑CKG		Esp5II	CTGCAG
	MspA1I	CMG↑CKG		Esp141I	CTGCAG

Table 2. Continued

Enzyme	Isoschizomers	Recognition sequence[b]	Enzyme	Isoschizomers	Recognition sequence[b]
	*Gse*II	CTGCAG	*Pvu*II		CAG↑CTG
	*Hal*B6II	CTGCA↑G		*Bav*I	CAG↑CTG
	*Kpn*12I	CTGCAG		*Bav*AI	CAG↑CTG
	*Mau*I	CTGCAG		*Cfr*6I	CAG↑CTG
	*Mkr*I	CTGCAG		*Ecl*I	CAG↑CTG
	*Nas*I	CTGCAG		*Gsp*I	CAGCTG
	*Ngb*I	CTGCAG		*Mzi*I	CAGCTG
	*Noc*I	CTGCAG		*Nme*RI	CAGCTG
	*Pma*I	CTGCAG		*Pvu*HKUI	CAGCTG
	*Pma*44I	CTGCAG		*Sba*I	CAGCTG
	*Pmy*I	CTGCAG		*Sci*AII	CAGCTG
	*Sal*PI	CTGCA↑G		*Sma*AIV	CAGCTG
	*Sfl*I	CTGCA↑G		*Sol*3335I	CAGCTG
	*Ska*II	CTGCAG		*Spl*AIV	CAGCTG
	*Xma*II	CTGCAG	*Rle*AI		CCCACA(12/9)
	*Xor*I	CTGCAG	*Rsa*I		GT↑AC
	*Xph*I	CTGCAG		*Afa*I	GT↑AC
	*Yen*I	CTGCA↑G		*Csp*6I	G↑TAC
	*Yen*AI	CTGCAG		*Cvi*QI	G↑TAC
	*Yen*BI	CTGCAG	*Rsr*II		CG↑GWCCG
	*Yen*CI	CTGCAG		*Cpo*I	CGGWCCG
	*Yen*DI	CTGCAG		*Csp*I	CGGWCCG
	*Yen*EI	CTGCAG	*Sac*I		GAGCT↑C
*Pvu*I		CGAT↑CG		*Ecl*136II	GAG↑CTC
	*Asp*22MI[d]	CGAT↑CG		*Ecl*137I	GAGCTC
	*Asp*16RI[d]	CGAT↑CG		*Eco*ICRI	GAGCTC
	*Bma*I	CGATCG		*Nas*SI	GAGCTC
	*Bma*AI	CGATCG		*Sco*I	GAGCTC
	*Bma*BI	CGATCG		*Sst*I	GAGCT↑C
	*Bma*CI	CGATCG	*Sac*II		CCGC↑GG
	*Bma*DI	CGATCG		*Aos*III	CCGCGG
	*Cfr*51I	CGATCG		*Bac*I	CCGCGG
	*Drd*III	CGATCG		*Bac*465I	CCGCGG
	*Ecl*JI	CGATCG		*Bsp*12I	CCGCGG
	*Erh*B9I	CGAT↑CG		*Cfr*37I	CCGCGG
	*Nbl*I	CGAT↑CG		*Cfr*41I	CCGCGG
	*Ple*19I	CGAT↑CG		*Cfr*42I	CCGC↑GG
	*Psu*161I	CGAT↑CG		*Cfr*43I	CCGCGG
	*Rsh*I	CGAT↑CG		*Cfr*45II	CCGCGG
	*Rsp*I	CGATCG		*Csc*I	CCGC↑GG
	*Sma*AIII	CGATCG		*Drd*AI	CCGCGG
	*Spl*AIII	CGATCG		*Drd*BI	CCGCGG
	*Xgl*3216I	CGATCG		*Drd*CI	CCGCGG
	*Xgl*3217I	CGATCG		*Drd*EI	CCGCGG
	*Xgl*3218I	CGATCG		*Drd*FI	CCGCGG
	*Xgl*3219I	CGATCG		*Dsp*1I	CCGCGG
	*Xgl*3220I	CGATCG		*Eae*46I	CCGC↑GG
	*Xml*I	CGATCG		*Ecc*I	CCGCGG
	*Xml*AI	CGATCG		*Ecl*28I	CCGCGG
	*Xni*I	CGATCG		*Ecl*37I	CCGCGG
	*Xor*II	CGAT↑CG		*Eco*55I	CCGCGG

Restriction/Methylation

Table 2. Continued

Enzyme	Isoschizomers	Recognition sequence[b]
	Eco92I	CCGCGG
	Eco96I	CCGCGG
	Eco99I	CCGCGG
	Eco100I	CCGCGG
	Eco104I	CCGCGG
	Eco134I	CCGCGG
	Eco135I	CCGCGG
	Eco158I	CCGCGG
	Eco182I	CCGCGG
	Eco196I	CCGCGG
	Eco208I	CCGCGG
	FscI	CCGCGG
	GalI	CCGC↑GG
	GceI	CCGC↑GG
	GceGLI	CCGC↑GG
	KspI	CCGC↑GG
	MraI	CCGCGG
	NgoIII	CCGCGG
	NgoAIII	CCGC↑GG
	NgoDI	CCGCGG
	NgoPIII	CCGC↑GG
	NlaDIII	CCGCGG
	NlaSI	CCGCGG
	PaeAI	CCGC↑GG
	SaaI	CCGCGG
	SabI	CCGCGG
	SakI	CCGCGG
	SboI	CCGCGG
	SfrI	CCGCGG
	Sfr303I	CCGC↑GG
	Sfr382I	CCGCGG
	ShyI	CCGCGG
	SseII	CCGCGG
	Ssp1725I	CCGCGG
	SstII	CCGC↑GG
	TglI	CCGCGG
SalI		G↑TCGAC
	HgiCIII	G↑TCGAC
	HgiDII	G↑TCGAC
	KoyI	GTCGAC
	NopI	G↑TCGAC
	RflFI	GTCGAC
	RheI	GTCGAC
	RhpI	GTCGAC
	RrhI	GTCGAC
	RroI	GTCGAC
	XamI	GTCGAC
	XciI	G↑TCGAC
SauI		CC↑TNAGG
	AocI	CC↑TNAGG
	AxyI	CC↑TNAGG
	Bse21I	CC↑TNAGG
	Bsu36I	CC↑TNAGG
	CvnI	CC↑TNAGG
	EciCI	CCTNAGG
	Eco76I	CCTNAGG
	Eco81I	CC↑TNAGG
	Eco115I	CCTNAGG
	Eco118I	CCTNAGG
	Lmu60I	CC↑TNAGG
	MstII	CC↑TNAGG
	OxaNI	CC↑TNAGG
	SecIII	CCTNAGG
	SshAI	CC↑TNAGG
ScaI		AGT↑ACT
	Asp763I	AGTACT
	BstMI	AGTACT
ScrFI		CC↑NGG
	Bsp53I	CCNGG
	Bsp73I	CCNGG
	DsaV	↑CCNGG
	Eco43I	CCNGG
	Eco51II	CCNGG
	Eco80I	CCNGG
	Eco85I	CCNGG
	Eco93I	CCNGG
	Eco153I	CCNGG
	Eco200I	CCNGG
	Msp67I	CC↑NGG
	SsoII	↑CCNGG
SduI		GDGCH↑C
	AocII	GDGCH↑C
	Bka1125I	GDGCHC
	BmyI	GDGCH↑C
	Bsp1286I	GDGCH↑C
	NspII	GDGCH↑C
SecI		C↑CNNGG
	BsaJI	C↑CNNGG
SfaNI		GCATC(5/9)
	BscAI	GCATC
SfeI		C↑TRYAG
	BdiSI	C↑TRYAG
SfiI		GGCCNNNN↑NGGCC
SgrAI		CR↑CCGGYG
SmaI		CCC↑GGG
	AhyI	C↑CCGGG
	Cfr9I	C↑CCGGG
	CfrJ4I	CCC↑GGG
	EclRI	CCCGGG
	PaeBI	CCC↑GGG
	XcyI	C↑CCGGG
	XmaI	C↑CCGGG

Table 2. Continued

Enzyme	Isoschizomers	Recognition sequence[b]	Enzyme	Isoschizomers	Recognition sequence[b]
SnaI		GTATAC	TspEI		AATT
	Uba1107I	GTA↑TAC	Tth111I		GACN↑NNGTC
	XcaI	GTA↑TAC		AspI	GACN↑NNGTC
SnaBI		TAC↑GTA		FsuI	GACNNNGTC
	EciAI	TACGTA		NtaI	GACNNNGTC
	Eco105I	TAC↑GTA		SmaAII	GACNNNGTC
	Eco158II	TACGTA		SplII	GACNNNGTC
	SspJI	TACGTA		SplAII	GACNNNGTC
	SspM1I	TACGTA		TspI	GACNNNGTC
	SspM2I	TACGTA		TteI	GACNNNGTC
SpeI		A↑CTAGT		TtrI	GACNNNGTC
SphI		GCATG↑C	Tth111II		CAARCA(11/9)
	BbuI	GCATG↑C	Uba105I		GACNNN↑NNGTC
	Bsp121I	GCATGC	Uba1108I		TCGTAG
	PaeI	GCATG↑C	VspI		AT↑TAAT
	SpaXI	GCATGC		AseI	AT↑TAAT
SpiI		CCGC		AsnI	AT↑TAAT
SplI		C↑GTACG	XbaI		T↑CTAGA
	PfuI	CGTACG	XcmI		CCANNNNN↑NNNNTGG
	SmaAI	CGTACG	XhoI		C↑TCGAG
	SplAI	CGTACG		AbrI	C↑TCGAG
	SspKI	CGTACG		Asp47I	CTCGAG
SspI		AAT↑ATT		Asp703I	CTCGAG
StuI		AGG↑CCT		BadI	CTCGAG
	AatI	AGG↑CCT		BbfI	CTCGAG
	Asp78I	AGGCCT		BbiIII	CTCGAG
	ChyI	AGGCCT		BluI	C↑TCGAG
	Eco147I	AGG↑CCT		Bsp92I	CTCGAG
	GdiI	AGG↑CCT		BssHI	CTCGAG
	NtaSI	AGGCCT		BstHI	CTCGAG
	Pme55I	AGG↑CCT		BstLI	CTCGAG
	SarI	AGGCCT		BstVI	C↑TCGAG
	SsvI	AGGCCT		BsuMI	CTCGAG
	SteI	AGGCCT		BthI	CTCGAG
StyI		C↑CWWGG		CcrI	C↑TCGAG
	BssT1I	C↑CWWGG		CjaI	CTCGAG
	Eco130I	C↑CWWGG		DdeII	CTCGAG
	Eco208II	CCWWGG		DrdDI	CTCGAG
	EcoT14I	C↑CWWGG		MavI	C↑TCGAG
	EcoT104I	CCWWGG		McaI	CTCGAG
	ErhB9II	C↑CWWGG		MecI	CTCGAG
	SblAI	CCWWGG		MlaAI	CT↑CGAG
	SblBI	CCWWGG		MpuI	CTCGAG
	SblCI	CCWWGG		MrhI	CTCGAG
TaqI		T↑CGA		MsiI	CTCGAG
	TflI	TCGA		OcoI	CTCGAG
	TthHB8I	T↑CGA		PaeR7I	C↑TCGAG
TaqII[e]		GACCGA(11/9)		PanI	C↑TCGAG
		CACCCA(11/9)		PflNI	CTCGAG
TfiI		GAWTC		PflWI	CTCGAG
Tsp45I		GTSAC		Sal1974I	CTCGAG

Table 2. Continued

Enzyme	Isoschizomers	Recognition sequence[b]
	Sau3239I	C↑TCGAG
	Sca1827I	CTCGAG
	SciI	CTC↑GAG
	Sci1831I	CTCGAG
	ScuI	CTCGAG
	SexI	CTCGAG
	Sfr274I	C↑TCGAG
	Sfu1762I	CTCGAG
	SgaI	CTCGAG
	SgoI	CTCGAG
	Sgr1841I	CTCGAG
	Shy1766I	CTCGAG
	SlaI	C↑TCGAG
	SluI	CTCGAG
	SpaI	CTCGAG
	Sph1719I	CTCGAG
	Ssp4I	CTCGAG
	Sve194I	CTCGAG
	XpaI	C↑TCGAG
XhoII		R↑GATCY
	AitII	RGATCY
	AitAI	RGATCY
	BstYI	R↑GATCY
	DsaIII	R↑GATCY
	MflI	R↑GATCY
	Tru201I	R↑GATCY
XmaIII		C↑GGCCG
	AaaI	C↑GGCCG
	BstZI	CGGCCG
	EagI	C↑GGCCG
	EclXI	C↑GGCCG
	Eco52I	C↑GGCCG
XmnI		GAANN↑NNTTC
	Asp700I	GAANN↑NNTTC
	BbvAI	GAANN↑NNTTC

[a] For the primary sources of data, see ref. 1.
[b] Abbreviations : R = G or A; Y = C or T; M = A or C; K = G or T; S = G or C; W = A or T; H = A, C or T; B = G, T or C; V = G, C or A; D = G, A or T; N = A, C, G or T.
[c] BcgI cuts on both sides of the recognition sequence, 10 bp to the 5' and 12 bp to the 3', releasing the recognition site as a 34 bp fragment with 2 nt 3'-overhangs at each end.
[d] References giving additional data: BsmAI (3); Asp22MI (4, 5); Asp16RI (5).
[e] TaqII has two distinct recognition sequences.

Table 3. Type I restriction enzymes[a]

Enzyme	Recognition sequence[b]
CfrA1	GCANNNNNNNNGTGG
EcoAI	GAGNNNNNNNGTCA
EcoBI	TGANNNNNNNNTGCT
EcoDI	TTANNNNNNNGTCY
EcoDXXI	TCANNNNNNNATTC
EcoEI	GAGNNNNNNNATGC
EcoKI	AACNNNNNNGTGC
EcoR124I	GAANNNNNNRTCG
EcoR124/3I	GAANNNNNNNRTCG
StySBI	GAGNNNNNNRTAYG
StySJI	GAGNNNNNNNGTRC
StySPI	AACNNNNNNGTRC
StySQI	AACNNNNNNRTAYG

[a] For the primary sources of data, see ref. 1.
[b] Abbreviations: see Table 2, footnote b.

Table 4. Type III restriction enzymes[a]

Enzyme	Isoschizomers	Recognition sequence
EcoPI		AGACC
EcoP15I		CAGCAG
HinfIII		CGAAT
	HineI	CGAAT
StyLTI		CAGAG

[a] For the primary sources of data see ref. 1.

1.2. Reaction conditions for restriction enzymes

Recommended reaction conditions for the commercially available restriction enzymes are given in *Table 5*. This information should not be considered sacrosanct; many enzymes are reasonably flexible in their requirements and the best buffer for one type of DNA may not give the maximum restriction with DNA from another source. In particular, the Mg^{2+} concentration can generally be varied from 5 to 10 mM without noticeable effect. Dithiothreitol and β-mercaptoethanol are often considered interchangeable but there may be circumstances in which one is more suitable than the other. The addition of 100 $\mu g \ ml^{-1}$ BSA is recommended for all enzymes, especially for lengthy digestions.

Most commercial suppliers provide standard buffers ('low salt', 'medium salt', 'high salt', etc.), each recommended for a range of restriction enzymes. At least one supplier has admitted in print that "these buffers are sub-optimal for many enzymes". A more universal buffer has been developed (6) by replacing NaCl with potassium glutamate, but restrictions produced with this buffer do not electrophorese cleanly under normal conditions. Several companies have reached a compromise with potassium acetate buffers (e.g. Stratagene Universal Buffer, Pharmacia One-Phor-All). Independent data on the efficiency of these buffers has not yet appeared.

Table 5 also gives inactivation temperatures, effects of salt concentration, star activities and activities on ssDNA. These data are incomplete and further information will be welcomed by the Editor.

The data in *Table 5* were obtained from a number of sources: commercial catalogs (though these are often contradictory), original descriptions of the enzymes, personal experience, and colleagues' reports.

Table 5. Reaction conditions for commercial restriction enzymes

Enzyme	Tris (mM)	pH	MgCl₂ (mM)	NaCl (mM)	DTT[a] (mM)	2ME[a] (mM)	Others[a]
*Aat*I	10	7.5	7	–	–	6	60 mM KCl
*Aat*II	10	7.5	10	–	1	–	50 mM KCl
*Acc*I	6	8.0	6	6	–	6	0.01% TX-100
*Acc*II	10	7.5	10	25	–	10	–
*Acc*III	10	7.7	10	100	1	–	–
*Acy*I	10	8.5	7	100	1	–	–
*Afl*I	10	8.0	10	50	–	10	–
*Afl*II	10	8.0	10	50	–	10	–
*Afl*III	10	8.0	10	150	–	6	–
*Aha*I	10	7.5	10	25	–	10	–
*Aha*II	10	8.0	10	100	–	10	–
*Aha*III	25	7.7	10	–	1	–	–
*Alu*I	10	7.5	6	50	–	6	–
*Alw*I	10	7.4	10	–	–	10	–
*Alw*NI	10	7.4	10	50	–	10	–
*Aoc*I	10	7.7	10	150	1	–	–
*Aoc*II	10	7.5	10	25	–	10	–
*Aos*I	10	7.5	10	–	1	–	50 mM KCl
*Apa*I	10	7.5	6	6	–	6	–
*Apa*LI	10	7.5	10	–	–	10	–
*Apy*I	50	7.5	10	100	1	–	–
*Aqu*I	25	7.7	10	50	–	10	–
*Ase*I	10	7.5	10	–	–	–	150 mM KCl
*Asn*I	50	7.5	10	100	1	–	–
*Asp*700I	50	7.5	10	100	1	–	–
*Asp*718I	50	7.5	10	100	1	–	–
*Asu*I	6	7.6	6	50	–	6	–
*Asu*II	10	7.5	10	–	1	–	–
*Ava*I	10	8.0	10	50	–	6	0.01% TX-100
*Ava*II	10	8.0	10	60	–	6	–
*Ava*III	10	8.5	10	75	1	–	–
*Avr*II	10	7.4	10	50	–	10	–
*Axy*I	50	7.5	10	–	1	–	100 mM KCl
*Bal*I	50	8.5	5	–	–	10	–
*Bam*HI	20	7.4	7	100	–	6	–
*Ban*I	10	8.0	7	–	–	6	–
*Ban*II	10	7.5	7	50	–	6	–
*Ban*III	10	7.5	7	–	–	6	80 mM KCl
*Bbe*I	10	7.5	10	–	1	–	–
*Bbi*I	10	7.5	10	–	–	–	–
*Bbu*I	6	7.5	6	6	–	6	–
*Bbv*I	10	8.0	10	50	–	10	–
*Bcl*I	10	8.0	10	–	–	6	75 mM KCl
*Bcn*I	10	7.5	10	50	1	–	–
*Bgl*I	100	8.0	10	60	–	–	–
*Bgl*II	100	8.0	5[b]	60	–	6	–

Temperature (°C)		NaCl effects[d]				KCl effect[e]				Star activity[f]	Cuts ssDNA?
Reactn[c]	Inactn[c]	0	50	100	150	0	50	100	150		
37	75	●	●	●		●	●	●			
37	60		○	○		●					
37	90	●	●								
37		○	○	●	●						
60											
37		●	●								
37											
37			●	○	○						
37											
37											
37			○	●	●						
37			●	●	●						
37	70		●	●	○						no
37	65	●	●	○							
37	65	●	●	●							
37											
37											
37											
37	80	●	●	○							
37	65	●	○								
37											
37											
37			●	●	●		●	●	●		
37											
37											
37											
37			●	●							
37											
37	100	●	●	●	●					ab	
37	65	●	●	○							
37											
37		●	●	●	○						
37											
37	65	●	○	○							
37	60		○	●	●						
50	70	●	●	○	○					abcd	
37	60	●	●	●	●						
37	70										
37											
37											
37		●	●								
37	65	●	●	●	●						no
50	100	○	○	○	○		●	●			
37											
37	65		●	●	●						
37	100	○	●	●	●						

Table 5. Continued

Enzyme	Tris		MgCl$_2$	NaCl	DTT[a]	2ME[a]	Others[a]
	(mM)	pH	(mM)	(mM)	(mM)	(mM)	
*Bsm*I	10	7.4	10	50	–	10	–
*Bsp*1286I	10	7.5	10	–	–	10	–
*Bsp*HI	10	7.4	10	–	–	–	100 mM KCl
*Bsp*MI	10	7.5	10	150	–	–	–
*Bsp*MII	10	7.5	10	150	–	–	–
*Bss*HII	10	7.5	10	25	–	10	–
*Bst*I	10	7.5	10	50	–	1	–
*Bst*BI	10	7.5	10	50	–	10	–
*Bst*EII	6	8.0	6	150	–	6	–
*Bst*NI	10	7.7	10	150	1	–	–
*Bst*UI	10	8.0	10	–	–	–	–
*Bst*XI	10	7.6	7	150	–	6	–
*Bst*YI	10	8.0	10	–	–	–	–
*Bsu*36I	10	7.4	10	100	–	–	–
*Ccr*II	10	7.7	10	100	1	–	–
*Cfo*I	6	7.6	6	50	–	6	–
*Cfr*10I	20	8.5	–	–	–	–	100 mM KCl
							3 mM MgSO$_4$
							0.02% TX-100
*Cfr*13I	10	8.5	5	50	–	–	–
*Cla*I	10	7.9	10	50	–	6	–
*Csp*45I	10	7.5	7	60	–	–	–
*Cvn*I	10	7.5	10	50	1	–	–
*Dde*I	10	7.5	6	150	–	6	–
*Dpn*I	10	7.5	7	150	–	10	–
*Dra*I	10	7.5	10	50	–	6	–
*Dra*II	10	7.5	10	25	–	10	–
*Dra*III	10	8.5	10	100	1	–	–
*Eae*I	50	7.4	10	–	–	10	50 mM KCl
*Eag*I	10	8.2	10	150	–	10	–
*Eco*47I	10	7.5	7	100	–	7	–
*Eco*47III	10	8.5	7	100	–	7	–
*Eco*52I	10	9.0	3	100	–	–	–
*Eco*81I	10	8.5	7	–	–	7	20 mM KCl
*Eco*105I	10	7.5	5	20	–	10	–
*Eco*NI	10	7.5	10	50	–	10	–
*Eco*O109I	40	8.0	10	–	–	10	–
*Eco*RI	100	7.5	10	50	–	6	–
*Eco*RII	25	7.7	10	50	–	10	–
*Eco*RV	10	8.0	6	100	–	6	–
*Eco*T14I	50	7.5	10	100	1	–	–
*Eco*T22I	10	7.5	7	125	–	7	–
*Ehe*I	10	8.0	10	–	1	–	–
*Esp*I	7	7.5	7	100	–	6	–
*Fdi*II	10	8.0	10	60	–	7	–
*Fnu*4HI	10	7.4	10	10	–	10	5 mM KPO$_4$

Temperature (°C)		NaCl effects[d]				KCl effect[e]				Star activity[f]	Cuts ssDNA?
Reactn[c]	Inactn[c]	0	50	100	150	0	50	100	150		
60		●	●	●	●						
37		●	●	○							
37	65	○	●	●	○	○	●	●	○		
37		○	○	●	●						
60		○	○	●	●						
50		●	●	●	●						
55										b	
65		●	●	○							
60	100		○	●	●						
60		○	○	●	●						slow
60		●	●	○							
55	65	○	●	●	●						
60		●	●	○	○						
37			○	●	○						
37											
37		●	●	○	○						
37	100		●	●	●		●	●	●		
37	100	●	●	●	○						
37	65	●	●	●	○						
37		●	●								
37	65										
37	70	○	●	●	●					ef	slow
37	65		○	●	●						no
37	65	○	●								
37											
37	65	○	●	●	○						
37	65	○	●	○		○	●	○			
37	65	○	●	●	●						
37	100		○	●	●						
37	100		○	●	●						
37	80	○	●	●	○						
37	90	●	●			●					
37	65	●	○								
37		●	●	●	●						
37	65	●	●	●	●						
37	70		●	●	●					abce	
37			●	●	●						
37	100			●						f	
37											
37			○	●	●						
37		●	●								
37											
50											
37		●	●	○							

Table 5. Continued

Enzyme	Tris		MgCl$_2$	NaCl	DTT[a]	2ME[a]	Others[a]
	(mM)	pH	(mM)	(mM)	(mM)	(mM)	
FnuDII	10	7.5	10	–	1	–	–
FokI	10	7.7	10	–	–	10	10 mM KCl
FspI	10	7.4	10	50	–	10	–
HaeII	50	7.5	6	50	–	6	–
HaeIII	50	7.5	6	50	1	–	–
HapII	10	7.5	10	–	1	–	–
HgaI	10	7.4	10	50	1	–	–
HgiAI	10	8.0	10	150	–	10	–
HhaI	10	8.0	6	100	–	6	–
Hin1I	10	8.5	5	25	–	–	–
HincII	10	7.5	7	100	–	6	–
HindII	10	7.5	10	50	1	–	–
HindIII	50	8.0	10	60	–	–	–
HinfI	10	7.5	7	60	–	6	–
HinPI	50	8.0	5	–	0.5	–	–
HpaI	10	7.4	10	–	1	–	20 mM KCl
HpaII	10	7.5	10	–	–	6	10 mM KCl
HphI	10	7.5	10	–	1	–	10 mM KCl
KpnI	10	7.5	10	10	–	6	–
MaeI	50	7.5	10	100	1	–	–
MaeII	50	7.5	10	100	1	–	–
MaeIII	50	7.5	10	100	1	–	–
MboI	10	7.5	10	100	1	–	–
MboII	10	7.4	10	–	1	–	10 mM KCl
MflI	10	7.5	10	–	1	–	–
MluI	10	7.5	7	50	–	6	–
MnlI	12	7.6	12	50	–	10	–
MroI	10	8.0	12	20	–	–	–
MseI	10	7.4	10	50	–	10	–
MspI	10	7.5	10	50	–	6	0.02% TX-100
MstI	10	7.7	10	150	1	–	–
MstII	10	7.7	10	150	1	–	–
MvaI	10	8.5	15	–	1	–	150 mM KCl
MvnI	10	7.5	10	50	1	–	–
NaeI	10	8.0	10	20	–	6	–
NarI	10	7.5	10	–	–	6	10 mM KCl
NciI	10	7.5	10	25	–	6	–
NcoI	10	7.5	10	150	–	–	–
NdeI	10	7.8	7	150	–	6	–
NdeII	50	7.5	10	100	1	–	–
NheI	10	8.0	10	50	–	10	–
NlaIII	10	7.6	10	–	–	10	50 mM (NH$_4$)$_2$SO$_4$
NlaIV	10	7.4	10	–	–	10	50 mM (NH$_4$)$_2$SO$_4$
NotI	10	7.5	10	150	–	6	0.01% TX-100
NruI	10	8.0	10	150[d]	–	6	–
NsiI	10	7.7	10	150	–	6	–

Temperature (°C)		NaCl effects[d]				KCl effect[e]				Star activity[f]	Cuts ssDNA?
Reactn[c]	Inactn[c]	0	50	100	150	0	50	100	150		
37		●									no
37	65	●	●	●	●	●	●	●	●		no
37	65		●	○	○						
37		●	●	●	○						
37	90	●	●	●	●					ab	yes
37											
37	65	●	●								slow
37	65			○	●						
37	90		○	●	●					abf	yes
37		●	●	○							
37	70	○	●	●	●						
37											
37	90	○	●	●	○					cf	
37	80	○	●	●	●						slow
37		●	●	●	●						yes
37	90		●	●		●	●	●		ab	
37	90	●	●	○		●	●	○			no
37	65	●	●	●		●	●	●			no
37	60	●									
45											
50											
45											
37	65	○	●	●	●						no
37	65	●	●	●	●	●	●	●	●		no
37											
37	100	○	●	●	○						
37	65	●	●	●	○						yes
37		●	●	○							
37	65	●	●	○							
37	90	●	●	●	●						no
37			●	●	●						
37	65		○	●	●						
37	100		○	●	●		○	●	●		
37											
37	100	●	●	●							
37	65	●	●			●	●				
37	80	●	●	○							
37	65		○	●	●						
37	65			○	●						
37											
37	65	●	●	●	●						
37	65										
37	65										
37	100		●	●	●						
37	80		●	●	●						
37		○	○	○	○						

Table 5. Continued

Enzyme	Tris (mM)	pH	MgCl₂ (mM)	NaCl (mM)	DTT[a] (mM)	2ME[a] (mM)	Others[a]
NspI	10	8.0	10	20	–	7	–
NspII	20	8.5	10	–	–	10	–
NspIII	10	8.0	10	–	–	10	25 mM KCl
NspIV	10	8.0	10	–	–	10	–
NspV	10	8.0	10	25	–	6	–
NspBII	10	7.5	10	–	1	–	50 mM KCl
NspHI	10	7.5	10	–	1	–	50 mM KCl
NunII	10	7.5	10	–	1	–	50 mM KCl
PaeR7I	10	7.4	10	–	–	10	–
PalI	10	7.5	10	–	–	1	–
PflMI	10	7.4	10	50	–	10	–
PleI	6	7.8	6	–	–	–	–
PpuMI	10	7.4	10	50	–	10	–
PssI	10	7.5	10	50	1	–	–
PstI	10	7.5	10	100	–	6	–
PvuI	10	7.4	7	–	–	6	150 mM KCl
PvuII	10	7.5	6	60	–	6	–
RsaI	10	8.0	10	50	–	6	–
RspXI	50	7.5	10	–	1	–	100 mM KCl
RsrI	25	7.7	10	–	1	–	–
RsrII	10	8.0	5	10	1	–	–
SacI	6	7.4	6	20	–	6	–
SacII	10	7.5	10	–	–	10	10 mM KCl
SalI	6	7.9	7[b]	150	–	6	–
SauI	25	7.7	10	–	1	–	–
Sau3AI	10	7.5	7	100	–	–	–
Sau96I	10	7.7	10	50	–	10	–
ScaI	10	7.4	6	100	–	6	–
ScrFI	10	7.6	10	50	–	6	–
SduI	10	7.5	10	–	1	–	–
SfaNI	10	7.5	10	150	–	–	–
SfiI	10	7.8	10	50	–	10	–
SinI	6	7.4	6	20	–	6	–
SmaI	10	8.0	6	–	–	6	20 mM KCl
SnaBI	10	7.7	10	50	–	10	–
SpeI	6	7.5	10	50	–	6	–
SphI	8	7.4	7	150	–	6	–
SplI	50	7.5	10	100	1	–	–
SpoI	15	7.5	7	–	–	6	50 mM KCl
SspI	10	7.4	10	100	1	–	–
SstI	10	7.5	10	50	1	–	–
SstII	10	7.5	10	50	1	–	–
StuI	10	8.0	10	50	–	6	–
StyI	10	8.5	100	10	1	–	–
TaqI	6	8.4	6	100	–	6	–
ThaI	10	7.5	10	–	1	–	–

Temperature (°C)		NaCl effects[d]				KCl effect[e]				Star activity[f]	Cuts ssDNA?
Reactn[c]	Inactn[c]	0	50	100	150	0	50	100	150		
50											
37											
37											
37											
50											
37											
37											
37		●	●	●							
37											
37	65	○	●	●	○						
37	65	●									
37		●	●	○							
37											
37	70	●	●	●	●					abf	
37	100		○	●	●	○	●	●			
37	95	●	●	●	●					abf	
37	100	●	●	●	●						
37											
37											
37	65	●	○								
37	60	●	○	○							
37	80	●					●				
37	80			○	●					abf	
37											
37	70	●	●	●	●					ab	no
37	65	●	●	●	●						
37	100		●	●	○						
37	65	○	●	●	●						
37											
37	65			●	●						no
50		●	●	●	●						
37	65	●									
30	60	●	○			●	●				
37		●	●	○							
37	65	○	●	●	●						
37	100			●	●						
55											
37			●	●	○	●	●	○			
37	65	○	●	●	○						
37		●	●								
37											
37		●	●	●	●						
37	65		○	●	●						
65	90	●	●	●	○						slow
60		●	●								

Table 5. Continued

Enzyme	Tris		MgCl$_2$	NaCl	DTT[a]	2ME[a]	Others[a]
	(mM)	pH	(mM)	(mM)	(mM)	(mM)	
*Tth*111I	10	7.5	10	50	–	10	–
*Tth*HB8I	50	7.5	10	100	1	–	–
*Xba*I	10	8.0	6	150	–	6	–
*Xcy*I	10	7.5	10	50	1	–	–
*Xho*I	10	8.0	6	150	–	6	–
*Xho*II	25	7.7	10	–	1	–	–
*Xma*I	10	7.5	10	25	–	10	–
*Xma*III	10	7.5	10	–	1	–	–
*Xmn*I	10	8.0	10	6	–	10	
*Xor*II	10	7.5	10	–	1	–	–

[a] Abbreviations: DTT, dithiothreitol; 2ME, β-mercaptoethanol; TX-100, Triton X-100.
[b] Use MgSO$_4$ not MgCl$_2$.
[c] Reactn = recommended reaction temperature. Inactn = temperature needed to achieve inactivation of the enzyme after 15 min. The data for inactivation are affected by the presence of trace amounts of compounds able to stabilize the enzyme. Use these data as a guide but if 100% inactivation is essential, either test the inactivation temperature with your system or use phenol extraction to remove the enzyme.
[d] Concentration (mM) results in: ●, 50–100% of maximal activity; ○, 10–50% of maximal activity; no entry, <10% of maximal activity.
[e] For those enzymes where KCl is the recommended salt. Symbols explained in footnote d.
[f] Conditions known to induce star activity. Key: a, high enzyme concentration; b, high glycerol; c, substitution of Mn^{2+} for Mg^{2+}; d, low salt; e, high pH; f, presence of DMSO.

1.3. Effect of DNA methylation on enzyme activity

dam and dcm methylases

The *dam* methylase of *E. coli* methylates at the N^6 position of adenine in the sequence 5'-GATC-3' (7). The *dcm* methylase adds methyl groups to the C^5 position of the internal cytosines in the sequences 5'-CCAGG-3' and 5'-CCTGG-3' (8, 9). The following restriction enzymes will not completely restrict DNA that has been *dam*- or *dcm*-methylated:

dam

*Acc*III	*Alw*I	*Ban*III	*Bcl*I	*Bsp*HI	*Bst*EII
*Cla*I	*Hph*I	*Mbo*I	*Mbo*II	*Nru*I	*Rsp*XI
*Taq*I	*Tth*HB8I	*Xba*I			

dcm

*Aat*I	*Aha*II	*Apa*I	*Asp*718I	*Ava*II	*Bal*I
*Bst*XI	*Cfr*13I	*Dra*II	*Eae*I	*Eco*47I	*Eco*O109I
*Eco*RII	*Pfl*MI	*Sau*96I	*Scr*FI	*Sfi*I	*Stu*I

The following enzymes are able to restrict DNA even when their recognition sites have been *dam*- or *dcm*-methylated:

dam

*Bam*HI	*Bgl*II	*Bsp*MII	*Bst*I	*Bst*YI	*Mro*I
*Pvu*I	*Sau*3AI	*Spo*I	*Xho*II	*Xor*II	

Temperature (°C)		NaCl effects[d]				KCl effect[e]				Star activity[f]	Cuts ssDNA?
Reactn[c]	Inactn[c]	0	50	100	150	0	50	100	150		
65		•	•	•							
65											
37	70	•	•	•						abd	
37											
37	80	○	•	•	•						
37											
37	65	•	•	○							
25		•	•	•							
37	65	•	•								
37		•									

dcm

AflI	ApyI	BamHI	BbeI	BglI	BstI
BstEII	BstNI	FokI	HaeIII	KpnI	MvaI
NarI	RsaI				

One enzyme – DpnI – is known to cut only if its recognition site is dam-methylated.

Note that these lists cover only commercially available enzymes. For comprehensive information on the methylation sensitivities of restriction enzymes see Table 6. For dam⁻ and dcm⁻ strains of E. coli see Chapter 1, Table 4.

Mammalian methylases

Mammalian DNA is often methylated at the C^5 position of the cytosine in 5'-CG-3'. The following enzymes will not restrict DNA modified in this way (10):

BssHII	BspMII	ClaI	CspI	EagI
Eco47III	FspI	MluI	NaeI	NarI
NotI	PvuI	RsrII	SalI	XhoI
XorII				

Only the following enzymes are known to be wholly unaffected:

*Acc*III   *Asu*II   *Cfr*9I   *Xma*I

Plant methylases

In plant DNA the cytosines in 5'-CG-3' and 5'-CNG-3' are often methylated at C^5. The restriction of plant DNA by the following enzymes is unaffected (10):

*Acc*III	*Afl*II	*Aha*III	*Ase*I	*Asu*II
*Bcl*I	*Bsp*HI	*Bsp*NI	*Bst*EII	*Bst*NI
*Cvi*QI	*Dpn*I	*Dra*I	*Eco*RV	*Hinc*II
*Hpa*I	*Kpn*I	*Mbo*II	*Mse*I	*Nde*I
*Nde*II	*Rsa*I	*Rsp*XI	*Sfi*I	*Spe*I
*Sph*I	*Ssp*I	*Taq*I	*Tth*HB8I	*Xmn*I

2. SITE-SPECIFIC METHYLATION

(Data kindly supplied by M. McClelland.)

Table 6 lists the sensitivities of restriction endonucleases to the site-specific DNA modifications ^{m4}C, ^{m5}C, ^{hm5}C and ^{m6}A which are commonly found in DNA from prokaryotes, eukaryotes and viruses. *Table 7* lists the DNA methyltransferases so far characterized together with their modification specificities. Several restriction endonuclease isoschizomers are known to differ in their sensitivity to methylation at particular modified sites. *Table 8* lists 19 isoschizomer pairs along with the modified recognition sites at which they differ.

Table 6. Methylation sensitivity of restriction endonucleases[a,c]

Restriction enzyme	Recognition sequence	Sites cut	Sites not cut
*Aac*I	CCWGG	$C^{m5}CWGG$	?
*Aat*I	AGGCCT	?	$AGG^{m5}CCT$
			$AGGC^{m5}CT$
			$AGGC^{m4}CT$
*Aat*II	GACGTC	?	$GACGT^{m5}C$
*Acc*I	GTMKAC	?	$GTMK^{m6}AC^{\#}$
			$GTMKA^{m5}C^{b}$
*Acc*II	CGCG	?	$^{m5}CGCG$
*Acc*III	TCCGGA	$T^{m5}CCGGA$	$TCCGG^{m6}A$
		$TC^{m5}CGGA$	
*Afl*I	GGWCC	$GGWC^{m5}C??$	?
		$GGWC^{m4}C^{b}$	
*Afl*II	CTTAAG	?	$^{m5}CTTAAG$
*Afl*III	ACRYGT	?	$A^{m5}CRYGT$
*Aha*II	$GRCGYC^{b}$	?	$GR^{m5}CGYC$
			$GRCGY^{m5}C$
*Alu*I	AGCT	?	$^{m6}AGCT$
			$AG^{m4}CT$
			$AG^{m5}CT^{\#}$
			$AG^{hm5}CT$
*Alw*I	GGATC	?	$GG^{m6}ATC$
			$GGAT^{m4}C$

Table 6. Continued

Restriction enzyme	Recognition sequence	Sites cut	Sites not cut
*Ama*I	TCGCGA	TCGCG^{m6}A	?
*Aos*II	GRCGYC	?	GR^{m5}CGYC
*Apa*I	GGGCCC	?	GGG^{m5}CCC$^{\#}$
			GGGCC^{m5}C
*Ape*I	ACGCGT	?	A^{m5}CGCGT
*Apa*LI	GTGCAC	GTGC^{m6}AC	?
*Apy*I	CCWGG	C^{m5}CWGG	^{m5}CCWGG
*Aqu*I	CYCGRG	?	^{m5}CYCGRG$^{\#}$
*Asp*700I	GAAN$_4$TTC	GA^{m6}AN$_4$TTC	G^{m6}AAN$_4$TTC
		GAAN$_4$TT^{m5}C	
*Asp*718I	GGTACC	GGT^{m6}A^{m5}CC^b	GGTAC^{m5}C
			GGTA^{m5}C^{m5}C^b
*Asu*II	TTCGAA	TT^{m5}CGAA	?
*Atu*CI	TGATCA	?	TG^{m6}ATCA
*Ava*I	CYCGRG	C^{m5}CCGGG	^{m5}CYCGRG
			CY^{m5}CGRG
			CTCG^{m6}AG^b
*Ava*II	GGWCC	GGWC^{m4}C^b	GGW^{m5}CC
			GGWC^{m5}C
			GGW^{hm5}C^{hm5}C
*Avi*II	AGCGCT	^{m6}AGCGCT	AG^{m5}CGCT
*Bal*I	TGGCCA	?	TGG^{m5}CCA$^{\#}$
			TGGC^{m5}CA^b
*Bam*FI	GGATCC	GG^{m6}ATCC	GGAT^{m4}CC
*Bam*HI	GGATCC	GGATC^{m5}C	GGAT^{m4}CC$^{\#}$
		GG^{m6}ATCC	GGAT^{m5}CC
		GG^{m6}ATC^{m5}C	GGAT^{hm5}C^{hm5}C
		GGATC^{m4}C	
*Bam*KI	GGATCC	GG^{m6}ATCC	GGAT^{m4}CC
*Ban*I	GGYRCC^b	GG^{m5}CGCC	?
		GGYRC^{m4}C	
*Ban*II	GRGCYC	?	GRG^{m5}CYC
*Ban*III	ATCGAT	?	ATCG^{m6}AT
*Bbe*I	GGCGCC	GGCG^{m5}CC	GG^{m5}CGCC
		GGCGC^{m5}C	
*Bbv*I	GCWGC	?	G^{m5}CWGC$^{\#}$
*Bcl*I	TGATCA	TGAT^{m5}CA	TG^{m6}ATCA
*Bcn*I	CCSGG	^{m5}CCSGG	C^{m4}CSGG$^{\#}$
*Bep*I	CGCG	?	^{m5}CGCG
*Bgl*I	GCCN$_5$GGC	GC^{m5}CN$_5$GGC^b	G^{m5}CCN$_5$GGC
			GCCN$_5$GG^{m5}C^b
			GC^{m4}CN$_5$GGC^b
*Bgl*II	AGATCT	AG^{m6}ATCT	AGAT^{m5}CT
			AGAT^{hm5}CT
*Bin*I	GGATC	?	GG^{m6}ATC
*Bma*DI	CGATCG	CG^{m6}ATCG	CGAT^{m6}CG
*Bme*216I	GGWCC	?	GGWC^{m5}C

Table 6. Continued

Restriction enzyme	Recognition sequence	Sites cut	Sites not cut
*Bsm*I	GAATGC	?	G^{m6}AATGC
*Bsm*AI	GTCTC	?	GT^{m5}CT^{m5}C
*Bsp*106I	ATCGAT	?	ATCG^{m5}AT$^{\#}$
*Bsp*1286I	GDGCHC	?	GDG^{m5}CHC
*Bsp*HI	TCATGA	?	TC^{m6}ATGA
			TCATG^{m6}A
*Bsp*MI	ACCTGC	?	ACCTG^{m5}C
*Bsp*MII	TCCGGA	TCCGG^{m6}A	T^{m5}CCGGA
			TC^{m5}CGGA
*Bsp*NI	CCWGG	^{m5}CCWGG	?
		C^{m5}CWGG	
*Bsp*XI	ATCGAT	?	ATCG^{m6}AT
*Bsp*XII	TGATCA	?	TG^{m6}ATCA
*Bss*HII	GCGCGC^{b}	?	G^{m5}CGCGC
*Bst*I	GGATCC	GG^{m6}ATCC	GGAT^{m4}CC
		GGATC^{m5}C	GGAT^{m5}CC
			GGATC^{m4}C
*Bst*BI	TTCGAA	?	TTCG^{m6}AA
*Bst*EII	GGTNACC	GGTNA^{m5}C^{m5}C^{b}	GGTNA^{hm5}C^{hm5}C
		GGTNAC^{m4}C	
*Bst*EIII	GATC	?	G^{m6}ATC
*Bst*GI	TGATCA	?	TG^{m6}ATCA
*Bst*NI	CCWGG^{b}	^{m5}CCWGG^{b}	^{hm5}C^{hm5}CWGG
		C^{m5}CWGG	C^{m4}CWGG
		^{m5}C^{m5}CWGG^{b}	
*Bst*UI	CGCG	?	^{m5}CGCG
*Bst*XI	CCAN$_{6}$TGG	?	^{m5}CCAN$_{6}$TGG
*Bst*YI	RGATCY	RG^{m6}ATCY	RGAT^{m4}CY
		RGAT^{m5}CY	
*Bsu*EI	CGCG	?	^{m5}CGCG$^{\#}$
*Bsu*FI	CCGG	?	^{m5}CCGG$^{\#}$
*Bsu*MI	CTCGAG	?	CT^{m5}CGAG$^{\#}$
*Bsu*QI	CCGG	?	^{m}CCGG
*Bsu*RI	GGCC	?	GG^{m5}CC$^{\#b}$
*Bsu*RII	CTCGAG		CT^{m}CGAG$^{\#}$
*Ccr*I	CTCGAG	?	CTCG^{m6}AG
*Cfo*I	GCGC	?	G^{m5}CGC
			G^{hm5}CG^{hm5}C
*Cfr*I	YGGCCR	?	YGG^{m5}CCR$^{\#}$
*Cfr*6I	CAGCTG	?	CAG^{m4}CTG$^{\#}$
			CAG^{m5}CTG
*Cfr*9I	CCCGGG	C^{m5}CCGGG	^{m4}CCCGGG
		CC^{m5}CGGG	^{m5}CCCGGG
			C^{m4}CCGGG$^{\#}$
			CC^{m4}CGGG
*Cfr*10I	RCCGGY	?	R^{m5}CCGGY$^{\#}$
*Cfr*13I	GGNCC	?	GGN^{m5}CC$^{\#}$

Table 6. Continued

Restriction enzyme	Recognition sequence	Sites cut	Sites not cut
*Cla*I	ATCGAT	?	m6ATCGAT
			ATm5CGAT
			ATCGm6AT[#]
*Cpe*I	TGATCA	?	TGm6ATCA
*Csp*I	CGGWCCG	CGGWCm5CG	CGGWm5CCG
			m5CGGWCCG
*Csp*45I	TTCGAA	?	TTCGm6AA
*Cvi*AI	GATC	?	Gm6ATC
*Cvi*BI	GANTC	?	Gm6ANTC[#]
*Cvi*JI	RGCY	?	RGm5CY[#]
*Cvi*NYI	CC	Cm5C	m5CC[#]
*Cvi*QI	GTAC	GTAm5C	GTm6AC[#]
*Dde*I	CTNAG	?	m5CTNAG[#]
			hm5CTNAG
*Dpn*I	Gm6ATC[b]	Gm6ATC	GATC
		Gm6ATm5C[b]	GATm4C
		Gm6ATm4C	GATm5C
*Dpn*II	GATC	?	Gm6ATC[#]
*Dra*II	RGGNCCY	?	RGGNCm5CY
*Eae*I	YGGCCR	?	YGGm5CCR[#]
			YGGCm5CR
*Eag*I	CGGCCG	?	CGGm5CCG
			m5CGGCm5CG
*Ear*I	GAAGAG	?	Gm6AAGAG
			GAAGm6AG
			m5CTm5CTTm5C
*Eca*I	GGTANm6ACC	?	GGTAm6ACC[#]
*Ecl*XI	CGGCCG	?	m5CGGCm5CG
			CGGm5CCG
*Eco*47I	GGWCC	?	GGWCm5C
*Eco*47III	AGCGCT	m6AGCGCT	AGm5CGCT
*Eco*A	GAGN₇GTCA[b]	?	Gm6AGN₇GmTCA[#b]
*Eco*B	TGAN₈TGCT[b]	?	TGm6ANₐmTGCT[#b]
*Eco*DXXI	TCAN₇AATC[b]	?	TCAN₇m6AAmTC[#b]
*Eco*E	GAGN₇ATGC	?	Gm6AGN₇ATGC
*Eco*K	AACN₆GTGC[b]	?	Am6ACN₆GmTGC[#b]
*Eco*O109I	RGGNCCY	?	RGGNCm5CY
*Eco*PI	AGACC[b]	AGAhm5Chm5C	AGm6ACC[#]
*Eco*P15	CAGCAG[b]	?	Cm6AGCAG[#]
*Eco*RI	GAATTC	GAATThm5C	Gm6AATTC[b]
			GAm6ATTC[#]
			GAATTm5C[b]
*Eco*RII	CCWGG[b]	m5CCWGG	m4CCWGG
			Cm4CWGG
			Cm5CWGG[#]
			CCm6AGG
			hm5Chm5CWGG

Restriction/Methylation

Table 6. Continued

Restriction enzyme	Recognition sequence	Sites cut	Sites not cut
*Eco*RV	GATATC	GATAT^{m5}C^b	G^{m6}ATATC$^{\#}$
			GAT^{m6}ATC
*Eco*R124	GAAN$_6$RTCG^b	?	GA^{m6}AN$_6$RTCG
			GAAN$_6$R^{m}TCG
*Eco*R124/3	GAAN$_7$RTCG^b	?	^{m6}A
*Esp*I	GCTNAGC	GCTNAG^{m5}C	G^{m5}CTNAGC
*Fnu*4HI	GCNGC	?	G^{m5}CNGC
			GCNG^{m5}C
*Fnu*DII	CGCG	?	^{m5}CGCG
			CG^{m5}CG
*Fnu*EI	GATC	G^{m6}ATC	?
*Fok*I	CATCC	CAT^{m5}CC	GG^{m6}ATG
		CATC^{m5}C^b	C^{m6}ATCC
		CATC^{m4}C	
*Fse*I	GGCCGGCC	?	GG^{m5}CCGGCC
			GGC^{m5}CGGCC
			GGCCGG^{m5}CC
*Fsp*I	TGCGCA	?	TG^{m5}CGCA
*Hae*II	RGCGCY^b	?	RG^{m5}CGCY
			RG^{hm5}CG^{hm5}CY
*Hae*III	GGCC	GGC^{m5}C	GG^{m5}CC$^{\#b}$
			GG^{hm5}C^{hm5}C
*Hap*II	CCGG	?	C^{m5}CGG$^{\#}$
*Hga*I	GACGC	?	GA^{m5}CGC
			GACG^{m5}C
*Hgi*AI	GRGCYC	?	GRG^{m5}CYC
*Hgi*JII	GGYRCC	?	GGYRC^{m5}C
*Hha*I	GCGC	?	G^{m5}CGC$^{\#}$
			GCG^{m5}C
			G^{hm5}CG^{hm5}C
*Hha*II	GANTC	?	G^{m6}ANTC$^{\#}$
*Hinc*II	GTYRAC	GTYRA^{m5}C	GTYR^{m6}AC
			GTYRA^{hm5}C
*Hind*II	GTYRAC	?	GTYR^{m6}AC$^{\#}$
*Hinf*I	GANTC	GANT^{m5}C^b	G^{m6}ANTC
			GANT^{hm5}C
*Hind*III	AAGCTT	?	^{m6}AAGCTT$^{\#}$
			AAG^{m5}CTT
			AAG^{hm5}CTT
*Hin*PI	GCGC	?	G^{m5}CGC
*Hpa*I	GTTAAC	GTTAA^{m5}C	GTTA^{m6}AC$^{\#}$
			GTTAA^{hm5}C
*Hpa*II	CCGG	?	^{m4}CCGG
			^{m5}CCGG^b
			C^{m4}CGG^b
			C^{m5}CGG$^{\#}$
			^{hm5}C^{hm5}CGG

Table 6. Continued

Restriction enzyme	Recognition sequence	Sites cut	Sites not cut
_Hph_I	TCACC	?	$T^{m5}CACC^{\#}$ $GGTG^{m6}A$
_Kpn_I	$GGTACC^{b}$	$GGTA^{m5}CC$ $GGTAC^{m5}C$ $GGTA^{m5}C^{m5}C^{b}$	$GGT^{m6}A^{m5}CC$ $GGTAC^{m4}C$
_Kpn_2I	TCCGGA	$TCCGG^{m6}A$	$T^{m5}CCGGA$ $TC^{m5}CGGA$
_Ksp_I	CCGCGG	?	$^{m5}CCGCGG$ $C^{m5}CGCGG$
_Mae_II	ACGT	?	$A^{m5}CGT^{b}$
_Mbo_I	GATC	$GAT^{m4}C$ $GAT^{m5}C$	$G^{m6}ATC^{\#b}$ $GAT^{hm5}C$
_Mbo_II	GAAGA	$T^{m5}CTT^{m5}C^{b}$	$GAAG^{m6}A^{\#}$
_Mfl_I	$RGATCY^{b}$	?	$RG^{m6}ATCY$ $RGAT^{m4}CY$ $RGAT^{m5}CY$
_Mlu_I	ACGCGT	$^{m6}ACGCGT$	$A^{m5}CGCGT$
_Mlu_9273I	TCGCGA	?	$T^{m5}CGCGA$
_Mlu_9273II	GCCGGC		$G^{m5}CCGGC$ $GC^{m5}CGGC$
_Mme_II	GATC	?	$G^{m6}ATC$
_Mnl_I	CCTC	?	$^{m5}CCTC$ $^{m5}C^{m5}CT^{m5}C$
_Mph_I	CCWGG	?	$C^{m5}CWGG$
_Mro_I	TCCGGA	$TCCGG^{m6}A$	$T^{m5}CCGGA$ $TC^{m5}CGGA$
_Mse_I	TTAA	$TT^{m6}AA$	?
_Msp_I	$CCGG^{b}$	$^{m4}CCGG$ $C^{m4}CGG$ $C^{m5}CGG$	$^{m5}CCGG^{\#}$ $^{hm5}C^{hm5}CGG$
_Mst_II	CCTNAGG	$^{m5}CCTNAGG$	?
_Mva_I	CCWGG	$C^{m5}CWGG^{b}$ $^{m5}CCWGG$	$C^{m4}CWGG^{\#}$ $CC^{m6}AGG$ $^{m4}CCWGG^{b}$ $^{m5}C^{m5}CWGG^{b}$
_Mvn_I	CGCG	?	$^{m5}CGCG$
_Nae_I	GCCGGC	?	$G^{m5}CCGGC$ $GC^{m5}CGGC$ $GCCGG^{m5}C$
_Nan_II	$G^{m6}ATC^{b}$	$G^{m6}ATC$ $G^{m6}AT^{m5}C^{b}$	GATC $GAT^{m5}C$
_Nar_I	GGCGCC	$GGCGC^{m5}C$	$GG^{m5}CGCC$ $GGCGC^{m4}C$
_Nci_I	CCSGG	$^{m5}CCSGG$	$C^{m4}CSGG$ $C^{m5}CSGG^{b}$
_Nco_I	CCATGG	$CC^{m6}ATGG$	$^{m4}CCATGG^{b}$ $^{m5}CCATGG$

Table 6. Continued

Restriction enzyme	Recognition sequence	Sites cut	Sites not cut
*Ncr*I	AGATCT	AG^{m6}ATCT^b	?
*Ncu*I	GAAGA	GAAG^{m6}A	?
*Nde*I	CATATG	^{m5}CATATG^b	^{m6}A
*Nde*II	GATC	GAT^{m5}C^b	G^{m6}ATC
*Ngo*I	RGCGCY	?	RG^{m5}CGCY
*Ngo*II	GGCC	?	GG^{m5}CC$^#$
*Ngo*BI	TCACC	?	T^{m5}CACC
*Ngo*PII	GGCC	?	GGC^{m5}C^b
			GG^{m5}CC$^#$
*Nhe*I	GCTAGC	?	GCTAG^{m5}C
*Nla*III	CATG	?	C^{m6}ATG
*Nmu*DI	G^{m6}ATC^b	G^{m6}ATC	GATC
*Nmu*EI	G^{m6}ATC^b	G^{m6}ATC	GATC
*Not*I	GCGGCCGC	GCGGCCG^{m5}C	GCGG^{m5}CCGC
			GCGGC^{m5}CGC
*Nru*I	TCGCGA	?	T^{m5}CGCGA
			TCGCG^{m6}A
*Nsi*I	ATGCAT	?	ATGC^{m6}AT
*Nsp*BII	CMGCKG	C^{m5}CGCKG	?
*Pfa*I	GATC	G^{m6}ATC	?
*Pfl*MI	CCAN$_5$TGG	?	C^{m4}CAN$_5$TGG
			C^{m5}CAN$_5$TGG
*Pfu*I	CGTACG	?	CGTA^{m5}CG
*Pae*R7I	CTCGAG	?	CTCG^{m6}AG$^#$
			CTGC^{m6}AG$^#$
*Pvu*I	CGATCG	CG^{m6}ATCG	CGAT^{m4}CG
			CGAT^{m5}CG
*Pvu*II	CAGCTG	?	CAG^{m4}CTG$^#$
			CAG^{m5}CTG
*Rsa*I	GTAC^b	GTA^{m5}C^b	GT^{m6}A^{m5}C
*Rsh*I	CGATCG	CG^{m6}ATCG	?
*Rsp*XI	TCATGA	?	TC^{m6}ATGA
			TCATG^{m6}A
*Rsr*I	GAATTC	?	G^{m6}AATTC
			GA^{m6}ATTC$^{#b}$
*Rsr*II	CGGWCCG	?	^{m5}CGGWCCG
			CGGW^{m5}CCG
			CGGWC^{m5}CG
*Sac*I	GAGCTC	G^{m6}AGCTC	GAG^{m5}CTC
*Sac*II	CCGCGG	?	^{m5}CCGCGG
*Sal*I	GTCGAC	GTCGA^{m5}C	GT^{m5}CGAC
			GTCG^{m6}AC$^#$
*Sal*DI	TCGCGA	TCGCG^{m6}A	T^{m5}CGCGA
*Sau*3AI	GATC^b	G^{m6}ATC	GAT^{m5}C^b
			GAT^{m4}C
			GAT^{hm5}C

Table 6. Continued

Restriction enzyme	Recognition sequence	Sites cut	Sites not cut
*Sau*96I	GGNCC	?	GGN^{m5}CC GGNC^{m5}C GGN^{hm5}C^{hm5}C
*Sbo*13I	TCGCGA	TCGCG^{m6}A	T^{m5}CGCGA
*Scr*FI	CCNGG	^{m5}CCNGG	C^{m5}CNGG C^{m4}CNGG
*Sfa*NI	GATGC	GATG^{m5}C	G^{m6}ATGC
*Sfi*I	GGCCN$_5$GGCC	GG^{m5}CCN$_5$GG^{m5}CC[b] GGCCN$_5$GGC^{m5}C	GGC^{m5}CN$_5$GGCC
*Sfl*I	CTGCAG	?	CTGC^{m6}AG
*Sin*I	GGWCC	?	GGW^{m5}CC[#]
*Sma*I	CCCGGG	C^{m5}CCGGG	^{m4}CCCGGG ^{m5}CCCGGG[b] C^{m4}CCGGG[b] CC^{m4}CGGG CC^{m5}CGGG[b]
*Spe*I	ACTAGT	?	^{m6}ACTAGT
*Sph*I	GCATGC	GCATG^{m5}C G^{hm5}CATG^{hm5}C	GC^{m6}ATGC
*Spl*I	CGTACG	CGT^{m6}ACG[b]	?
*Spo*I	TCGCGA	TCGCG^{m6}A	T^{m5}CGCGA
*Sso*II	CCNGG	?	C^{m5}CNGG ^{m5}CCNGG
*Sso*47I	GAATTC	?	G^{m6}AATTC[#]
*Sst*I	GAGCTC	?	GAG^{m5}CTC GAG^{hm5}CT^{hm5}C
*Stu*I	AGGCCT	?	AGG^{m5}CCT AGGC^{m5}CT AGGC^{m4}CT
*Sty*SBI	GAGN$_6$RTAYG[b]	?	G^{m6}AGN$_6$R^{m}TAYG[#b]
*Sty*SPI	AACN$_6$GTRC[b]	?	A^{m6}ACN$_6$G^{m}TRC[#b]
*Taq*I	TCGA	T^{m5}CGA[b] T^{hm5}CGA[b]	TCG^{m6}A[#]
*Taq*II	GACCGA CACCCA	?	G^{m6}ACCGA
*Taq*XI	CCWGG	^{m5}CCWGG C^{m5}CWGG	?
*Tfi*I	TCGA	?	TCG^{m6}A
*Tha*I	CGCG	?	^{m5}CGCG ^{hm5}CG^{hm5}CG
*Tth*HB8I	TCGA	T^{m5}CGA	TCG^{m6}A[#]
*Xba*I	TCTAGA	?	TCTAG^{m6}A# T^{m5}CTAGA[b] T^{hm5}CTAGA
*Xho*I	CTCGAG	?	CT^{m5}CGAG CTCG^{m6}AG ^{m5}CTCGAG

Table 6. Continued

Restriction enzyme	Recognition sequence	Sites cut	Sites not cut
*Xho*II	RGATCY	RG^{m6}ATCY	RGAT^{m5}CY^b
*Xma*I	CCCGGG	CC^{m5}CGGG^b	^{m4}CCCGGG
			^{m5}CCCGGG
			C^{m4}CCGGG
			CC^{m4}CGGG
*Xma*III	CGGCCG	?	CGG^{m5}CCG
*Xmn*I	GAAN$_4$TTC	GA^{m6}AN$_4$TTC	G^{m6}AAN$_4$TTC
			GAAN$_4$TT^{m5}C^b
*Xor*II	CGATCG	CG^{m6}ATCG	CGAT^{m5}CG
			^{hm5}CGAT^{hm5}CG

^{a} $^\#$ denotes canonical modification methyltransferase specificity. M = A or C; K = G or T; N = A, C, G, or T; R = A or G; Y = C or T; W = A or T; S = G or C; D = A, G or T; H = A, C or T. Sequences are in 5'-3' order. ^{m4}C = N4-methylcytosine; ^{m5}C = C5-methylcytosine; ^{hm5}C = hydroxymethylcytosine; ^{m}C = methylcytosine, N4 or C5-methylcytosine unspecified; ^{m6}A = N6-methyladenine.

^{b} *Acc*I nicking occurs slowly in the unmethylated strand of the hemi-methylated sequence GTMKA^{m5}C. *Afl*II cuts slowly at GGWC^{m4}C.

*Aha*II (GRCGYC) will cut GRCGCC *faster* if these sites are methylated at GRCG^{m5}CC, but will not cut GRCGY^{m5}C sites.

*Asp*718I cuts M·*Cvi*QI-modified (GT^{m6}AC) *Chlorella* virus NY2A DNA. *Asp*718I does not cut GGTAC^{m5}CWGG overlapping *dcm* sites or ^{m5}C-substituted phage XP12 DNA, whereas *Kpn*I cuts MP12 readily.

*Ava*I nicking occurs slowly in the unmethylated strand of the hemi-methylated sequence CTCG^{m6}AG/CTCGAG.

*Ava*II cuts slowly at GGWC^{m4}C.

*Bal*I sites overlapping *dcm* sites (TGGC^{m5}CAGG) are 50-fold slower than unmethylated sites.

*Ban*I gives various rate effects when its recognition sequence is ^{m4}C- or ^{m5}C-methylated at different positions.

*Bgl*I cleavage rate at certain GC^{m5}CN$_5$GGC, GC^{m4}CN$_5$GGC, and GCCN$_5$GG^{m5}C hemi-methylated sites is extremely slow. However, ^{m5}C bi-methylated M·*Hae*III–*Bgl*I sites are completely refractory to *Bgl*I.

*Bss*HII does not cut M·*Hha*I-modified DNA, in which two different cytosine positions are hemi-methylated, G^{m5}CGCGC/GCG^{m5}CGC.

M·*Bst*I modifies the internal cytosine GGAT^{m}CC, but it is not known whether this modification is ^{m5}C or ^{m4}C.

*Bst*EII cuts the fully ^{m5}C-substituted phage XP12 DNA.

*Bst*NI cuts C^{m5}CWGG, ^{m5}CCWGG and ^{m5}C^{m5}CWGG. *Bst*NI isoschizomers that are insensitive to C^{m5}CWGG include *Aor*I, *Apy*I, *Bsp*NI, *Mva*I and *Taq*XI.

*Bsu*RI nicking occurs in the unmethylated strand of the hemi-methylated sequence GG^{m5}CC/GGCC.

M·*Cre*I is from the unicellular eukaryote *Chlamydomonas reinhardtii*.

*Dpn*I requires adenine methylation on both DNA strands. Isoschizomers of *Dpn*I include *Cfu*I, *Nan*II, *Nmu*EI, *Nmu*DI and *Nsu*DI. *Dpn*I cuts *dam*-modified XP12 DNA.

M·*Eca*I GGT^{m6}ACC is cloned.

M·*Eco dam* modifies GAT^{m5}C at a reduced rate.

*Eco*A is a Type I restriction endonuclease. ^{m}T represents a 6-methyladenine in the complementary strand.

*Eco*B is a Type I restriction endonuclease. ^{m}T represents a 6-methyladenine in the complementary strand.

*Eco*DXXI is a Type I restriction endonuclease. ^{m}T represents a 6-methyladenine in the complementary strand.

*Eco*K is a Type I restriction endonuclease. ^{m}T represents a 6-methyladenine in the complementary strand.

*Eco*PI is a Type III restriction endonuclease.

*Eco*P15 is a Type III restriction endonuclease.

*Eco*RI cannot cut hemi-methylated G^{m6}AATTC/GAATTC sites. Bi-methylated GA^{m6}ATTC/GA^{m6}ATTC sites are not cut by *Eco*RI or *Rsr*I. *Eco*RI shows a reduced rate of cleavage at hemi-methylated GAATT^{m5}C and does not cut an oligonucleotide that contains GAATT^{m5}C in both strands.

*Eco*RII isoschizomers that are sensitive to C^{m5}CWGG include *Atu*BI, *Atu*II, *Bst*GII, *Bin*SI, *Cfr*5I, *Cfr*III, *Eca*II, *Ecl*II, *Eco*27I, *Eco*38I and *Mph*I. *Eco*RII shows reduced rate of cleavage at hemi-methylated ^{m5}CCWGG/CCWGG sites.

*Eco*RV cuts the fully ^{m5}C-substituted phage XP12 DNA.

*Eco*R124 is a Type I restriction endonuclease. ^{m}T represents a 6-methyladenine in the complementary strand.

*Eco*R124/3 is a Type I restriction endonuclease.

*Fok*I cuts about 2- to 4-fold more slowly at CATC^{m5}C than at unmodified sites.

*Hae*II show a reduction in rate of cleavage when its recognition sequence is modified at RGCG^{m5}CY.

*Hae*III nicking occurs in the unmethylated strand of the hemi-methylated sequence GG^{m5}CC/GGCC.

*Hin*fI cuts GANT^{m5}C, however, detectable rate differences are observed between unmethylated, hemi-methylated (GANT^{m5}C/GANTC) and bi-methylated (GANT^{m5}C/GANT^{m5}C) target sequences.

*Hin*fI does cut phage XP12 DNA, although at a reduced rate. *Hin*fI cuts unmethylated GANTC faster than hemi-methylated GANT^{m5}C/GANTC, which is cut faster than GANT^{m5}C/GANT^{m5}C. However, the rate difference between unmethylated and fully methylated *Hin*fI sites is only about 10-fold.

*Hpa*II nicking occurs in the unmethylated strand of the hemi-methylated sequence ^{m5}CCGG/CCGG.

*Kpn*I sensitivity to hemi-methylated GGTA^{m5}CC and GGTAC^{m5}C sites has been reported. *Kpn*I efficiently cuts ^{m5}C-substituted phage XP12 DNA, but not *Chlorella* virus NY2A DNA, which carries both GT^{m6}AC and ^{m5}CC modifications.

*Mae*II nicks slowly in the unmethylated strand of hemi-methylated A^{m5}CGT/ACGT.

*Mbo*I isoschizomers that are sensitive to G^{m6}ATC include, *Bsa*PI, *Bsp*74I, *Bsp*76I, *Bsp*105I, *Bss*GII, *Bst*EIII, *Bst*XII, *Cpa*I, *Cty*I, *Cvi*AI, *Cvi*BII, *Cvi*HI, *Dpn*II, *Fnu*AII, *Fnu*CI, *Hac*I, *Meu*I, *Mkr*AI, *Mme*II, *Mno*III, *Mos*I, *Msp*67II, *Mth*I, *Mth*AI, *Nde*II, *Nfl*AII, *Nfl*BI, *Nfl*I, *Nla*DI, *Nla*II, *Nme*CI, *Nph*I, *Nsi*AI, *Nsp*AI, *Nsu*I, *Pfa*I, *Rlu*1I, *Sal*AI, *Sal*HI, *Sau*6782I, *Sin*MI, *Tru*II.

*Mbo*I cuts the fully ^{m5}C-substituted phage XP12 DNA, although certain hemi-methylated ^{m5}C-containing substrates are reported not to be cut.

*Mfl*I cuts slowly at ^{m6}AGATCY sites.

M·*Mmu*I is the mammalian ^{m5}CG methyltransferase from *Mus musculus* (mouse).

*Msp*I cuts the unmethylated strand and methylated strand of C^{m5}CGG/CCGG and C^{m4}CGG/CCGG duplexes. *Msp*I cuts very slowly at GGC^{m5}CGG. An M·*Msp*I clone methylates ^{m5}CCGG. However, there is a report that *Moraxella* sp. chromosomal DNA is methylated at ^{m5}C^{m5}CGG.

*Mva*I nicking occurs in the unmethylated strand of the hemi-methylated sequence C^{m4}CWGG/CCWGG.

*Mva*I cuts XP12 DNA very slowly at ^{m5}C^{m5}CWGG.

*Nan*II requires adenine methylation on both DNA strands. *Nan*II cuts M·*Eco dam* modified XP12 DNA.

*Nci*I may cut ^{m5}C^{m5}CGG methylated DNA. Possibly the second methylation negates the effect of C^{m5}CGG.

*Nco*I is blocked by M·*Sec*I (CCNNGG).

*Ncr*I is a *Bgl*II isoschizomer from *Nocardia carnia* Beijing.

*Nde*I cuts the fully ^{m5}C-substituted phage XP12 DNA.

*Nde*II cuts the fully ^{m5}C-substituted phage XP12 DNA.

*Ngo*PII does not cut overlapping *dcm* sites.

*Nmu*DI requires adenine methylation on both DNA strands.

*Nmu*EI requires adenine methylation on both DNA strands.

*Rsa*I cuts the fully ^{m5}C-substituted phage XP12 DNA, but does not cut *Chlorella* virus NY2A DNA, which is modified at GT^{m6}AC. DNA from *Rhodopseudomonas sphaeroides* species Kaplan is cut by *Asp*718I, but not by *Rsa*I or *Kpn*I. It is likely that M·*Rsa*I specifies GTA^{m4}C; and high levels of ^{m4}C are present in *R. sphaeroides* DNA.

*Rsr*I cannot cut hemi-methylated G^{m6}AATTC/GAATTC sites.

*Sau*3AI nicking occurs in the unmethylated strand of the hemi-methylated sequence GAT^{m5}C/GATC.

*Sau*3AI cuts at a reduced rate at ^{m6}AGATC. *Sau*3AI isoschizomers that are insensitive to G^{m6}ATC include *Bce*243I, *Bsp*49I, *Bsp*51I, *Bsp*52I, *Bsp*54I, *Bsp*57I, *Bsp*58I, *Bsp*59I, *Bsp*60I, *Bsp*61I, *Bsp*64I, *Bsp*65I, *Bsp*66I, *Bsp*67I, *Bsp*72I, *Bsp*AI, *Bsp*91I, *Bsr*PII, *Cpe*I, *Cpf*I, *Csp*5I, *Fnu*EI, *Msp*BI, *Sau*CI, *Sau*DI, *Sau*EI, *Sau*FI, *Sau*GI and *Sau*MI.

*Sfi*I cannot cut M·*Bgl*I-modified DNA.

*Sma*I nicking occurs in the unmethylated strand of the hemi-methylated sequence CC^m5CGGG/CCCGGG. *Sma*I may cut C^m5C^m5CGGG-methylated DNA. Possibly the second methylation negates the effect of CC^m5CGGG. There are conflicting results regarding *Sma*I: ^m5CCCGGG is not cut when modified by M·*Aqu*I methyltransferase or at overlapping M·*Hae*III–*Sma*I sites (GG^m5CCCGGG). Other investigators have reported that *Sma*I cuts at a reduced rate at hemi-methylated ^m5CCCGGGG sites.

*Spl*I cuts GT^m6AC-modified *Chlorella* virus NY2A DNA, but does not cut *Kpn*I-digested XP12 DNA.

*Sty*SBI is a Type I restriction endonuclease. ^mT represents a 6-methyladenine in the complementary strand.

*Sty*SPI is a Type I restriction endonuclease. ^mT represents a 6-methyladenine in the complementary strand.

*Taq*I cuts very slowly at T^hm5CGA. *Taq*I cuts the fully ^m5C-substituted phage XP12 DNA.

M·*Taq*I methylates T^m5CGA at a rate at least 20-fold slower than unmodified TCGA.

*Xba*I will cut T^m5CTAGA/TCTAGA hemi-methylated DNA at high enzyme levels (>100 U *Xba*I μg^{-1}), but will not cut this sequence in 20- to 40-fold overdigestions.

*Xho*II nicking occurs slowly in the unmethylated strand of the hemi-methylated sequence RGAT^m5CY/RGATCY.

*Xma*I is claimed not to cut CC^m5CGGG in one report.

*Xmn*I cuts the fully ^m5C-substituted phage XP12 DNA. *Xmn*I cuts slowly at some sites in DNA methylated on *both* strands at GAAN₄TT^m5C.

[c] This compilation is based upon that given in ref. 10 updated by M. McClelland. Data from ref. 10 reproduced with permission from Oxford University Press.

Table 7. DNA methyltransferases and their modification specificities[a,c]

Methylase	Specificity	Methylase	Specificity
M·*Acc*I	GTMK^m6AC	M·*Bsu*P11I	GG^m5CC and G^m5CNGC
M·*Afl*II	CTTAAG(^m6A)	M·*Bsu*P11s	GGCC and GDGCHC
M·*Ala*K21	GAT^m5C	M·*Bsu*QI	^mCCGG
M·*Alu*I	AG^m5CT	M·*Bsu*RI	GG^m5CC
M·*Alw*26I	GT^m5CTC and G^m6AGAC	M·*Bsu*RII	CT^mCGAG
M·*Apa*I	GGG^m5CCC	M·*Bsu*SPB	GG^m5CC and G^m5CNGC
M·*Aqu*I	^m5CYCGRG	M·*Bsu*SPRI	GG^m5CC and ^m5C^m5CGG and C^m5CWGG
M·*Bal*I	TGG^m5CCA		
M·*Bam*HI	GGAT^m4CC	M·*Bsu*SPR191	^m5C^m5CGG and C^mCWGG
M·*Bam*HII	G^mCWGC	M·*Bsu*SPR83I	GG^m5CC and C^m5CWGG
M·*Bbv*I	G^m5CWGC	M·*Cfr*I	YGG^m5CCR
M·*Bbv*SI	G^mCWGC	M·*Cfr*6I	CAG^m4CTG
M·*Bbv*SII	G^m6AT	M·*Cfr*9I	C^m4CCGGG
M·*Bbv*SIII	A^m6AG	M·*Cfr*10I	R^m5CCGGY
M·*Bcn*I	C^m4CSGG	M·*Cfr*13I	GGN^m5CC
M·*Bep*I	^m5CGCG	M·*Cla*I	ATCG^m6AT
M·*Bme*216I	GGWC^mC	M·*Cre*I[b]	T^m5CR
M·*Bsp*RI	GG^m5CC	M·*Cvi*BI	G^m6ANTC
M·*Bsp*JL14I	GGN^mCC	M·*Cvi*BIII	TCG^m6A
M·*Bsp*106I	ATCG^m6AT	M·*Cvi*JI	RG^m5CY
M·*Bst*I[b]	GGAT^mCC	M·*Cvi*NYI	^m5CC
M·*Bst*VI	CTCG^m6AG	M·*Cvi*QI	GT^m6AC
M·*Bst*YI	RGAT^mCY	M·*Dde*I	^m5CTNAG
M·*Bsu*EI	^m5CGCG	M·*Dpn*II	G^m6ATC
M·*Bsu*FI	^m5CCGG	M·*Eae*I	YGG^m5CCR
M·*Bsu*MI	YT^m5CGAR	M·*Eca*I[b]	GGT^m6ACC
M·*Bsu*Phi3T	GG^m5CC and G^m5CNGC	M·*Eco dam*[b]	G^m6ATC

Table 7. Continued

Methylase	Specificity	Methylase	Specificity
M·Eco dcmI	C^mCWGG	M·MboII	$GAAG^{m6}A$
M·Eco dcmII	R^mCCGG	M·MmuI[b]	^{m5}CG
M·Eco dcmIII	^mCCWGG	M·MspI	$^{m5}CCGG$[b]
M·Eco dcmIV	$GGWC^mC$	M·MvaI	$C^{m4}CWGG$
M·EcoA	$G^{m6}AGN_7G^mTCA$	M·NcoI	$CCATGG(^mC)$
M·EcoB	$TG^{m6}AN_8{}^mTGCT$	M·NdeI	$CATATG(^{m6}A)$
M·EcoE	$G^{m6}AGN_7ATGC$	M·NgoII	GG^mCC
M·EcoK	$A^{m6}ACN_6G^mTGC$	M·NgoIV	$G^{m5}CCGGC$
M·EcoPI	$AG^{m6}ACC$	M·NgoV	$GGNN^{m5}CC$
M·EcoP1 dam	$G^{m6}ATC$	M·NgoVI	$G^{m6}ATC$
M·EcoP15	$C^{m6}AGCAG$	M·NgoVII	G^mCWGC
M·EcoRI	$GA^{m6}ATTC$	M·NgoAI	$GG^{m5}CC$
M·EcoRII	$C^{m5}CWGG$	M·NgoBI	$T^{m5}CACC$
M·EcoRV	$G^{m6}ATATC$	M·NgoBII	$GTAN_5{}^{m5}CTC$
M·EcoR124	$GAAN_6RTCG(^{m6}A)$	M·NgoPI	RG^mCGCY
M·EcoR124/3	$GAAN_7RTCG(^{m6}A)$	M·NgoPII	GG^mCC
M·EcoT1 dam	$G^{m6}ATC$	M·NlaIII	$C^{m6}ATG$
M·EcoT2 dam	$G^{m6}AT$	M·PaeR7I	$CTCG^{m6}AG$
M·EcoT4 dam	$G^{m6}ATC$	M·PstI	$CTGC^{m6}AG$
M·Eco31I	$GGT^{m5}CTC$ and $G^{m6}AGACC$	M·PvuII	$CAG^{m4}CTG$
M·Eco57I	$CTGAAG\ (^{m6}A)$	M·RsrI	$GA^{m6}ATTC$
M·Eco72I	$CACGTG\ (^{m5}C)$	M·SalI	$GTCG^{m6}AC$
M·FnuDII	$^{m5}CGCG$	M·Sau96I	GGN^mCC
M·FokI	$GG^{m6}ATG$ and $C^{m6}ATCC$	M·SinI	$GGW^{m5}CC^{\#}$
M·HaeII	$RGCGCY$	M·SmaI	CC^mCGGG
M·HaeIII	$GG^{m5}CC$	M·Sso47I	$G^{m6}AATTC$
M·HapII	C^mCGG	M·Sso47II	C^mCNGG
M·HgaI	$GACGC(^mC)$	M·SspMQI	^{m5}CG
M·HhaI	$G^{m5}CGC$	M·StySBI	$G^{m6}AGN_6R^mTYG$
M·HhaII	$G^{m6}ANTC$	M·StySPI	$A^{m6}ACN_6G^mTRC$
M·HincII	$GTYR^{m6}AC$	M·StySQ	$A^{m6}ACN_6R^mTAYG$
M·HindII	$GTYR^{m6}AC$	M·StySJ	$G^{m6}AGN_6G^mTRC$
M·HindIII	$^{m6}AAGCTT$	M·TaqI[b]	$TCG^{m6}A$
M·HinfI	$G^{m6}ANTC$	M·TthHB8I	$TCG^{m6}A$
M·HpaI	$GTTA^{m6}AC$	M·TflI	$TCG^{m6}A$
M·HpaII	$C^{m5}CGG$	M·XbaI	$TCTAG^{m6}A$
M·HphI	$T^{m5}CACC$	M·XmaIII	CGG^mCCG
M·MboI	$G^{m6}ATC$		

[a]See footnote a of *Table 6*.
[b]See footnote b of *Table 6*.
[c]See footnote c of *Table 6*.

Table 8. Isoschizomer/isomethylator pairs that differ in their sensitivity to sequence-specific methylation[a]

| Methylated sequence[d] | Restriction isoschizomer pairs[b,c] | |
	cut by	not cut by
m^4CCGG	*Msp*I	*Hpa*II
Cm^5CGG	*Msp*I	*Hpa*II (*Hap*II)
Cm^4CGG	*Msp*I	*Hpa*II
CCm^5CGGG	*Xma*I (*Cfr*9I)	*Sma*I
Cm^5CWGG	*Bst*NI (*Mva*I)	*Eco*RII
Gm^6ATC	*Sau*3AI (*Fnu*EI)	*Mbo*I (*Nde*II)
GATm^5C	*Mbo*I	*Sau*3A
GATm^4C	*Mbo*I	*Sau*3A
GGCm^5C	*Hae*III	*Ngo*PII
GGTACm^5C	*Kpn*I	*Asp*718I
GGTAm^5Cm^5C	*Kpn*I	*Asp*718I
GGWCm^5C	*Afl*I	*Ava*II (*Eco*47I)
RGm^6ATCY	*Xho*II (*Bst*YI)	*Mfl*I
Tm^5CCGGA	*Acc*III	*Bsp*MII (*Mro*I)
TCm^5CGGA	*Acc*III	*Bsp*MII (*Mro*I)
TCCGGm^6A	*Bsp*MII (*Mro*I)	*Acc*III
TCGCGm^6A	*Sbo*13I (*Sal*DI)	*Nru*I
CGGWCm^5CG	*Csp*I	*Rsr*II

[a] In each row the first column lists a methylated sequence, the second column lists an isoschizomer that cuts this sequence, and the third column lists an isoschizomer that does not cut this sequence.

[b] An enzyme is classified as insensitive to methylation if it cuts the methylated sequence at a rate that is at least one tenth the rate at which it cuts the unmethylated sequence. An enzyme is classified as sensitive to methylation if it is inhibited at least 20-fold by methylation relative to the unmethylated sequence.

[c] See footnote c of *Table 6*.

[d] See footnote a of *Table 6*.

3. REFERENCES

1. Roberts, R.J. (1990) *Nucleic Acids Res.*, **18**, 2331.

2. Nomenclature Committee of the International Union of Biochemistry (1985) *Eur. J. Biochem.*, **150**, 1.

3. Kong, H., Morgan, R.D. and Chen, Z. (1990) *Nucleic Acids Res.*, **18**, 686.

4. Renbaum, P., Abrahamove, D., Fainsod, A., Wilson, G.G., Rottem, S. and Razin, A. (1990) *Nucleic Acids Res.*, **18**, 1145.

5. Inagaki, K., Dou, D.X., Kita, K., Hiraoka, N., Kishimoto, N., Sugio, T. and Tano, T. (1990) *J. Ferment. Bioeng.*, **69**, 60.

6. McClelland, M., Hanish, J., Nelson, M. and Patel, Y. (1988) *Nucleic Acids Res.*, **16**, 364.

7. Hattman, S., Brooks, J.E. and Masrekar, M. (1978) *J. Mol. Biol.*, **126**, 367.

8. Marinus, M.G. and Morris, N.R. (1973) *J. Bacteriol.*, **114**, 1143.

9. May, M.S. and Hattman, S. (1975) *J. Bacteriol.*, **123**, 768.

10. Nelson, M. and McClelland, M. (1989) *Nucleic Acids Res.*, **17**, r389.

CHAPTER 5
DNA AND RNA MODIFYING ENZYMES

This chapter covers those commercially available enzymes and proteins that are used to modify DNA and/or RNA molecules, with the exception of restriction endonucleases (Chapter 4) and RNases primarily used in enzymatic RNA sequencing (Chapter 2, *Table 11*). Enzymes relevant to molecular biology but not acting on DNA or RNA (glycosidases and proteases) are described in Chapter 2, *Tables 9* and *10*.

The enzymes and proteins described in this chapter are grouped as follows:

If in doubt, consult the main index to find the enzyme you are interested in.

Modifying Enzymes

1. DNA POLYMERASES

1.1. DNA POLYMERASE I (Kornberg polymerase)

Source
E. coli. Most commercial preparations are obtained from an *E. coli* lysogen carrying a λ*polA* transducing phage, e.g. *E. coli* NM964 (1) or CM5199 (2).

Description
A single polypeptide chain, 109 kd, with one polymerase and two exonuclease activities (3–5). With dsDNA and excess dNTPs the 3' to 5' exonuclease is usually masked by the 5' to 3' polymerase.

Properties
(i)   5' to 3' DNA-dependent DNA polymerase, requiring a ssDNA template and a DNA or RNA primer with a 3'-OH terminus:

(ii)   5' to 3' exonuclease, degrading dsDNA or a DNA–RNA hybrid (including the RNA component) from a 5'-P terminus:

(iii)   3' to 5' exonuclease, degrading ssDNA or dsDNA from a 3'-OH terminus:

Applications
(i) Labeling DNA by nick translation (1, 6–8).
(ii) Second strand cDNA synthesis in conjunction with RNase H (9).

Reaction conditions
The pH optimum is around 7.4 (5). Requires Mg^{2+}. Reactions are usually carried out at room temperature or 25°C.

$10 \times$ Nick Translation Buffer: 500 mM Tris–HCl (pH 7.4), 100 mM $MgCl_2$, 10 mM DTT, 500 µg ml^{-1} BSA.

1.2. DNA POLYMERASE I (Klenow fragment)

Source
E. coli, originally by subtilisin proteolysis of the Kornberg polymerase (10), now usually from a recombinant *E. coli* expressing a truncated *polA* gene (11).

Description
The large or Klenow fragment of DNA polymerase I has a molecular mass of 75 kd and lacks the 5' to 3' exonuclease activity of the Kornberg enzyme (10, 12–15).

Properties
(cf. DNA polymerase I — Kornberg polymerase.)

(i) 5' to 3' DNA-dependent DNA polymerase, requiring a ssDNA template and a DNA or RNA primer with a 3'-OH terminus.
(ii) 3' to 5' exonuclease, degrading ssDNA or dsDNA from a 3'-OH terminus.

Applications
(i) End-filling and end-labeling dsDNA with a 5' overhang (16–19).
(ii) Labeling ssDNA by random priming (20, 21).
(iii) Chain termination DNA sequencing (22).
(iv) Second strand synthesis of cDNA (23–26).
(v) Second strand synthesis in site-directed mutagenesis (27–31).
(vi) Production of ssDNA probes by primer extension (7, 32, 33).

Reaction conditions
Requirements are similar to the Kornberg enzyme but the exact conditions depend on the application. Reactions are usually carried out at room temperature or 25°C.

10 × End-labeling Buffer: 500 mM Tris–HCl (pH 7.5), 100 mM $MgCl_2$, 10 mM DTT, 500 µg ml^{-1} BSA.

10 × Random Priming Buffer: 1500 mM HEPES–NaOH (pH 7.6), 40 mM Tris–HCl (pH 7.6), 4 mM $MgCl_2$, 8 mM β-mercaptoethanol.

10 × TM Buffer (DNA sequencing): 100 mM Tris–HCl (pH 7.6), 600 mM NaCl, 66 mM $MgCl_2$.

2 × Second Strand cDNA Buffer: 200 mM HEPES–NaOH (pH 6.9), 140 mM KCl, 20 mM $MgCl_2$, 5 mM DTT.

Modified versions
There is a second engineered form (United States Biochemical Corp.) that lacks the 3' to 5' exonuclease activity (34) and is reputably better for random primer labeling (application ii above).

1.3. T4 DNA POLYMERASE

Source
T4-infected *E. coli* (35, 36) or a recombinant *E. coli* expressing a cloned T4 polymerase gene.

Description
A single polypeptide, 114 kd. Lacks the 5' to 3' exonuclease of the Kornberg polymerase and so is functionally similar to the Klenow fragment (37), though the T4 polymerase possesses a more active 3' to 5' exonuclease.

Properties
(cf. DNA polymerase I — Kornberg polymerase.)

(i)   5' to 3' DNA-dependent DNA polymerase, requiring a ssDNA template and a DNA or RNA primer with a 3'-OH terminus.
(ii) 3' to 5' exonuclease, degrading ssDNA (very active) or dsDNA from a 3'-OH terminus.

The active 3' to 5' exonuclease of T4 DNA polymerase enables the enzyme to carry out a replacement reaction with dsDNA molecules possessing blunt ends or 3' overhangs:

Applications
(i)   Labeling dsDNA by the replacement reaction (38, 39).
(ii)   End-filling and end-labeling dsDNA with a 5' overhang (40).
(iii) Gap-filling in site-directed mutagenesis with short mismatch oligonucleotides (41, 42).
(iv)   Generation of overlapping subclones for DNA sequencing (43).
(v)   Detection of thymine dimers (44).

Reaction conditions
The pH optimum is 8.0–9.0 with only 50% maximal activity at pH 7.5. Requires Mg^{2+}, optimum 6 mM; 37°C.

$10 \times$ T4 Buffer: 330 mM Tris–acetate (pH 7.9), 660 mM KAc, 100 mM MgAc, 5 mM DTT, 1 mg ml^{-1} BSA.

1.4. T7 DNA POLYMERASE

Source
E. coli infected with T7 phage.

Description
A dimer comprising the T7 gene 5 product (84 kd) and *E. coli* thioredoxin (12 kd) (45, 46). The T7 gene 5 product is the polymerase/exonuclease and the *E. coli* protein binds the complex tightly to the template, preventing early dissociation during complementary strand synthesis. Functionally similar to the Klenow polymerase but has a faster rate of DNA synthesis (300 nucleotides per sec).

Properties
(cf. DNA polymerase I — Kornberg polymerase.)

(i)   5' to 3' DNA-dependent DNA polymerase, requiring a ssDNA template and a DNA or RNA primer with a 3'-OH terminus.
(ii)   3' to 5' exonuclease, degrading ssDNA or dsDNA from a 3'-OH terminus.

Applications
T7 DNA polymerase is used as an alternative to Klenow polymerase in those applications where the rapidity and template affinity of the T7 enzyme are an advantage.

(i)   Chain termination DNA sequencing (47–49).
(ii)   Second strand synthesis in site-directed mutagenesis (50, 51).

Reaction conditions
The supplier's recommendations should be followed as the conditions for the unmodified enzyme must be set up to minimize the exonuclease activity.

Modified versions
United States Biochemical Corp. have patented an engineered form of T7 DNA polymerase that lacks the 3' to 5' exonuclease (Sequenase) and is therefore ideal for DNA sequencing (52). Other companies market unmodified T7 DNA polymerase but specify precise reaction conditions aimed at suppressing the exonuclease.

1.5. *Taq* DNA POLYMERASE

Source
Thermus aquaticus YT1 (53, 54) or from recombinant *E. coli* (55).

Description
This enzyme is functionally similar to the *E. coli* DNA polymerase I but is thermostable up to 94°C and has an optimum working temperature of 80°C.

Properties
(cf. DNA polymerase I — Kornberg polymerase.)

(i)   5' to 3' DNA-dependent DNA polymerase, requiring a ssDNA template and a DNA or RNA primer with a 3'-OH terminus.
(ii) 5' to 3' exonuclease, degrading dsDNA from a 5'-P terminus.

Applications
(i)   DNA amplification by the polymerase chain reaction (56, 57).
(ii) Chain termination DNA sequencing at elevated temperatures to minimize problems with secondary structure (58).
(iii) Genomic footprinting (59).

Reaction conditions
Although the requirements are not stringent (Mg^{2+}, pH 8.0–8.5) the quality of commercial *Taq* polymerase is variable. The supplier's recommendations should be followed when setting up reaction conditions. Use the supplier's own buffer if one is provided.

Related enzymes
Bio-Rad market a different thermostable DNA polymerase, the *Bst* polymerase (from *Bacillus stearothermophilus*). This enzyme is functionally identical to the *E. coli* Klenow polymerase but will function at temperatures up to 70°C (60). United States Biochemical Corp. market *Tth* DNA polymerase (*Thermus thermophilus*) which has properties similar to *Taq* polymerase but is claimed to lack detectable exonuclease activities (61). Other thermostable DNA polymerases are commercially available but their identities have not been disclosed. These include Pyrostase (Molecular Genetic Resources), Vent DNA polymerase (New England Biolabs), and Hot Tub (Amersham).

1.6. MICROCOCCAL DNA POLYMERASE

Source
Micrococcus luteus.

Description
Functionally identical to DNA polymerase I of *E. coli* but has been preferred for certain specific applications.

Properties
(cf. DNA polymerase I — Kornberg polymerase.)

(i)   5' to 3' DNA-dependent DNA polymerase, requiring a ssDNA template and a DNA or RNA primer with a 3'-OH terminus.
(ii)   5' to 3' exonuclease, degrading dsDNA from a 5'-P terminus.
(iii) 3' to 5' exonuclease, degrading ssDNA or dsDNA from a 3'-OH terminus.

Applications
(i)   Mutagenesis procedures (62, 63).
(ii)   Synthesis of artificial copolymers.

Reaction conditions
Activity is reduced in Tris buffers. Requires Mg^{2+}; 37°C.

$10 \times$ MDP Buffer: 400 mM K phosphate (pH 7.0), 80 mM $MgCl_2$, 10 mM β-mercaptoethanol.

1.7. ALPHA DNA POLYMERASE

Source
Calf thymus.

Description
Alpha DNA polymerase is the most active eukaryotic DNA polymerase and the one most clearly linked to chromosomal DNA replication (64). The enzyme lacks the exonuclease activities of prokaryotic DNA polymerases.

Properties
(cf. DNA polymerase I — Kornberg polymerase.)

5' to 3' DNA-dependent DNA polymerase, requiring a ssDNA template and a DNA or RNA primer with a 3'-OH terminus.

Applications
As a standard eukaryotic DNA polymerase.

Reaction conditions
Requirements are similar to prokaryotic DNA polymerases; 37°C.

$10 \times$ Alpha Buffer: 600 mM Tris–HCl (pH 7.5), 60 mM $MgCl_2$, 50 mM β-mercaptoethanol, 5 mg ml^{-1} BSA.

1.8. AMV REVERSE TRANSCRIPTASE

Source
Chicks infected with avian myeloblastosis virus (AMV), or a recombinant *E. coli* expressing the AMV *pol* gene.

Description
A dimer of 62 kd and 94 kd subunits, with the small subunit being a proteolytic product of the large one (65). The enzyme possesses a complex array of activities (66, 67).

Properties
(i)   5' to 3' DNA- or RNA-dependent DNA polymerase, requiring a ssDNA or ssRNA template and a DNA or RNA primer with a 3'-OH terminus (68, 69):

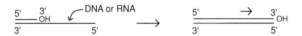

(ii) 5' to 3' and 3' to 5' exoribonuclease, specific for the RNA component of a DNA–RNA hybrid (70):

(iii) DNA endonuclease (66).
(iv) DNA unwinding activity.

Applications
(i)   First strand cDNA synthesis (9, 67, 71).
(ii) Chain termination RNA sequencing (72).
(iii) Modified chain termination DNA sequencing, especially to read GC-rich regions intractable with Klenow polymerase (73).
(iv) Probing RNA secondary structure (74).
(v)   3'-end labeling DNA fragments (75).

Reaction conditions
Requires Mg^{2+}. AMV reverse transcriptase can be incubated at 42°C. Actinomycin D (50 µg ml^{-1}) can be added to the reaction mix to inhibit the DNA-dependent DNA polymerase activity (76).

10 × AMV RT Buffer: 500 mM Tris–HCl (pH 8.3), 400 mM KCl, 100 mM $MgCl_2$, 50 mM DTT.

1.9. M-MuLV REVERSE TRANSCRIPTASE

Source
Recombinant *E. coli* containing plasmid pB6B15.23, which carries the *pol* gene of Moloney murine leukemia virus (77).

Description
The enzyme is obtained as a fusion protein comprising the first six amino acids of *trpE* followed by the M-MuLV *pol* gene, lacking the last 48 amino acids (77). Total molecular mass is 71 kd.

Properties
As for AMV reverse transcriptase except that:

(i)   the DNA endonuclease activity is absent;
(ii) the exoribonuclease activities are much reduced (78).

Applications
As for AMV reverse transcriptase. The reduced RNase activities of the M-MuLV enzyme are reputed to be an advantage for cDNA synthesis (77, 79–81), but see comparisons in refs 82 and 83.

Reaction conditions
Similar to AMV reverse transcriptase though the M-MuLV enzyme should be incubated at 37°C. Actinomycin D can be used to suppress the DNA-dependent DNA polymerase. Some suppliers recommend including 4 mM $MnCl_2$ in the reaction mixture.

$10 \times$ M-MuLV RT Buffer: 500 mM Tris–HCl (pH 8.3), 750 mM KCl, 30 mM $MgCl_2$, 100 mM DTT, 1 mg ml^{-1} BSA.

Modified versions
GIBCO-BRL market an engineered form of M-MuLV reverse transcriptase (Superscript) which lacks the exoribonuclease activity and so is ideal for cDNA synthesis.

2. RNA POLYMERASES

2.1. *E. coli* RNA POLYMERASE

Source
E. coli.

Description
The holoenzyme that can transcribe but not initiate efficiently has the subunit structure $\alpha_2\beta\beta'$. For initiation of transcription a fifth subunit, σ, is required (84).

Properties
DNA-dependent RNA polymerase, able to recognize a variety of promoter sequences related to the *E. coli* consensus:

Applications
(i)   *In vitro* transcription of genes with suitable promoters.
(ii) Synthesis of labeled RNA (85).

Reaction conditions
Requires Mg^{2+}; 37°C. EDTA is recommended.

$10 \times$ RNA Pol Buffer: 400 mM Tris–HCl (pH 7.9), 1500 mM KCl, 100 mM $MgCl_2$, 1 mM Na_2. EDTA, 1 mM DTT, 5 mg ml^{-1} BSA.

Modifying Enzymes

2.2. SP6 RNA POLYMERASE

Source
Salmonella typhimurium LT2 infected with SP6 phage (86) or a recombinant *E. coli.*

Description
A single polypeptide, 96 kd. The SP6 promoter is highly efficient and recognized specifically by the SP6 polymerase (87–90). Vectors containing the SP6 promoter can direct synthesis of microgram amounts of RNA transcript *in vitro* (8 mol RNA per mol DNA template, ref. 91). The SP6 polymerase does not terminate at nicks in the template.

Properties
(cf. *E. coli* RNA polymerase.)

DNA-dependent RNA polymerase, highly specific for the SP6 promoter.

Applications
Synthesis of RNA transcripts from sequences ligated downstream of the SP6 promoter, for:

(a)   studies of RNA processing (89, 91–93);
(b)   production of antisense RNA (94);
(c)   synthesis of mRNA for *in vitro* translation (95);
(d)   obtaining highly labeled RNA probes (91, 96–99);
(e)   RNA sequencing (100).

Reaction conditions
Requires Mg^{2+}, 37°C. Salt is not required but may be included in the reaction mix at up to 100 mM NaCl to decrease non-specific transcript initiation. The polymerization will, however, be reduced by about 50%. The activity is greatly stimulated by spermidine (86).

$5 \times$ SP6 Buffer: 200 mM Tris–HCl (pH 7.9), 30 mM $MgCl_2$, 10 mM spermidine, 50 mM DTT.

2.3. T3 RNA POLYMERASE

Source
Recombinant *E. coli* containing plasmid pCM56, which carries the T3 polymerase gene downstream of the *lac* UV5 promoter (101).

Description
A single polypeptide, approximately 100 kd. Highly specific for the T3 promoter (102). Vectors carrying this promoter are available.

Properties
(cf. *E. coli* RNA polymerase.)

DNA-dependent RNA polymerase, highly specific for the T3 promoter.

Applications
Synthesis of RNA transcripts from sequences ligated downstream of the T3 promoter, with applications similar to the SP6 RNA polymerase (97, 103).

Reaction conditions
The same requirements as SP6 RNA polymerase. The pH can be as low as 7.2.

$5 \times$ T3 Buffer: 200 mM Tris–HCl (pH 8.0), 250 mM NaCl, 40 mM $MgCl_2$, 10 mM spermidine, 150 mM DTT.

2.4. T7 RNA POLYMERASE

Source
Recombinant *E. coli* containing plasmid pAR1219, which carries the T7 polymerase gene downstream of the *lac* UV5 promoter (104).

Description
A single polypeptide, 98 kd (105). Highly specific for the T7 promoter (106). Vectors carrying the T7 promoter can be used to obtain 30 µg of transcript per µg of DNA in 30 min.

Properties
(cf. *E. coli* RNA polymerase.)

DNA-dependent RNA polymerase, highly specific for the T7 promoter.

Applications
Synthesis of RNA transcripts from sequences ligated downstream of the T3 promoter, with applications similar to the SP6 RNA polymerase (90, 91, 97–100, 107–111).

Reaction conditions
Similar to SP6 RNA polymerase. The pH can be as low as 7.2.

$5 \times$ T7 Buffer: 200 mM Tris–HCl (pH 8.0), 125 mM NaCl, 40 mM $MgCl_2$, 10 mM spermidine, 25 mM DTT.

2.5. RNA POLYMERASE II

Source
Wheat germ.

Description
The eukaryotic RNA polymerase responsible for transcribing most protein-coding genes.

Properties
(cf. *E. coli* RNA polymerase.)

DNA-dependent RNA polymerase, able to recognize a variety of promoter sequences.

Applications
Eukaryotic transcription studies (112).

Reaction conditions
The enzyme has unusual requirements and works best at 25°C.

10 × Pol II Buffer: 100 mM Tris–HCl (pH 7.9), 10 mM $MnCl_2$, 200 mM $MgCl_2$, 500 mM $(NH_4)_2SO_4$, 1.25 mg ml^{-1} BSA.

Modifying Enzymes

3. NUCLEASES

3.1. Bal31 NUCLEASE

Source
Alteromonas espejiana Bal31.

Description
An extracellular nuclease; Bal31 is highly sequence dependent, so the rate at which it shortens a DNA molecule can vary considerably. At extreme dilutions only the ssDNA endonuclease activity is evident.

Properties (113–119)
(i) 3' to 5' exonuclease that progressively removes nucleotides from the 3'-OH termini of ssDNA or dsDNA molecules having either blunt or sticky ends:

(ii) Endodeoxyribonuclease, with high specificity for ssDNA:

(iii) The combination of (i) and (ii) enables Bal31 to shorten dsDNA molecules progressively:

(iv) Inefficient ribonuclease (120).

Applications
(i) Controlled trimming of dsDNA molecules to remove unwanted sequences prior to cloning (121).
(ii) Generating overlapping subclones for DNA sequencing.
(iii) Restriction mapping (116).
(iv) Mapping DNA secondary structures (114, 119).

Reaction conditions
Bal31 has an absolute requirement for Mg^{2+} and Ca^{2+}. Reactions can therefore be stopped with EGTA. Incubate at 30°C. Some suppliers warn against freezing stocks of Bal31.

2 × Bal31 Buffer: 40 mM Tris–HCl (pH 7.4), 600 mM NaCl, 25 mM $MgCl_2$, 25 mM $CaCl_2$, 2 mM Na_2.EDTA.

3.2. S1 NUCLEASE

Source
Aspergillus oryzae.

Description
A Zn metalloprotein, glycosylated, 32 kd (122, 123). S1 nuclease is relatively thermostable.

Properties
Single-strand specific endonuclease, more active on ssDNA than ssRNA (122, 124):

Applications
(i)   Analysis of DNA–RNA hybrids (S1 nuclease mapping) to locate introns (125, 126) and transcript termini (127–129).
(ii) Removal of overhangs to produce blunt-ended molecules (130, 131).
(iii) Opening hairpins formed during cDNA synthesis (132).

Reaction conditions
S1 nuclease requires Zn^{2+} and a low pH (4.0–4.5), being inactive at pH 6.0 (133, 134). Reactions can be stopped with EDTA (133, 134). High NaCl is used to prevent nicking of dsDNA (135); 37°C.

5 × S1 Buffer: 250 mM NaAc (pH 4.5), 1000 mM NaCl, 5 mM $ZnSO_4$, 2.5% (v/v) glycerol.

3.3. MUNG BEAN NUCLEASE

Source
Mung bean (*Vigna radiata*) sprouts.

Description
Functionally identical to S1 nuclease (136–140) but is less likely to cause nicks in dsDNA, and is also less likely to cut the intact strand opposite a nick.

Properties
(cf. S1 nuclease.)

Single-strand specific endonuclease, more active on ssDNA than ssRNA:

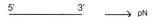

Applications
Mung bean nuclease is useful as an alternative to S1 nuclease when cleavage of double-stranded molecules is a problem (124, 139, 141–145).

Reaction conditions
Requirements are similar to S1 nuclease, except that the optimal pH is a little higher (pH 5.0) and the optimal temperature is 30°C. Mung bean nuclease is very 'sticky' and glycerol is added to prevent the enzyme adhering to the surfaces of the reaction tube.

$10 \times$ Mung Bean Buffer: 300 mM NaAc (pH 5.0), 500 mM NaCl, 10 mM $ZnCl_2$, 50% (v/v) glycerol.

3.4. P1 NUCLEASE

Source
Penicillium citrinum.

Description
A Zn metalloprotein, highly glycosylated, 24 kd.

Properties
Non-specific phosphodiesterase that acts both as an endonuclease and an exonuclease. P1 is most active with ssDNA and ssRNA but also has substantial activity with dsDNA and dsRNA (146):

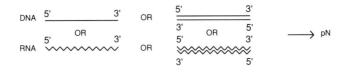

Applications
(i)   Mobility shift analysis in RNA sequencing.
(ii) Possibly as an alternative to S1 and mung bean nucleases for ssDNA or ssRNA cleavage, though the high dsDNA and dsRNA activities make P1 nuclease a poor choice.

Reaction conditions
The optimal pH is between 4.5 and 8.0, depending on the substrate (146, 147). Zn^{2+} is not essential. The optimal temperature is in fact 70°C, though 37°C is generally used. A suitable 10 × buffer is 500 mM NaAc pH 5.5.

3.5. RIBONUCLEASE A

Source
Bovine pancreas.

Description
A very active and stable enzyme, 13.7 kd (148). Most preparations are contaminated with a separate DNase that can be inactivated by heating to 100°C. RNase A is the non-glycosylated form of the standard pancreatic ribonuclease.

Properties
Endoribonuclease, cleaving phosphodiester bonds 3' to a pyrimidine (149). No activity with DNA:

5' ⌇⌇⌇⌇⌇⌇⌇⌇⌇⌇ 3' ⟶ OLIGONUCLEOTIDES WITH A TERMINAL
PYRIMIDINE 2', 3' - CYCLIC PHOSPHATE

Applications
(i)   Removing RNA from DNA preparations.
(ii) Mapping single bp mutations in DNA and RNA (150, 151).

Reaction conditions
Ribonuclease A is active over a wide pH range (though the optimum is pH 7.0–7.5) and has no cofactor requirements. Normally a DNA solution is not adjusted specifically to optimize conditions for treatment with RNase A.

DNase-free RNase A
Commercial preparations of RNase A are often contaminated with DNases. To prepare DNase-free RNase A the solid enzyme is dissolved in 10 mM Tris–HCl (pH 7.5), 15 mM NaCl, heated at 100°C for 15 min and then cooled slowly to room temperature. Aliquots can be stored at –20°C for months. A stock solution of 10 mg ml^{-1} is suitable for most applications.

3.6. RIBONUCLEASE H

Source
E. coli, in many cases a recombinant strain carrying the RNase H gene on a pBR322 plasmid (152).

Description
Monomer, 17.5 kd.

Properties
Endoribonuclease, specific for the RNA component of a DNA–RNA hybrid (153):

Applications
(i)   Second strand cDNA synthesis (9, 71).
(ii) Removing poly(A) tails from mRNA, possibly to improve band resolution in gel electrophoresis (154, 155).
(iii) Detection of DNA–RNA hybrids (156, 157).
(iv) Studying RNA priming of DNA synthesis (158).
(v)   Site-specific cleavage of RNA hybridized to a short oligonucleotide (159).

Reaction conditions
Requires Mg^{2+}, 37°C. The pH optimum is 7.5–9.1 (153). RNase H is relatively insensitive to salt, being 50% active in 0.3 M NaCl.

10 × RNase H Buffer: 400 mM Tris–HCl (pH 7.5), 40 mM $MgCl_2$, 10 mM DTT.

Modifying Enzymes

3.7. DEOXYRIBONUCLEASE I

Source
Bovine pancreas.

Description
A glycoprotein, 31 kd, usually obtained as a mixture of several isoenzymes. Most suppliers market RNase-free versions.

Properties
Endodeoxyribonuclease, degrading ssDNA and dsDNA by cutting preferentially 5' to pyrimidines, resulting in mono- and oligonucleotides with an average length of 4 nt (160, 161).

In the absence of Mn^{2+}:

In the presence of Mn^{2+}:

Applications
(i)   Introducing nicks into dsDNA prior to labeling by nick translation (8).
(ii)   DNase I protection and footprinting to locate proteins bound to DNA (163, 164).
(iii) Removal of DNA during RNA preparation.
(iv)   Removal of the DNA template after *in vitro* transcription (165).
(v)   Detection of transcriptionally active regions of chromatin (166).
(vi)   Producing overlapping subclones for DNA sequencing (167).
(vii) Nicking circular DNA prior to bisulfite mutagenesis (168).

Reaction conditions
Requires Mg^{2+}, but this can be replaced by Mn^{2+} to cause dsDNA breaks (162). 37°C. Some suppliers recommend addition of 20 mM $CaCl_2$ to the reaction mixture.

10 × DNase Buffer: 500 mM Tris–HCl (pH 7.5), 10 mM $MgCl_2$, 1 mg ml^{-1} BSA.

RNase-free DNase I
Pancreatic DNase I is often contaminated with RNases. Commercial RNase-free preparations are available but the enzyme should be used with a suitable RNase inhibitor (see Chapter 2, *Table 19*) if RNase activity will be a problem. DNase I can be rendered RNase-free by dissolving it at a concentration of 1 mg ml^{-1} in 0.1 M iodoacetic acid, 0.15 M sodium acetate (pH 5.2), heating to 55°C for 45 min, cooling on ice, and then adding $CaCl_2$ to a concentration of 5 mM. Aliquots can be stored at –20°C for months.

3.8. S7 NUCLEASE (microccoccal nuclease)

Source
Staphylococcus aureus.

Description
S7 nuclease is active on DNA and RNA. With DNA it displays a preference for ssDNA and AT-rich regions in dsDNA.

Properties
Endonuclease, active on DNA and RNA, both single- and double-stranded:

Applications
(i)   Preparation of an mRNA-dependent protein synthesis system from rabbit reticulocyte lysates (169, 170).
(ii) Digestion of chromatin to prepare nucleosomes (171).

Reaction conditions
Requires Ca^{2+}, so reactions can be stopped with EGTA; 37°C.

$10 \times$ S7 Buffer: 500 mM Na borate (pH 8.8), 50 mM NaCl, 25 mM $CaCl_2$.

3.9. T7 ENDONUCLEASE

Source
E. coli infected with T7 phage, or a recombinant *E. coli* strain expressing T7 gene 3.

Description
The only ssDNA specific endonuclease that is active at neutral and alkaline pH (172). T7 endonuclease has a weak activity with dsDNA [1% of the ssDNA activity (173, 174)] with a preference for branched structures, probably because of short single-stranded regions within them (172).

Properties
Endodeoxyribonuclease, primarily specific for ssDNA:

Applications
(i)   Studying complex structures in dsDNA (174, 175).
(ii) Footprinting to locate proteins bound to DNA (176, 177).
(iii) Studying the unwinding of DNA caused by DNA-binding proteins (176).
(iv) Activation of the SOS response in *E. coli recA* (178).

Reaction conditions
Requires Mg^{2+}; 37°C.

$10 \times$ T7 Nuclease Buffer: 500 mM Tris–HCl (pH 8.0), 200 mM NaCl, 60 mM $MgCl_2$, 50 mM DTT, 1 mg ml^{-1} BSA.

3.10. EXONUCLEASE I

Source
E. coli.

Description
Monomer, 55 kd.

Properties
3' to 5' exodeoxyribonuclease, specific for ssDNA:

$$5' \underline{\hspace{3cm}} 3'_{\text{OH}} \longrightarrow 5' \underline{\underset{\leftarrow}{\hspace{2cm}}} 3'_{\text{OH}} + \text{pN}$$

Applications
(i) Studying DNA helicase activity (179).
(ii) Assaying endonucleolytic cleavage of ssDNA (180).

Reaction conditions
10 × ExoI buffer: 670 mM glycine–KOH (pH 9.5), 67 mM MgCl$_2$, 100 mM β-mercaptoethanol; 37°C.

3.11. EXONUCLEASE III

Source
E. coli BE 257 or SR80, both of which contain a thermo-inducible gene on the plasmid pSGR3 (181, 182).

Description
Monomer, 28 kd (183), with a number of activities. The exonuclease is inactive on ssDNA or dsDNA with a 3' overhang of 4 nt or more.

Properties
(i) 3' to 5' exodeoxyribonuclease, specific for dsDNA with 3'-OH termini (183):

$$5' \underline{\hspace{3cm}} 3'_{OH}$$
$$HO\underline{\hspace{3cm}}5' \qquad \longrightarrow \qquad 5'\underline{\hspace{2cm}}3'_{OH} \quad HO\underline{\hspace{1.5cm}}5'$$
$$3'$$

(ii) 3' phosphatase, removing phosphate groups from the 3' termini of ssDNA or dsDNA (182, 184):

$$5'\underline{\hspace{3cm}}3'_{P}$$
OR
$$5'\underline{\hspace{3cm}}3'_{P}$$
$$P\underline{\hspace{3cm}}5' \qquad \longrightarrow \qquad 5'\underline{\hspace{3cm}}3'_{OH}$$
$$3' \qquad\qquad HO\underline{\hspace{2cm}}5'$$

(iii) Endodeoxyribonuclease, specific for nucleotides from which the purine or pyrimidine base has been cleaved (156, 182, 184).

(iv) Endoribonuclease, specific for the RNA component of an RNA–DNA hybrid (182, 184):

$$5' \xrightarrow{\text{DNA}} 3' \qquad \longrightarrow \qquad 5'\underline{\hspace{3cm}}3' + \text{OLIGORIBONUCLEOTIDES}$$
$$3' \underset{\text{RNA}}{\wedge\wedge\wedge} 5'$$

Applications
(i) Generation of overlapping subclones for DNA sequencing (130, 185–188).
(ii) End-labeling of blunt-ended dsDNA, in conjunction with Klenow polymerase (189, 190).
(iii) Deletion of sequences at the termini of dsDNA fragments, in conjunction with a single-strand specific endonuclease (130, 191).
(iv) Mutagenesis techniques (51).
(v) Protein protection studies (192–194).

Reaction conditions
The exo- and endonucleases have pH optima of 7.6–8.5. The optimum for the phosphatase is pH 6.8–7.4. Requires Mg^{2+}. Salt may be added to the reaction mixture for certain applications.

$10 \times$ Basic Exo III Buffer: 500 mM Tris–HCl (pH 8.0), 50 mM $MgCl_2$, 100 mM β-mercaptoethanol.

3.12. EXONUCLEASE VII

Source
E. coli HMS137.

Description
An exonuclease that can attack both the 5' and 3' termini of ssDNA. Exonuclease VII produces oligonucleotides, which means it may not leave blunt ends when used with dsDNA with overhangs.

Properties
3' to 5' and 5' to 3' exodeoxyribonuclease, specific for ssDNA (126, 195, 196):

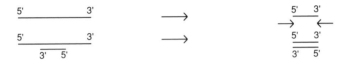

Applications
Exon mapping (127, 197). Exonuclease VII has no endonucleolytic activity, so can be used in transcript mapping to distinguish ssDNA loops from ssDNA overhangs in DNA–RNA hybrids.

Note that exonuclease VII is not suitable for blunt-ending dsDNA, as its mode of action may result in single nucleotide overhangs remaining after treatment.

Reaction conditions
Does not require Mg^{2+} and is active in 8 mM EDTA. The optimum temperature is 37°C, but exonuclease VII retains most of its activity at 42°C. The higher temperature plus low salt (up to 30 mM KCl) may be used to destabilize secondary structures in ssDNA.

$10 \times$ ExoVII Buffer: 670 mM K phosphate (pH 7.9), 83 mM Na_2.EDTA, 100 mM β-mercaptoethanol.

3.13. LAMBDA EXONUCLEASE

Source
E. coli λ lysogen SG5519.

Description
This exonuclease is 10–100 times more active with dsDNA (blunt-ended or with a 3' overhang) than with ssDNA (198, 199), but is inefficient with dsDNA with a 5' overhang. Nicks and gaps are not usually recognized as starting points.

Properties
5' to 3' exodeoxyribonuclease, specific for dsDNA preferably with a terminal 5'-P (200):

Applications
(i) Removing 5' overhangs on dsDNA, especially prior to labeling with terminal transferase (201).
(ii) Restriction mapping.

Reaction conditions
Lambda exonuclease requires Mg^{2+} and a relatively high pH. 37°C is recommended though the enzyme is active at room temperature.

$10 \times$ LE Buffer: 670 mM glycine–KOH (pH 9.4), 25 mM $MgCl_2$, 500 μg ml^{-1} BSA.

3.14. *N. crassa* NUCLEASE

Source
Neurospora crassa.

Description
Acts as an endonuclease on ssRNA and ssDNA, and as an exonuclease on ssDNA and dsDNA (202). The substrate can therefore be either linear or circular. At low enzyme concentration nicks can be made in supercoiled DNA (203). This is the only enzyme known to produce RNA fragments and ribonucleotides with 5'-terminal phosphates (204, 205).

Properties
(i) Endonuclease active on ssRNA and dsRNA, and ssDNA:

(ii) Exonuclease active on ssDNA and dsDNA:

Applications
Degrading non-paired single-stranded regions in DNA molecules, for example during hybridization experiments (206).

Reaction conditions
The optimum $1 \times$ buffer is 100 mM Tris–HCl (pH 7.5–8.5) with the salt and Mg^{2+} concentrations dependent on the substrate. Mg^{2+} is not essential but 10 mM $MgCl_2$ stimulates the activity on DNA and represses that on RNA. The exonuclease is inhibited by Zn^{2+}.

3.15. PHOSPHODIESTERASES I and II

Source
Phosphodiesterase I: *Crotalus durissus* venom (207).
Phosphodiesterase II: calf spleen.

Description
Both enzymes act on DNA and RNA with the appropriate termini.

Properties
Phosphodiesterase I: exonuclease, specific for DNA and RNA with a 3'-OH terminus, yielding 5'-dNMPs and NMPs:

E.g.
$$
\begin{array}{c}
5' \quad\quad\quad\quad 3' \\
\text{HO} \overline{\quad\quad\quad\quad} \text{OH} \\
3' \quad\quad\quad\quad 5'
\end{array}
\longrightarrow \text{pN}
$$

Phosphodiesterase II: exonuclease, specific for DNA and RNA with a 5'-OH terminus, yielding 3'-dNMPs and NMPs:

E.g.
$$
\begin{array}{c}
5' \quad\quad\quad\quad 3' \\
\text{HO} \overline{\quad\quad\quad\quad} \text{OH} \\
3' \quad\quad\quad\quad 5'
\end{array}
\longrightarrow \text{Np}
$$

Applications
DNA and RNA structural studies.

Reaction conditions
Phosphodiesterase II does not require Mg^{2+}.

$10 \times$ PI Buffer: 1000 mM Tris–HCl (pH 8.9), 1000 mM NaCl, 140 mM $MgCl_2$, incubate at 25°C.

$2 \times$ PII Buffer: 250 mM succinate–HCl (pH 6.5), incubate at 37°C.

4. LIGASES

4.1. T4 DNA LIGASE

Source
E. coli λ lysogens NM969 (208, 209) or 60 p*c*1857 p*PLc*28lig, or a recombinant *E. coli*.

Description
Monomer, 68 kd.

Properties
DNA ligase, catalyzing the formation of a phosphodiester bond between adjacent 5'-P and 3'-OH termini (210).

(i)   With cohesive-ended molecules:

(ii)   With blunt-ended molecules:

(iii)   With nicked molecules:

Applications
Joining DNA molecules (211–215).

Reaction conditions
Requires Mg^{2+} and ATP. For blunt-end ligations about 50 times as much enzyme is needed, compared with cohesive-end ligation. Blunt-end ligation is stimulated by T4 RNA ligase (216) or PEG (217) and inhibited by more than 50 mM NaCl.

'Ideal' reaction conditions have been worked out (218) and are as follows:

For blunt-end ligation (vector:insert ratio = 3:1), 10× buffer = 250 mM Tris–HCl (pH 7.5), 50 mM $MgCl_2$, 25% (w/v) PEG8000, 5 mM DTT, 4 mM ATP. Use 1 unit enzyme, incubate for 4 h at 23°C in 20 µl total volume.

For cohesive-end ligation (vector:insert ratio = 1:2), 10× buffer = 250 mM Tris–HCl (pH 7.5), 100 mM $MgCl_2$, 100 mM DTT, 4 mM ATP. Use 0.01 units enzyme, incubate overnight at 8°C in 20 µl total volume.

For ligating linkers to blunt-ended DNA, 10 × buffer = 300 mM Tris–HCl (pH 7.5), 300 mM NaCl, 80 mM $MgCl_2$, 20 mM DTT, 2 mM Na_2.EDTA, 70 mM spermidine, 1 mg ml^{-1} BSA, 2.5 mM ATP. Use 2 units enzyme, 5 µg DNA, 500 ng (10–12 mer linker) or 12.5 µg (8 mer linker), incubate overnight at 4°C in 20 µl total volume.

Modifying Enzymes

4.2. *E. coli* DNA LIGASE

Source
E. coli 594 (*su⁻*) lysogenic for λgt4-*lop*-11 *lig* Sam7 (219, 220).

Description
Monomer, 74 kd (221), functionally similar to T4 DNA ligase, but requires NAD instead of ATP as the cofactor and is more stringent in its activity, for instance being inactive with RNA-primed DNA (222, 223).

Properties
As for T4 DNA ligase (224).

Applications
Joining DNA molecules, though generally T4 DNA ligase is preferred.

Reaction conditions
Requires Mg^{2+}. As with T4 DNA ligase, the reaction conditions should be optimized for different substrates. However, the ideal conditions for the *E. coli* enzyme have not yet been reported.

$10 \times$ Basic EcL Buffer: 300 mM Tris–HCl (pH 8.0), 40 mM $MgCl_2$, 12 mM Na_2.EDTA, 10 mM DTT, 0.26 mM NAD, 500 μg ml^{-1} BSA, incubate at 12°C.

4.3. T4 RNA LIGASE

Source
T4-infected *E. coli* or recombinant *E. coli* containing the plasmid pRF-E35 (225).

Description
Although called an RNA ligase this enzyme works on ssDNA as well as ssRNA. ATP is needed as a cofactor (226–228).

Properties
Ligase, catalyzing the formation of a phosphodiester bond between 5'-P and 3'-OH groups on ssDNA and ssRNA molecules:

Applications
(i)   Joining ssRNA or ssDNA molecules (229–231).
(ii) Increasing the efficiency of T4 DNA ligase in blunt-end ligation (216).
(iii) 3'-end labeling of RNA with 3'-5'-CDP (226, 228, 232, 233).
(iv) Stimulating misacylation of tRNAs (234).
(v)   Modifying tRNA anticodons (235).

Reaction conditions
Requires Mg^{2+} and is generally used at 37°C.

$10 \times$ RL Buffer: 500 mM Tris–HCl (pH 7.5), 100 mM $MgCl_2$, 100 mM DTT, 10 mM ATP, 100 µg ml^{-1} BSA.

5. END-MODIFICATION ENZYMES

5.1. ALKALINE PHOSPHATASE

Source
Calf intestine [calf intestine phophatase/alkaline phosphatase (CIP/CIAP)] or *E. coli* [bacterial alkaline phosphatase (BAP)] (236, 237).

Description
Dimeric glycoprotein, two identical 69 kd subunits. Contains four atoms of zinc per molecule (237). CIP is inactivated by heating to 68°C whereas BAP is stable at this temperature. BAP is resistant to phenol extraction. Preparations of BAP frequently contain a contaminating exodeoxyribonuclease.

Properties
5' phosphatase, removing 5' phosphate groups from ssDNA and dsDNA, ssRNA and dsRNA, dNTPs and NTPs (238, 239):

E.g.

$$P \xrightarrow[\ 3'\ \quad\quad\quad 5'\ ]{\ 5'\ \quad\quad\quad 3'\ } P \longrightarrow HO \xrightarrow[\ 3'\ \quad\quad\quad 5'\ ]{\ 5'\ \quad\quad\quad 3'\ } OH$$

Applications
(i)   Dephosphorylation of dsDNA to prevent self-ligation (240, 241).
(ii) Dephosphorylation of DNA and RNA to allow 5'-end labeling with T4 polynucleotide kinase (242).
(iii) As a component of a conjugated antibody immunodetection system (243, 244).

Reaction conditions
Alkaline phosphatases do not have particularly stringent buffer requirements:

$5 \times$ CIP Buffer: 250 mM Tris (pH 9.0), 50 mM $MgCl_2$, 5 mM $ZnCl_2$, 50 mM spermidine.

$10 \times$ BAP Buffer: 500 mM Tris–HCl (pH 8.0).

Temperature is more important:

For CIP use 37°C for RNA and dsDNA with 5' overhangs, and 56°C for blunt-ended dsDNA and dsDNA with 3' overhangs.

With BAP use 65°C for DNA (to suppress the contaminating exonuclease) and 45°C for RNA. If your sample of BAP has no significant exonuclease activity then with dsDNA use 37°C for blunt ends or 5' overhangs and 60°C for 3' overhangs.

5.2. T4 POLYNUCLEOTIDE KINASE

Source
T4 infected *E. coli* (245, 246) or a recombinant *E. coli* (247).

Description
Tetramer of identical 33 kd subunits (36, 248). The phosphatase is active on ssDNA, dsDNA, ssRNA and dsRNA and also with 3'-dNMPs and 3'-NMPs.

Properties
(i)　5' kinase, transferring the γ-phosphate of ATP to a 5'-OH terminus on a ssDNA, dsDNA, ssRNA or dsRNA molecule (36):

E.g.　HO—5'————3'　$\xrightarrow{\text{ATP}}$　P—5'————3' + ADP

With excess ADP an exchange reaction occurs:

A-P-P* + P—5'————3'　$\longrightarrow$　A-P-P + *P—5'————3'

(ii)　3' phosphatase, removing phosphate groups from the 3'-termini of ssDNA, dsDNA, ssRNA or dsRNA molecules:

E.g.　5'————3'P　$\longrightarrow$　5'————3'OH + P$_i$

Applications
(i)　Labeling the 5' termini of DNA or RNA by the kinase or exchange reactions (245, 249, 250), especially prior to Maxam–Gilbert sequencing (251–253), enzymatic RNA sequencing (250, 254), restriction mapping (255, 256) and footprinting (163, 257).
(ii)　Adding 5' phosphates to synthetic oligonucleotides.
(iii)　Removing 3' phosphates (258).

Reaction conditions
Requires Mg^{2+}. The kinase and exchange reactions require different conditions, though both are carried out at 37°C. The kinase activity is maximal at pH 7.6 with at least 1 µM ATP and a 5:1 ratio of ATP to 5'-OH ends. The phosphatase has a pH optimum of 5.0–6.0 (259).

10 × Kinase Buffer: 500 mM Tris–HCl (pH 7.6), 70 mM $MgCl_2$, 50 mM DTT.

10 × Exchange Buffer: 500 mM imidazole–HCl (pH 6.6), 100 mM $MgCl_2$, 50 mM DTT, 3 mM ADP.

Up to 10 mM sperimidine may be added to the reaction mixture to stabilize the enzyme. For kinasing blunt-ended molecules the 10× buffer should contain 50% (v/v) glycerol.

5.3. T4 POLYNUCLEOTIDE KINASE (phosphatase-free)

Source
E. coli infected with a mutant T4 phage.

Description
An altered version of the standard T4 polynucleotide kinase but still a tetramer of identical 33 kd subunits (260).

Properties
5' kinase, transferring the γ-phosphate of ATP to a 5'-OH terminus on a ssDNA, dsDNA, ssRNA or dsRNA molecule.

Applications
Addition of a 5'-P group to 3'-CMP to produce labeled 5'-3'-CDP, used to 3'-end label RNA (261).

Reaction conditions
As for T4 polynucleotide kinase.

5.4. TERMINAL DEOXYNUCLEOTIDYL TRANSFERASE (TdT)

Source
Calf thymus (262).

Description
Dimer of 80 kd + 26 kd. TdT can polymerize ribonucleotides with limited efficiency. The 3'-OH acceptor can be an overhang, blunt or recessed end, though an overhang is most efficient (263, 264).

Properties
Template-independent DNA polymerase, adding deoxynucleotides to a 3'-OH terminus on ssDNA or dsDNA (263, 265).

(i)   With ssDNA or dsDNA with a 3' overhang of at least 2 nt:

(ii)   With dsDNA with a blunt or recessed 3'-OH terminus (266):

Applications
(i)   Homopolymer tailing (19, 71, 267).
(ii)   Labeling 3' termini with labeled 3'-dNTP (268), 3'-NTP (269), cordycepin triphosphate (268) or a dideoxy-NTP (270).

Reaction conditions
The choice of buffer is critical for the type of acceptor molecule being used, the type of nucleotides being polymerized, and the purpose of the experiment. See refs 271 and 272.

5.5. POLYNUCLEOTIDE PHOSPHORYLASE

Source
Micrococcus luteus or *E. coli* B.

Description
Tetramer, total molecular mass 260 kd. The polymerization reaction is reversible with excess inorganic phosphate. The enzyme will use modified NDPs as substrates, especially if Mg^{2+} is replaced by Mn^{2+}.

Properties
Template-independent ribonucleotide polymerase, able to add ribonucleotides to the 3'-OH terminus of an RNA acceptor:

$$5'\text{wwwwww}\,3'\,OH \xrightarrow[\text{NDPs}]{} 5'\text{wwwwww}\,pNpNpN\,3' + P_i$$

The enzyme also catalyzes an exchange reaction between inorganic phosphate and the β-phosphate of an NDP:

$$^*P_i + N\text{-}P\text{-}P \longrightarrow P_i + N\text{-}P\text{-}P^*$$

Applications
Synthesis of artificial RNA molecules.

Reaction conditions
Requires Mg^{2+}; 37°C.

10 × PP Buffer: 1000 mM Tris–HCl (pH 9.0), 500 mM $MgCl_2$, 4 mM Na_2.EDTA, 200 mM NDP.

5.6. POLY(A) POLYMERASE

Source
E. coli.

Description
Monomer, 58 kd. As well as polymerizing ATP the enzyme uses CTP and UTP at 5% maximal activity, but does not work with GTP. It is specific for ssRNA, having no activity with ssDNA or dsDNA, and negligible activity with dsRNA. Short ssRNAs (e.g. trinucleotides) are poor acceptors. Up to 2000 residues can be added to a good acceptor.

Properties
Template-independent 3' poly(A) polymerase, specific for ssRNA with a 3'-OH terminus (273):

Applications
(i)   Labeling RNA with a tail of labeled nucleotides (273) or a single cordycepin monophosphate (274).
(ii) Providing a poly(A) tail for priming cDNA synthesis with oligo(dT) (275).

Reaction conditions
Requires Mg^{2+} and a high concentration of monovalent cations. Stimulated by Mn^{2+}. 37°C.

$10 \times$ Poly(A) Buffer: 500 mM Tris–HCl (pH 7.9), 100 mM $MgCl_2$, 25 mM $MnCl_2$, 2500 mM NaCl, 500 mg ml^{-1} BSA. Use 0.25 mM ATP in the final mix for a complete reaction with 250 µg RNA.

5.7. mRNA GUANYLTRANSFERASE

Source
Vaccinia virus or *Saccharomyces cerevisiae pep4*.

Description
An enzyme complex with a very high specificity for RNA molecules with a 5' triphosphate terminus.

Properties
A three-stage enzymatic activity that attaches the 5'-cap structure to a suitable acceptor RNA.

(i)   RNA triphosphatase:

$$5'\text{ppp} \sim\sim\sim\sim\sim\sim \longrightarrow 5'\text{ppp} \sim\sim\sim\sim\sim\sim + P_i$$

(ii)   Guanyltransferase:

$$\text{G-P-P-P} + \text{pp} \sim\sim\sim\sim\sim\sim \longrightarrow \text{Gppp} \sim\sim\sim\sim\sim\sim + PP_i$$

(iii)   RNA (guanine-7) methyltransferase:

$$\text{Gppp} \sim\sim\sim\sim\sim\sim \xrightarrow{\text{SAM}} \text{m}^7\text{Gppp} \sim\sim\sim\sim\sim\sim$$

Applications
In transcript mapping, to distinguish a 5' terminus created by transcription initiation from one resulting from RNA processing (276–280).

Reaction conditions
Requires Mg^{2+}; 37°C.

$10 \times$ GT Buffer: 500 mM Tris–HCl (pH 7.9), 60 mM KCl, 12.5 mM $MgCl_2$, 25 mM DTT, 400 µM GTP.

6. TOPOISOMERASES

6.1. DNA TOPOISOMERASE I

Source
Calf thymus.

Description
Single polypeptide, 105 kd (281).

Properties
Type I DNA topoisomerase, able to relax supercoiled DNA by introducing transient single-stranded breaks in the sugar–phosphate backbone. After the intact strand has been passed through the break the free polynucleotide ends are rejoined (282–284):

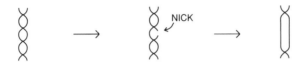

This activity also enables topoisomerase I to:

(i)   catenate and decatenate nicked circular dsDNA;
(ii) attach a ssDNA molecule to a 5'-OH terminus on a second ssDNA or dsDNA molecule (285–287).

Applications
(ii) Studying nucleosome assembly (288, 289).
(ii) Studying DNA tertiary structure (290, 291).

Reaction conditions
Requires Mg^{2+}. The pH optimum is 7.5 and the enzyme is active in 200 mM NaCl. 37°C.

$10 \times$ TopI Buffer: 500 mM Tris–HCl (pH 7.5), 500 mM KCl, 100 mM $MgCl_2$, 5 mM DTT, 1 mM Na_2.EDTA, 300 µg ml^{-1} BSA.

Modifying Enzymes

6.2. DNA TOPOISOMERASE II (DNA gyrase)

Source
Micrococcus luteus (292).

Description
Multimer probably of identical subunits. ATP is needed as a cofactor.

Properties
Type II DNA topoisomerase, able to introduce negative superhelical turns into cccDNA by double-stranded breakage followed by rejoining of the phosphodiester bonds (82, 282):

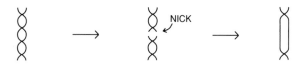

This activity enables DNA gyrase to reversibly knot and catenate cccDNA.

Applications
Introducing supercoils into plasmids and other cccDNA.

Reaction conditions
Requires Mg^{2+} and ATP. 37°C. The addition of 10% (v/v) glycerol to the reaction mixture may improve DNA gyrase activity with some substrates.

5 × Gyrase Buffer: 175 mM Tris–HCl (pH 7.5), 100 mM KCl, 100 mM $MgCl_2$, 0.5 mM Na_2.EDTA, 10 mM spermidine, 5 mM ATP, 50 mM β-mercaptoethanol, 200 μg ml^{-1} BSA.

7. DNA BINDING PROTEINS

7.1. *E. coli* SINGLE-STRAND BINDING PROTEIN

Source
E. coli, including recombinant *E. coli*.

Description
Tetramer of four identical 18.9 kd subunits. Binds cooperatively to ssDNA but not dsDNA.

Properties
Single-strand DNA binding protein.

Applications
(i)   D-loop mutagenesis procedure (62, 293).
(ii)   Visualizing ssDNA by electron microscopy.
(iii)   Studying DNA helicase activity, in conjunction with exonuclease I (179).
(iv)   Stimulating the transition of phage genomes from ssDNA to dsDNA (294).
(v)   Stimulating DNA polymerase activity (295).
(vi)   Sequencing through regions of strong secondary structure (61).

Reaction conditions
Optimal attachment requires high salt (1 M NaCl), but this can subsequently be dialyzed away.

Modifying Enzymes

7.2. T4 GENE 32 PROTEIN

Source
E. coli infected with a T4 triple mutant am*N*134/am*BL*292/am*E*219 (defective for genes 33, 35 and 58–61), or an overproducing recombinant *E. coli* (296).

Description
The gene 32 protein is a ssDNA-binding protein that stimulates the activity of T4 DNA polymerase 5- to 10-fold (297–301). The protein binds to ssRNA with a 10- to 1000-fold lower affinity.

Properties
Single-strand DNA binding protein.

Applications
(i)   Stimulating the activity of T4 DNA polymerase in replicating a ssDNA template (299, 302).
(ii) Visualizing ssDNA by electron microscopy (303–306).
(iii) Site-directed mutagenesis protocols (41).
(iv) Improving the efficiency of restriction for large scale DNA preparations (307).
(v)   Recognition of damaged DNA (308).

Reaction conditions
For optimum attachment to most substrates use the following 10× buffer: 500 mM Tris–HCl (pH 7.5), 100 mM $MgCl_2$, 1 mM Na_2.EDTA, 70 mM β-mercaptoethanol.

7.3. RecA PROTEIN

Source
E. coli, including recombinant *E. coli*.

Description
Required for homologous genetic recombination and DNA repair (309, 310). The protein promotes DNA strand exchange *in vitro* (311).

Properties
(i)   Single-strand DNA binding protein.
(ii) DNA-dependent ATPase.

Applications
(i)   D-loop mutagenesis procedure (62, 293).
(ii) Visualizing ssDNA by electron microscopy (312).

Reaction conditions
These depend on the application (62, 312). A suitable 10× buffer for D-loop formation is 200 mM Tris–HCl (pH 8.0), 100 mM $MgCl_2$, 10 mM DTT, 13 mM ATP.

Modifying Enzymes

8. RIBOZYMES

8.1. RNAzyme TET 1.0

Source
Engineered version of the *Tetrahymena* self-splicing intron (313).

Description
This is a modified version of the autocatalytic RNA contained within the intron of the *Tetrahymena* pre-LrRNA. It catalyzes a sequence-specific cleavage in ssRNA (314). A RNAzyme kit is marketed by United States Biochemical Corp.

Properties
Sequence-specific endoribonuclease, cutting ssRNA specifically 3' of the sequence 5'-CUCU-3':

Applications (315, 316)
(i)   Direct restriction-type mapping of ssRNA.
(ii) Enzymatic RNA sequencing.
(iii) Analysis of higher order structure in ssRNA.

Reaction conditions
GTP is required as a cofactor and reactions are carried out at 50°C. Use the supplier's recommendations when setting up reactions.

9. REFERENCES

1. Maniatis, T., Jeffrey, A. and Kleid, D.G. (1975) *Proc. Natl. Acad. Sci. USA*, **72**, 1184.

2. Murray, N.E. and Kelley, W.S. (1979) *Mol. Gen. Genet.*, **175**, 77.

3. Lehman, I.R., Bessman, M.J., Simms, E.S. and Kornberg, A. (1958) *J. Biol. Chem.*, **233**, 163.

4. Jovin, T.M., Englund, P.T. and Bertsch, L.L. (1969) *J. Biol. Chem.*, **244**, 2996.

5. Richardson, C.C., Schildkraut, C., Aposhian, H.V. and Kornberg, A. (1964) *J. Biol. Chem.*, **239**, 222.

6. Kelly, R.B., Cozzarelli, N.R., Deutscher, M.P., Lehman, I.R. and Kornberg, A. (1970) *J. Biol. Chem.*, **245**, 39.

7. Meinkoth, J. and Wahl, G. (1984) *Anal. Biochem.*, **138**, 267.

8. Rigby, P.W.J., Dieckmann, M., Rhodes, C. and Berg, P. (1977) *J. Mol. Biol.*, **113**, 237.

9. Gubler, U. and Hoffman, B.J. (1982) *Gene*, **25**, 263.

10. Jacobsen, H., Klenow, H. and Overgaard-Hansen, K. (1974) *Eur. J. Biochem.*, **45**, 623.

11. Joyce, C.M. and Grindley, N.D.F. (1983) *Proc. Natl. Acad. Sci. USA*, **80**, 1830.

12. Brutlag, D., Atkinson, M.R., Setlow, P. and Kornberg, A. (1969) *Biochem. Biophys. Res. Commun.*, **37**, 982.

13. Klenow, H. and Henningsen, I. (1970) *Proc. Natl. Acad. Sci. USA*, **65**, 168.

14. Setlow, P., Brutlag, D. and Kornberg, A. (1972) *J. Biol. Chem.*, **247**, 224.

15. Setlow, P. and Kornberg, A. (1972) *J. Biol. Chem.*, **247**, 232.

16. Telford, J.L., Kressman, A., Koski, R.A., Grosschedl, R., Muller, F., Clarkson, S.G. and Birnstiel, M.L. (1979) *Proc. Natl. Acad. Sci. USA*, **76**, 2590.

17. Kramer, W., Drutsa, V., Jansen, H.-W., Kramer, B., Pflugfelder, M. and Fritz, H.-J. (1984) *Nucleic Acids Res.*, **12**, 9441.

18. Carter, P., Bedouelle, H. and Winter, G. (1985) *Nucleic Acids Res.*, **13**, 4431.

19. Anderson, S., Gait, M.J., Mayol, L. and Young, I.G. (1980) *Nucleic Acids Res.*, **8**, 1731.

20. Feinberg, A.P. and Vogelstein, B. (1983) *Anal. Biochem.*, **132**, 6.

21. Feinberg, A.P. and Vogelstein, B. (1984) *Anal. Biochem.*, **137**, 266.

22. Sanger, F., Nicklen, S. and Coulson, A.R. (1977) *Proc. Natl. Acad. Sci. USA*, **74**, 5463.

23. Houdebiene, L.-M. (1976) *Nucleic Acids Res.*, **3**, 615.

24. Goeddel, D.V., Shepherd, H.M., Yelverton, E., Leung, D. and Crea, R. (1980) *Nucleic Acids Res.*, **8**, 4057.

25. Rougeon, F., Kourilsky, P. and Mach, B. (1975) *Nucleic Acids Res.*, **2**, 2365.

26. Wickens, M.P., Buell, G.N. and Schimke, R.T. (1978) *J. Biol. Chem.*, **253**, 2483.

27. Wallace, R.B., Johnson, P.F., Tanaka, S., Schold, M., Itakura, K. and Abelson, J. (1980) *Science*, **209**, 1396.

28. Messing, J. and Vieira, J. (1982) *Gene*, **19**, 269.

29. Wallace, R.B., Johnson, M.J., Suggs, S.V., Miyoshi, K., Bhatt, R. and Itakura, K. (1981) *Gene*, **16**, 21.

30. Carter, P.J., Winter, G., Wilkinson, A.J. and Ferscht, A.R. (1984) *Cell*, **38**, 835.

31. Zoller, M.J. and Smith, M. (1982) *Nucleic Acids Res.*, **10**, 6487.

32. Seiler-Tuyns, A., Eldridge, J.D. and Paterson, B.M. (1984) *Proc. Natl. Acad. Sci. USA*, **81**, 2980.

33. Studencki, A.B. and Wallace, R.B. (1984) *DNA*, **3**, 7.

34. Derbyshire, V., Freemont, P.S., Sanderson, M.R., Beese, L., Friedman, J.M., Joyce, C.M. and Steitz, T.A. (1988) *Science*, **240**, 199.

35. Challberg, M.D. and Englund, P.T. (1979) *J. Biol. Chem.*, **254**, 7820.

36. Panet, A., van de Sande, J.H., Loewen, P.C., Khorana, H.G., Raae, A.J., Lillehaug, J.R. and Kleppe, K. (1973) *Biochemistry*, **12**, 5045.

37. Englund, P.T. (1971) *J. Biol. Chem.*, **246**, 3269.

38. O'Farrell, P.H., Kutter, E. and Nakanishi, M. (1980) *Mol. Gen. Genet.*, **179**, 421.

39. Morris, C.F., Hama-Inaba, H., Mace, D., Sinha, N.K. and Alberts, B. (1979) *J. Biol. Chem.*, **254**, 6787.

40. Burd, J.F. and Wells, R.D. (1974) *J. Biol. Chem.*, **249**, 7094.

41. Craik, C.S., Largman, C., Fletcher, T., Roczniak, S., Barr, P.J., Fletterick, R. and Rutter, W.J. (1985) *Science*, **228**, 291.

42. Nossal, N.G. (1974) *J. Biol. Chem.*, **249**, 5668.

43. Dale, R.M.K., McClure, B.A. and Houchins, J.P. (1985) *Plasmid*, **12**, 31.

44. Doetsch, P.W., Chan, G.L. and Haseltine, W.A. (1985) *Nucleic Acids Res.*, **13**, 3285.

45. Modrich, P. and Richardson, C.C. (1975) *J. Biol. Chem.*, **250**, 5515.

46. Mark, D.F. and Richardson, C.C. (1976) *Proc. Natl. Acad. Sci. USA*, **73**, 780.

47. Tabor, S. (1987) *PhD Dissertation*. Harvard University, Boston.

48. Huber, H.E., Tabor, S. and Richardson, C.C. (1987) *J. Biol. Chem.*, **262**, 16224.

49. Tabor, S., Huber, H.E. and Richardson, C.C. (1987) *J. Biol. Chem.*, **262**, 16212.

50. Lechner, R.L., Engler, M.J. and Richardson, C.C. (1983) *J. Biol. Chem.*, **258**, 11174.

51. Vandeyar, M.A., Weiner, M.P., Hutton, C.J. and Batt, C.A. (1988) *Gene*, **65**, 129.

52. Tabor, S. and Richardson, C.C. (1987) *Proc. Natl. Acad. Sci. USA*, **84**, 4767.

53. Chien, A., Edgar, D.B. and Trela, J.M. (1976) *J. Bacteriol.*, **127**, 1550.

54. Saiki, R.K. and Gelfand, D.H. (1989) *Amplifications*, **1**, 4.

55. Lawyer, F.C., Stoffel, S., Saiki, R.K., Myambo, K., Drummond, R. and Gelfand, D.H. (1989) *J. Biol. Chem.*, **264**, 6427.

56. Saiki, R.K., Scharf, S., Faloona, F., Mullis, K.B., Horn, G.T., Erlich, H.A. and Arnheim, N. (1985) *Science*, **230**, 1350.

57. Saiki, R.K., Gelfand, D.H., Stoffel, S., Scharf, S.J., Higuchi, R., Horn, G.T., Mullis, K.B. and Erlich, H.A. (1988) *Science*, **239**, 487.

58. Innis, M.A., Myambo, K.B., Gelfand, D.H. and Brow, M.A.D. (1988) *Proc. Natl. Acad. Sci. USA*, **85**, 9436.

59. Saluz, H.P. and Jost, J.P. (1989) *Nature*, **338**, 277.

60. Ye, S.Y. and Hong, G.F. (1987) *Scientia Sinica*, **30**, 503.

61. Anon (1990) *Molecular Biology Reagents Catalogue*. United States Biochemical Corp., Cleveland.

62. Shortle, D., Koshland, D., Wienstock, G.M. and Botstein, D. (1980) *Proc. Natl. Acad. Sci. USA*, **77**, 5375.

63. Shortle, D., Grisafi, P., Benkovic, S.J. and Botstein, D. (1982) *Proc. Natl. Acad. Sci. USA*, **79**, 1588.

64. Kornberg, A. (1980) *DNA Replication*. Freeman, San Francisco.

65. Grandgenett, D.P., Gerard, G.F. and Green, M. (1973) *Proc. Natl. Acad. Sci. USA*, **70**, 230.

66. Leis, J.P., Duyk, G., Johnson, S., Longiaru, M. and Skalka, A. (1983) *J. Virol.*, **45**, 727.

67. Verma, I.M. (1977) *Biochim. Biophys. Acta*, **473**, 1.

68. Breathnach, R., Mandel, J.L. and Chambon, P. (1977) *Nature*, **270**, 314.

69. Tilghman, S.M., Curtis, P.J., Tiemeier, D.C., Leder, P. and Weissman, C. (1978) *Proc. Natl. Acad. Sci. USA*, **75**, 1309.

70. Champoux, J.J., Gilboa, E. and Baltimore, D. (1984) *J. Virol.*, **49**, 686.

71. Okayama, H. and Berg, P. (1982) *Mol. Cell. Biol.*, **2**, 161.

72. Kaartinen, M., Griffiths, G.M., Hamlyn, P.M., Markham, A.F., Karjlainen, K., Pelkonen, J.L.T., Makela, O. and Milstein, C. (1983) *J. Immunol.*, **130**, 937.

73. Chen, E.Y. and Seeburg, P.H. (1985) *DNA*, **4**, 165.

74. Inoue, T. and Cech, T.R. (1985) *Proc. Natl. Acad. Sci. USA*, **82**, 648.

75. Oyama, F., Kikuchi, R., Omori, A. and Uchida, T. (1988) *Anal. Biochem.*, **172**, 444.

76. Fujinaga, K., Parsons, T., Beard, J.W., Beard, D. and Green, M. (1970) *Proc. Natl. Acad. Sci. USA*, **67**, 1432.

77. Roth, M.J., Tanese, N. and Goff, S.P. (1985) *J. Biol. Chem.*, **260**, 9326.

78. Moelling, K. (1974) *Virology*, **62**, 46.

79. Kotewicz, M.L., D'Alessio, J.M., Driftmier, K.M., Blodgett, K.P. and Gerard, G.F. (1985) *Gene*, **35**, 249.

80. Tanese, N., Roth, M. and Goff, S.P. (1985) *Proc. Natl. Acad. Sci. USA*, **82**, 4944.

81. Gerard, G.F. (1986) *Focus*, 7(1), 1.

82. Perbal, B. (1988) *A Practical Guide to Molecular Cloning (2nd edn)*. Wiley, New York.

83. Krug, M.S. and Berger, S.L. (1987) *Methods Enzymol.*, **152**, 316.

84. Travers, A.A. and Burgess, R.R. (1969) *Nature*, **222**, 537.

85. Calva, E., Rosenvold, E.C., Szybalski, W. and Burgess, R.R. (1980) *J. Biol. Chem.*, **255**, 11011.

86. Butler, E.T. and Chamberlin, M.J. (1982) *J. Biol. Chem.*, **257**, 5772.

87. Melton, D.A., Krieg, P.A., Rebagliati, M.R., Maniatis, T., Zinn, K. and Green, M.R. (1984) *Nucleic Acids Res.*, **12**, 7035.

88. Green, M.R., Maniatis, T. and Melton, D.A. (1983) *Cell*, **32**, 681.

89. Kassavetis, G.A., Butler, E.T., Roulland, D. and Chamberlin, M.J. (1982) *J. Biol. Chem.*, **257**, 5779.

90. Kotani, H., Ishizaki, Y., Hiraoka, N. and Obayashi, A. (1987) *Nucleic Acids Res.*, **15**, 2653.

91. Johnson, M.T. and Johnson, B. (1984) *BioTechniques*, **2**, 156.

92. Krainer, A.F., Maniatis, T., Ruskin, B. and Green, M.R. (1984) *Cell*, **36**, 993.

93. Krieg, P.A. and Melton, D.A. (1984) *Nature*, **308**, 203.

94. Melton, D.A. (1985) *Proc. Natl. Acad. Sci. USA*, **82**, 144.

95. Krieg, P.A. and Melton, D.A. (1984) *Nucleic Acids Res.*, **12**, 7057.

96. Zinn, K., DiMaio, D. and Maniatis, T. (1983) *Cell*, **34**, 865.

97. Church, G.M. and Gilbert, W. (1984) *Proc. Natl. Acad. Sci. USA*, **81**, 1991.

98. Cox, K.H., DeLeon, D.V., Angerer, L.M. and Angerer, R.C. (1984) *Dev. Biol.*, **101**, 485.

99. Looney, J.E., Han, J.H. and Harding, J.D. (1984) *Gene*, **27**, 67.

100. Parvin, J.D., Smith, F.I. and Palese, P. (1986) *DNA*, **5**, 167.

101. Morris, C.E., Klement, J.F. and McAllister, W.T. (1986) *Gene*, **41**, 193.

102. Brown, J.E., Klement, J.F. and McAllister, W.T. (1986) *Nucleic Acids Res.*, **14**, 3521.

103. Peebles, C.L. (1980) *Gene Anal. Tech.*, **3**, 59.

104. Davanloo, P., Rosenberg, A.H., Dunn, J.J. and Studier, F.W. (1984) *Proc. Natl. Acad. Sci. USA*, **81**, 2035.

105. Stahl, S.J. and Zinn, K. (1981) *J. Mol. Biol.*, **148**, 481.

106. Dunn, J.J. and Studier, F.W. (1983) *J. Mol. Biol.*, **166**, 477.

107. Tabor, S. and Richardson, C.C. (1985) *Proc. Natl. Acad. Sci. USA*, **82**, 1074.

108. Southern, E.M. (1975) *J. Mol. Biol.*, **98**, 503.

109. Smith, G.E. and Summers, M.D. (1980) *Anal. Biochem.*, **109**, 123.

110. Schenborn, E.T. and Meirendorf, R.C. (1985) *Nucleic Acids Res.*, **13**, 6223.

111. Langer, P.R., Waldrop, A.A. and Ward, D.C. (1981) *Proc. Natl. Acad. Sci. USA*, **78**, 6633.

112. Lewis, M.K. and Burgess, R.R. (1980) *J. Biol. Chem.*, **255**, 4928.

113. Lau, P.P. and Gray, H.B. (1979) *Nucleic Acids Res.*, **6**, 331.

114. Gray, H.B., Winston, T.P., Hodnett, J.L., Legerski, R.J., Nees, D.W., Wei, C.-F. and Robberson, D.L. (1981) in *Gene Amplification and Analysis: Structural Analysis of Nucleic Acids.* (J.G.

Chirikjian and T.S. Papas eds) Elsevier, New York, Vol. 2, p. 169.

115. Gray, H.B., Ostrander, D.A., Hodnett, J.L., Legerski, R.J. and Robberson, D.L. (1975) *Nucleic Acids Res.*, **2**, 1459.

116. Legerski, R.J., Hodnett, J.L. and Gray, H.B. (1978) *Nucleic Acids Res.*, **5**, 1445.

117. Talmadge, K., Stahl, S. and Gilbert, W. (1980) *Proc. Natl. Acad. Sci. USA*, 77, 3369.

118. Shishido, K. and Ando, T. (1982) in *Nucleases.* (S. Linn and R. Roberts eds) Cold Spring Harbor Laboratory Press, New York, p. 155.

119. Wei, C.-F., Alianell, G.A., Bencen, G.H. and Gray, H.B. (1983) *J. Biol. Chem.*, **258**, 13506.

120. Miele, E.A., Mills, D.R. and Kramer, F.R. (1983) *J. Mol. Biol.*, **171**, 281.

121. Sambrook, J., Fritsch, E.F. and Maniatis, T. (1989) *Molecular Cloning: A Laboratory Manual. (2nd edn)*. Cold Spring Harbor Laboratory Press, New York.

122. Vogt, V.M. (1973) *Eur. J. Biochem.*, **33**, 192.

123. Oleson, A.E. and Sasakuma, M. (1980) *Arch. Biochem. Biophys.*, **204**, 361.

124. Kroeker, W.D. and Kowalski, D. (1978) *Biochemistry*, **17**, 3236.

125. Gannon, F., Jeltsch, J.M. and Perrin, F. (1980) *Nucleic Acids Res.*, **8**, 4405.

126. Chase, J.W. and Richardson, C.C. (1974) *J. Biol. Chem.*, **249**, 4545.

127. Berk, A.J. and Sharp, P.A. (1978) *Proc. Natl. Acad. Sci. USA*, 75, 1274.

128. Berk, A.J. and Sharp, P.A. (1978) *Cell*, **14**, 695.

129. Grosschedl, R. and Birnstiel, M.L. (1980) *Proc. Natl. Acad. Sci. USA*, 77, 1432.

130. Roberts, T.M., Kacich, R. and Ptashne, M. (1979) *Proc. Natl. Acad. Sci. USA*, 76, 760.

131. Weaver, R.F. and Weissman, C. (1979) *Nucleic Acids Res.*, 7, 1175.

132. Maniatis, T., Gee, S.G., Efstratiadis, A. and Kafatos, F.C. (1976) *Cell*, **8**, 163.

133. Zechel, K. and Weber, K. (1977) *Eur. J. Biochem.*, 77, 133.

134. Kedzierski, W., Laskowski, M. and Mandel, M. (1973) *J. Biol. Chem.*, **248**, 1277.

135. Vogt, V.M. (1980) *Methods Enzymol.*, **65**, 248.

136. Laskowski, M. (1980) *Methods Enzymol.*, **65**, 263.

137. Sung, S. and Laskowski, M. (1962) *J. Biol. Chem.*, **237**, 506.

138. Johnson, P.H. and Laskowski, M. (1968) *J. Biol. Chem.*, **243**, 3421.

Modifying Enzymes

DNA AND RNA MODIFYING ENZYMES

139. Johnson, P.H. and Laskowski, M. (1970) *J. Biol. Chem.*, **245**, 891.

140. Mikulski, A.A. and Laskowski, M. (1970) *J. Biol. Chem.*, **245**, 5026.

141. Green, M.R. and Roeder, R.G. (1980) *Cell*, **22**, 231.

142. Ghangas, G.S. and Wu, R. (1975) *J. Biol. Chem.*, **250**, 4601.

143. Hammond, A.W. and D'Alessio, J. (1986) *Focus*, **8**(4), 4.

144. Weisink, P.C., Finnegan, D.J., Donelson, J.E. and Hogness, D.S. (1974) *Cell*, **3**, 315.

145. Gubler, U. (1987) *Methods Enzymol.*, **152**, 330.

146. Furuichi, Y. and Miura, K. (1975) *Nature*, **253**, 374.

147. Kihara, K., Nomiyama, H., Yukuhiro, M. and Mukai, J.-I. (1976) *Anal. Biochem.*, **75**, 672.

148. Hirs, C.H.W., Moore, S. and Stein, W.H. (1956) *J. Biol. Chem.*, **219**, 623.

149. Davidson, J.N. (1972) *The Biochemistry of the Nucleic Acids. (7th edn)* Academic, New York.

150. Myers, R.M., Larin, Z. and Maniatis, T. (1985) *Science*, **230**, 1242.

151. Winter, E., Yamamoto, F., Almoguera, C. and Perucho, M. (1985) *Proc. Natl. Acad. Sci. USA*, **82**, 7575.

152. Kanaya, S. and Crouch, R.J. (1983) *J. Biol. Chem.*, **258**, 1276.

153. Berkower, I., Leis, J. and Hurwitz, J. (1973) *J. Biol. Chem.*, **248**, 5914.

154. Vournakis, J.N., Efstratiadis, A. and Kafatos, F.C. (1975) *Proc. Natl. Acad. Sci. USA*, **72**, 2959.

155. Stavrianopoulos, J.G., Gambino-Giuffrida, A. and Chargaff, E. (1976) *Proc. Natl. Acad. Sci. USA*, **73**, 1081.

156. Keller, W. and Crouch, R. (1972) *Proc. Natl. Acad. Sci. USA*, **69**, 3360.

157. Grossman, L.I., Watson, R. and Vinograd, J. (1973) *Proc. Natl. Acad. Sci. USA*, **70**, 3339.

158. Hillenbrand, G. and Staudenbauer, W.L. (1982) *Nucleic Acids Res.*, **10**, 833.

159. Donis-Keller, H. (1979) *Nucleic Acids Res.*, **7**, 179.

160. Liao, T.-H. (1974) *J. Biol. Chem.*, **249**, 2354.

161. Matsuda, M. and Ogoshi, H. (1966) *J. Biochem.*, **59**, 230.

162. Melgar, E. and Goldthwait, D.A. (1968) *J. Biol. Chem.*, **243**, 4409.

163. Galas, D.J. and Schmitz, A. (1978) *Nucleic Acids Res.*, **5**, 3157.

164. Jackson, P.D. and Felsenfeld, G. (1987) *Methods Enzymol.*, **152**, 735.

165. Krieg, P.A. and Melton, D.A. (1987) *Methods Enzymol.*, **155**, 397.

166. Weintraub, H. and Groudine, M. (1976) *Science*, **193**, 848.

167. Hong, G.-F. (1987) *Methods Enzymol.*, **155**, 93.

168. Greenfield, L., Simpson, L. and Kaplan, D. (1975) *Biochim. Biophys. Acta*, **407**, 365.

169. Pelham, H.R.B. and Jackson, R.J. (1976) *Eur. J. Biochem.*, **67**, 247.

170. Jackson, R.J. and Hunt, T. (1983) *Methods Enzymol.*, **96**, 50.

171. Noll, M. (1974) *Nature*, **251**, 249.

172. Pham, T.T. and Coleman, J.E. (1985) *Biochemistry*, **24**, 5672.

173. Center, M.S., Studier, F.W. and Richardson, C.C. (1970) *Proc. Natl. Acad. Sci. USA*, **65**, 242.

174. de Massy, B., Weisberg, R.A. and Studier, F.W. (1987) *J. Mol. Biol.*, **193**, 359.

175. Panayotatos, N. and Wells, R.D. (1981) *Nature*, **289**, 466.

176. Straney, D.C. and Crothers, D.M. (1987) *J. Mol. Biol.*, **193**, 279.

177. Osterman, H.L. and Coleman, J.E. (1981) *Biochemistry*, **20**, 4884.

178. Panayotatos, N. and Fontaine, A. (1985) *J. Biol. Chem.*, **260**, 3173.

179. Rosamond, J., Telander, K.M. and Linn, S. (1979) *J. Biol. Chem.*, 254, 8646.

180. Goldmark, P.J. and Linn, S. (1972) *J. Biol. Chem.*, **247**, 1849.

181. Rao, R.N. and Rogers, S.G. (1978) *Gene*, **3**, 247.

182. Rogers, S.G. and Weiss, B. (1980) *Gene*, **11**, 187.

183. Weiss, B. (1976) *J. Biol. Chem.*, **251**, 1896.

184. Rogers, S.G. and Weiss, B. (1980) *Methods Enzymol.*, **65**, 201.

185. Guo, L.-H. and Wu, R. (1982) *Nucleic Acids Res.*, **10**, 2065.

186. Yang, C., Guo, L. and Wu, R. (1983) in *Frontiers in Biochemical and Biophysical Studies of Proteins and Membranes.* (T. Lie, S. Sakakibara, A. Schechter, K. Yagi, H. Yajima and K. Tasanobu eds) Elsevier, New York, p. 5.

187. Ozkaynak, E. and Putney, S.D. (1987) *BioTechniques*, **5**, 770.

188. Henikoff, S. (1987) *Methods Enzymol.*, **155**, 156.

189. Smith, A.J.H. (1979) *Nucleic Acids Res.*, **6**, 831.

190. Donelson, J.E. and Wu, R. (1972) *J. Biol. Chem.*, **247**, 4661.

191. Guo, L.-H., Yang, R.C.A. and Wu, R. (1983) *Nucleic Acids Res.*, **11**, 5521.

192. Shalloway, D., Kleinberger, T. and Livingston, D.M. (1980) *Cell*, **20**, 411.

193. von der Ahe, D., Janich, S., Scheidereit, C., Renkawitz, R., Schutz, G. and Beato, M. (1985) *Nature*, **313**, 706.

194. Wu, C. (1985) *Nature*, **317**, 84.

195. Chase, J.W. and Richardson, C.C. (1974) *J. Biol. Chem.*, **249**, 4553.

196. Chade, J.W. and Vales, L.D. (1981) in *Gene Amplification and Analysis: Structural Analysis of Nucleic Acids.* (J.G. Chirikjian and T.S. Papas eds) Elsevier, New York, Vol. 2, p. 148.

197. Berk, A.J. and Sharp, P.A. (1977) *Cell*, **12**, 721.

198. Little, J.W. (1981) in *Gene Amplification and Analysis: Structural Analysis of Nucleic Acids.* (J.G. Chirikjian and T.S. Papas eds) Elsevier, New York, Vol. 2, p. 136.

199. Radding, C.M. (1966) *J. Mol. Biol.*, **18**, 235.

200. Little, J.W., Lehman, I.R. and Kaiser, A.D. (1967) *J. Biol. Chem.*, **242**, 672.

201. Jackson, D.A., Symons, R.H. and Berg, P. (1972) *Proc. Natl. Acad. Sci. USA*, **69**, 2904.

202. Fraser, M.J., Tjeerde, R. and Matsumoto, K. (1976) *Can. J. Biochem.*, **54**, 971.

203. Fraser, M.J. (1980) *Methods Enzymol.*, **65**, 255.

204. Krupp, G. and Gross, H.J. (1979) *Nucleic Acids Res.*, **6**, 3481.

205. Brederode, F.T., Koper-Zwarthoff, E.C. and Bol, J.F. (1980) *Nucleic Acids Res.*, **8**, 2213.

206. Linn, S. and Lehman, I.R. (1965) *J. Biol. Chem.*, **240**, 1287.

207. Bjork, W. (1963) *J. Biol. Chem.*, **238**, 2487.

208. Wilson, G.G. and Murray, N.E. (1979) *J. Mol. Biol.*, **132**, 471.

209. Murray, N.E., Bruce, S.A. and Murray, K. (1979) *J. Mol. Biol.*, **132**, 493.

210. Nilsson, S.V. and Magnusson, G. (1982) *Nucleic Acids Res.*, **10**, 1425.

211. Ferretti, L. and Sgaramella, V. (1981) *Nucleic Acids Res.*, **9**, 85.

212. Pohl, F.M., Thomae, R. and Karst, A. (1982) *Eur. J. Biochem.*, **123**, 141.

213. Dugaiczyk, A., Boyer, H.W. and Goodman, H.M. (1975) *J. Mol. Biol.*, **96**, 171.

214. Graf, H. (1979) *Biochim. Biophys. Acta*, **564**, 225.

215. Sgaramella, V. (1972) *Proc. Natl. Acad. Sci. USA*, **69**, 3389.

216. Sugino, A., Goodman, H.M., Heyneker, H.L., Shine, J., Boyer, H.W. and Cozzarelli, N.Z. (1977) *J. Biol. Chem.*, **252**, 3987.

217. Pheiffer, B.H. and Zimmerman, S.B. **(1983)** *Nucleic Acids Res.*, **11**, 7853.

218. Cobianchi, F. and Wilson, S.H. (1987) *Methods Enzymol.*, **152**, 94.

219. Panasenko, S.M., Cameron, J.R., Davis, R.W. and Lehman, I.R. (1977) *Science*, **196**, 188.

220. Panasenko, S.M., Alazard, R.J. and Lehman, I.R. (1978) *J. Biol. Chem.*, **253**, 4590.

221. Higgins, N.P. and Cozzarelli, N.R. (1979) *Methods Enzymol.*, **68**, 50.

222. Sgaramella, V. and Khorana, H.G. (1972) *J. Mol. Biol.*, **72**, 493.

223. Sano, H. and Feix, G. (1974) *Biochemistry*, **13**, 5110.

224. Zimmerman, S.B. and Pheiffer, B.H. (1983) *Proc. Natl. Acad. Sci. USA*, **80**, 5852.

225. Rand, K.N. and Gait, M.J. (1984) *EMBO J.*, **3**, 397.

226. Sugino, A., Snopek, T.J. and Cozzarelli, N.R. (1977) *J. Biol. Chem.*, **252**, 1732.

227. England, T.E., Gumport, R.I. and Uhlenbeck, O.C. (1977) *Proc. Natl. Acad. Sci. USA*, **74**, 4839.

228. Walker, G.L., Uhlenbeck, O.C., Bedows, E. and Gumport, R.I. (1975) *Proc. Natl. Acad. Sci. USA*, **72**, 122.

229. Tessier, D.C., Brousseau, R. and Vernet, T. (1986) *Anal. Biochem.*, **158**, 171.

230. Romaniuk, P.J. and Uhlenbeck, O.C. (1983) *Methods Enzymol.*, **100**, 52.

231. Middleton, T., Herlihy, W.C., Schimmel, P.R. and Munro, H.N. (1985) *Anal. Biochem.*, **144**, 110.

232. Bruce, A.G. and Uhlenbeck, O.C. (1978) *Nucleic Acids Res.*, **5**, 3665.

233. England, T.E. and Uhlenbeck, O.C. (1978) *Nature*, **275**, 560.

234. Noren, C.J., Anthony-Cahill, S.J., Griffith, M.C. and Schultz, P.G. (1989) *Science*, **244**, 182.

235. Bruce, A.G. and Uhlenbeck, O.C. (1982) *Biochemistry*, **21**, 855.

236. Chaconas, G. and van de Sande, J.H. (1980) *Methods Enzymol.*, **65**, 75.

237. Efstratiadis, A., Vournakis, J.N., Donis-Keller, H., Chaconas, G., Dougall, D.K. and Kafatos, F.C. (1977) *Nucleic Acids Res.*, **4**, 4165.

Modifying Enzymes

238. Garen, A. and Garen, S. (1963) *J. Mol. Biol.*, **6**, 433.

239. McCracken, S. and Meighen, E. (1980) *J. Biol. Chem.*, **255**, 2396.

240. Ullrich, A., Shine, J., Chirgwin, J., Pictet, R., Tischer, E., Rutter, W.J. and Goodman, H.M. (1977) *Science*, **196**, 1313.

241. Ish-Horowitz, D. and Burke, J.F. (1981) *Nucleic Acids Res.*, **9**, 2989.

242. Fosset, M., Chappelet-Tordo, D. and Lazdunski, M. (1974) *Biochemistry*, **13**, 1783.

243. McGadey, J. (1970) *Histochemie*, **23**, 180.

244. Blake, M.S., Johnston, K.H., Russell-Jones, G.J. and Gotschlich, E.C. (1984) *Anal. Biochem.*, **136**, 175.

245. Richardson, C.C. (1965) *Proc. Natl. Acad. Sci. USA*, **54**, 158.

246. Depew, R.E., Snopek, T.J. and Cozzarelli, N.R. (1975) *Virology*, **64**, 144.

247. Midgley, C.A. and Murray, N.E. (1985) *EMBO J.*, **4**, 2695.

248. Lillehaug, J.R. (1977) *Eur. J. Biochem.*, **73**, 499.

249. Berkner, K.L. and Folk, W.R. (1977) *J. Biol. Chem.*, **252**, 3176.

250. Simoncsits, A., Brownlee, G.G., Brown, R.S., Rubin, J.R. and Guilley, H. (1977) *Nature*, **269**, 833.

251. Maxam, A.M. and Gilbert, W. (1980) *Methods Enzymol.*, **65**, 499.

252. Reddy, M.V., Gupta, R.C. and Randerath, K. (1981) *Anal. Biochem.*, **117**, 271.

253. Randerath, K., Reddy, M.V. and Gupta, R.C. (1981) *Proc. Natl. Acad. Sci. USA*, **78**, 6126.

254. Gross, H.J., Krupp, G., Domdey, H., Raba, M., Jank, P., Lossow, C., Alberty, H., Ramm, K. and Sanger, H.L. (1981) *Eur. J. Biochem.*, **21**, 249.

255. Smith, H.O. and Birnstiel, M.L. (1976) *Nucleic Acids Res.*, **3**, 2387.

256. Maat, J. and Smith, A.J.H. (1978) *Nucleic Acids Res.*, **5**, 4537.

257. Johnsrud, L. (1978) *Proc. Natl. Acad. Sci. USA*, **75**, 5314.

258. Royer-Pokora, B., Gordon, L.K. and Haseltine, W.A. (1981) *Nucleic Acids Res.*, **9**, 4595.

259. Cameron, V. and Uhlenbeck, O.C. (1977) *Biochemistry*, **16**, 5120.

260. Soltis, D.A. and Uhlenbeck, O.C. (1982) *J. Biol. Chem.*, **257**, 11340.

261. Ohtsuka, E., Tanaka, S., Tanaka, T., Miyake, T., Markham, A.F., Nakagawa, E., Wakabayashi, T., Taniyama, Y., Nishikawa, S., Fukumoto, R., Uemura, H., Doi, T., Tokunaga, T. and Ikehara, M. (1981) *Proc. Natl. Acad. Sci. USA*, **78**, 5493.

262. Chang, L.M.S. and Bollum, F.J. (1971) *J. Biol. Chem.*, **246**, 909.

263. Deng, G. and Wu, R. (1981) *Nucleic Acids Res.*, **9**, 4173.

264. Michelson, A.M. and Orkin, S.H. (1982) *J. Biol. Chem.*, **257**, 14773.

265. Chang, L.M.S. and Bollum, F.J. (1986) *Crit. Rev. Biochem.*, **21**, 27.

266. Roychoudhury, R., Jay, E. and Wu, R. (1976) *Nucleic Acids Res.*, **3**, 863.

267. Heidecker, G. and Messing, J. (1983) *Nucleic Acids Res.*, **11**, 4891.

268. Tu, C.-P.D. and Cohen, S.N. (1980) Gene, **10**, 177.

269. Wu. R., Jay, E. and Roychoudhury, R. (1976) *Methods Cancer Res.*, **12**, 87.

270. Yousaf, S.I., Carroll, A.R. and Clarke, B.E. (1984) *Gene*, **27**, 309.

271. Deng, G. and Wu, R. (1983) *Methods Enzymol.*, **100**, 96.

272. Eschenfeldt, W.H., Puskas, R.S. and Berger, S.L. (1987) *Methods Enzymol.*, **152**, 337.

273. Sippel, A.E. (1973) *Eur. J. Biochem.*, **37**, 31.

274. Beltz, W.R. and Ashton, S.H. (1982) *Fed. Proc.*, **41**, 1450, Abstract 6896.

275. Gething, M.-J., Bye, J., Skehel, J. and Waterfield, M. (1980) *Nature*, **287**, 301.

276. Ahlquist, P., Dasgupta, R., Shih, D.S., Zimmern, D. and Kaesberg, P. (1979) *Nature*, **281**, 277.

277. Financsek, I., Mizumoto, K. and Muramatsu, M. (1982) *Gene*, **18**, 115.

278. Handa, H., Mizumoto, K., Oda, K., Okamoto, T. and Fukasawa, T. (1985) *Gene*, **33**, 159.

279. Itoh, N., Mizumoto, K. and Kaziro, Y. (1984) *J. Biol. Chem.*, **259**, 13923.

280. Itoh, N., Yamada, H., Kaziro, Y. and Mizumoto, K. (1987) *J. Biol. Chem.*, **262**, 1989.

281. Liu, L.F. and Miller, K.G. (1981) *Proc. Natl. Acad. Sci. USA*, **78**, 3487.

282. Gellert, M. (1981) *Annu. Rev. Biochem.*, **50**, 879.

283. Champoux, J.J. (1978) *Annu. Rev. Biochem.*, **47**, 449.

284. Wang, J.C. (1981) in *The Enzymes. (3rd edn)* (P.D. Boyer ed.) Academic, New York, Vol. 14, p. 331.

285. Been, M.D. and Champoux, J.J. (1981) *Proc. Natl. Acad. Sci. USA*, **78**, 2883.

286. Halligan, B.D., Davis, J.L., Edwards, K.A. and Liu, L.F. (1982) *J. Biol. Chem.*, **257**, 3995.

287. Prell, B. and Vosberg, H.P. (1980) *Eur. J. Biochem.*, **108**, 389.

288. Laskey, R.A., Mills, A.D. and Morris, N.R. (1977) *Cell*, **10**, 237.

289. Germond, J.-E., Vogt, V.M. and Hirt, B. (1974) *Eur. J. Biochem.*, **43**, 591.

290. Wang, J.C. (1979) *Proc. Natl. Acad. Sci. USA*, **76**, 200.

291. Peck, L.J. and Wang, J.C. (1981) *Nature*, **292**, 375.

292. Klevan, L. and Wang, J.C. (1980) *Biochemistry*, **19**, 5229.

293. McEntee, K., Weinstock, G.M. and Lehman, I.R. (1980) *Proc. Natl. Acad. Sci. USA*, 77, 857.

294. Christiansen, C. and Baldwin, R.L. (1977) *J. Mol. Biol.*, **115**, 441.

295. Kowalczykowski, S.C., Bear, D.G. and von Hippel, P.H. (1981) in *The Enzymes. (3rd edn)* (P.D. Boyer ed.) Academic, New York, Vol. 14, p. 373.

296. Shamoo, Y., Adari, H., Konigsberg, W.H., Williams, K.R. and Chase, J.W. (1986) *Proc. Natl. Acad. Sci. USA*, **83**, 8844.

297. Bittner, M., Burke, R.L. and Alberts, B.M. (1979) *J. Biol. Chem.*, **254**, 9565.

298. Alberts, B. and Sternglanz, R. (1977) *Nature*, **269**, 655.

299. Huberman, J.A., Kornberg, A. and Alberts, B.M. (1971) *J. Mol. Biol.*, **62**, 39.

300. Chase, J.W. and Williams, K.R. (1986) *Annu. Rev. Biochem.*, **55**, 103.

301. Williams, K.R., LoPresti, M.B. and Setoguchi, M. (1981) *J. Biol. Chem.*, **256**, 1754.

302. Topal, M.D. and Sinha, N.K. (1983) *J. Biol. Chem.*, **258**, 12274.

303. Morris, C.F., Sinha, N.K. and Alberts, B.M. (1975) *Proc. Natl. Acad. Sci. USA*, **72**, 4800.

304. Delius, H., Mantell, N.J. and Alberts, B.M. (1972) *J. Mol. Biol.*, **67**, 341.

305. Brack, C., Bickle, T.A. and Yuan, R. (1975) *J. Mol. Biol.*, **96**, 693.

306. Wu, M. and Davidson, N. (1975) *Proc. Natl. Acad. Sci. USA*, **72**, 4506.

307. Dombroski, D.F. and Morgan, A.R. (1985) *J. Biol. Chem.*, **260**, 415.

308. Toulme, J.J., Behmoaras, T., Guigues, M. and Helene, C. (1983) *EMBO J.*, **2**, 505.

309. Clark, A.J. (1973) *Annu. Rev. Genet.*, **7**, 67.

310. Radding, C.M. (1978) *Annu. Rev. Biochem.*, **47**, 847.

311. Soltis, D.A. and Lehman, I.R. (1983) *J. Biol. Chem.*, **258**, 6073.

312. Wasserman, S.A. and Cozzarelli, N.R. (1985) *Proc. Natl. Acad. Sci. USA*, **82**, 1079.

313. Zaug, A.J., Been, M.D. and Cech, T.R. (1986) *Nature*, **324**, 429.

314. Zaug, A.J., Grosshans, C.A. and Cech, T.R. (1988) *Biochemistry*, **27**, 8924.

315. Cech, T.R. (1989) *USB Comments*, **16**(2), 1.

316. Anon (1989) *USB Comments*, **16**(3), 13.

Modifying Enzymes

CHAPTER 6
CLONING VECTORS

This chapter covers most of the commercially available *E. coli* cloning vectors. A comprehensive encyclopedia of virtually all vectors for *E. coli* and other organisms has been published elsewhere (1).

1. *E. COLI* PLASMID CLONING VECTORS

The basic details of plasmid vectors for *E. coli* are listed in *Table 1* and important restriction sites for the vectors are given in *Table 2*. Vector maps are shown in *Figure 1*; to locate the map of any vector, refer to the 'Map page' column of either *Table 1* or *Table 2*. General information on the genetic markers and promoters carried by these vectors is provided in *Tables 3* and *4*.

Table 1. Basic details of *E. coli* plasmid vectors

Vector	Type	Size	Replicon	Markers[a]	Promoters[b]	References	Map page
c2RB	cosmid	6800	ColE1	amp, neo	–	2	200
pBluescript-II KS+/−, SK+/−	RNA expression, ssDNA	2950	ColE1, f1	amp, *lacZ'*	T3, T7	3,4	200
pBR322	general purpose	4363	pMB1	amp, tet	–	5,6	201
pBR325	general purpose	5995	pMB1	amp, cml, tet	–	7	201
pBS	RNA expression	2740	ColE1	amp, *lacZ'*	T3, T7	5,8,9	201
pBS+/−	RNA expression, ssDNA	3200	ColE1, f1	amp, *lacZ'*	T3, T7	4	201
pcos1EMBL	cosmid	6100	R6K	kan, tet	–	10	203
pDR540	protein expression	4063	pMB1	amp, *galK*	*tac*	11–14	202
pDR720	protein expression	4100	pMB1	amp, *galK*	*trp*	14	202
pEMBL8, 9	ssDNA	4000	pMB1, f1	amp, *lacZ'*	–	15	203
pFB69, I1, I2, I13, I14, I15	general purpose	3200	pMB1	amp, tet	–	16,17	203
pGEM series							
pGEM1	RNA expression	2865	pMB1	amp	SP6, T7	18	204
pGEM2	RNA expression	2869	pMB1	amp	SP6, T7	18	204
pGEM3	RNA expression	2867	pMB1	amp	SP6, T7	18	204
pGEM3Z	RNA expression	2743	pMB1	amp, *lacZ'*	SP6, T7	18	204
pGEM3Zf(+/−)	RNA expression, ssDNA	3199	pMB1, f1	amp, *lacZ'*	SP6, T7	18	205
pGEM4	RNA expression	2871	pMB1	amp	SP6, T7	18	205
pGEM4Z	RNA expression	2746	pMB1	amp, *lacZ'*	SP6, T7	18	205
pGEM5Zf(+/−)	RNA expression, ssDNA	3003	pMB1, f1	amp, *lacZ'*	SP6, T7	18	206
pGEM7Zf(+/−)	RNA expression, ssDNA	3000	pMB1, f1	amp, *lacZ'*	SP6, T7	18	206
pGEM9Zf(−)	RNA expression, ssDNA	2925	pMB1, f1	amp, *lacZ'*	SP6, T7	18	207
pGEM11Zf(+/−)	RNA expression, ssDNA	3223	pMB1, f1	amp, *lacZ'*	SP6, T7	18	207
pGEM13Zf(+)	RNA expression, ssDNA	3181	pMB1, f1	amp, *lacZ'*	SP6, T7	–	207

Table 1. Continued

Vector	Type	Size	Replicon	Markers[a]	Promoters[b]	References	Map page
pGEMEX-1	protein, RNA expression, ssDNA	4200	pMB1, f1	amp	SP6, T3, T7	18	208
pGEX2T	protein fusion	4948	pMB1	amp	tac	19	208
pGEX3X	protein fusion	4952	pMB1	amp	tac	19	208
pJB8	cosmid	5400	pMB1	amp	–	20	208
pKC30	protein expression	6442	pMB1	amp	λP_L	21–23	209
pKK175-6	promoter probe	4400	pMB1	amp, tet	–	24,25	209
pKK223-3	protein expression	4585	pMB1	amp	tac	24,26–28	209
pKK232-8	promoter probe	5096	pMB1	amp, cml	–	24	210
pKK233-2	protein expression	4593	pMB1	amp	trc	29–31	210
pKT279, 280, 287	protein secretion	4300	pMB1	tet	β-lactamase	32,33	210
pMC1871	protein fusion	7478	pMB1	$lacZ^c$, tet	–	34,35	211
pNH series							
pNH8a	protein expression	?	?	amp, $galK$	$tac–lac$	36	211
pNH16a	protein expression	?	?	amp, $galK$	lac	36	211
pNH18a	protein expression	?	?	amp	$tac–lac$	36	211
pNO1523	general purpose	5200	pMB1	amp, str^d	–	37	211
pRIT2T	protein fusion	4250	pMB1	amp	λP_R	38	212
pRIT5	protein fusion	6900	pMB1	amp, cml	Protein A	38	212
pSELECT-1	RNA expression, ssDNA	3422	pMB1,M13	tet, $lacZ$	SP6, T7	–	213
pSL1180, 1190	general purpose, ssDNA	5680	pMB1, f1	amp, $lacZ^e$	–	39	212,213

Cloning Vectors

Table 1. Continued

Vector	Type	Size	Replicon	Markers[a]	Promoters[b]	References	Map page
pSP series[g]							
pSP18	RNA expression	3004	pMB1	amp	SP6	–	214
pSP19	RNA expression	3010	pMB1	amp	SP6	–	214
pSP64	RNA expression	2999	pMB1	amp	SP6	9	214
pSP64(polyA)	poly(A)-RNA expression	3033	pMB1	amp	SP6	18	214
pSP65	RNA expression	3005	pMB1	amp	SP6	9	215
pSP70	RNA expression	2417	pMB1	amp	SP6, T7	18	215
pSP71	RNA expression	2419	pMB1	amp	SP6, T7	18	215
pSP72	RNA expression	2462	pMB1	amp	SP6, T7	18	215
pSP73	RNA expression	2464	pMB1	amp	SP6, T7	18	216
pSPORT1	RNA expression	4109	pMB1, f1	amp, *lacZ'*	SP6, T7	–	216
pSPT18, 19	RNA expression	3100	pMB1	amp	SP6, T7	9	216
pT3/T7 series							
pT3/T7-1, 2	RNA expression	2700	ColE1	amp, *lacZ'*	T3, T7	40	217
pT3/T7-3	RNA expression	2950	ColE1	amp, *lacZ'*	T3, T7	40	217
pT7/T3-18U, 19U	RNA expression, ssDNA	2890	pMB1, f1	amp, *lacZ'*	T3, T7	41	218
pT7-1, 2	RNA expression	2400	pMB1	amp	T7	42	218
pT712, 13	RNA expression	2818	pMB1	amp	T7	–	218
pTZ18R, 18U, 19R, 19U	RNA expression, ssDNA	2900	pMB1, f1	amp, *lacZ'*	T7	43	219
pUC7, 8, 9, 12, 13, 18[f], 19	general purpose	2700	pMB1	amp, *lacZ'*	–	44–46	220
pUC118, 119	ssDNA	3200	pMB1,M13	amp, *lacZ'*	–	47	219
pWE15	cosmid, RNA expression	8800	pMB1, SV40	amp, neo	T3, T7	48,49	220
pWE16	cosmid, RNA expression	8800	pMB1, SV40	amp, dhfr	T3, T7	48,49	220
pYEJ001	protein expression	4100	pMB1	amp, cml, tet	synthetic	50	220

Cloning Vectors

a Abbreviations (see also *Table 3*): amp, ampicillin resistance; cml, chloramphenicol resistance; dhfr, dihydrofolate reductase; *galK*, galactokinase; kan, kanamycin resistance; *lacZ*, β-galactosidase; *lacZ'*, β-galactosidase alpha-peptide; neo, neomycin phosphotransferase (confers kanamycin resistance); str, streptomycin sensitivity; tet, tetracycline resistance.

b Only those promoters relevant to protein or RNA expression are given.

c Note that in pMC1871 the marker is *lacZ* and not *lacZ'*.

d The str gene in pNO1523 confers streptomycin sensitivity. The vector must be used in a *rpsL* host so that insertional inactivation results in restoration of str[r].

e Although these vectors carry *lacZ'* the size of the polylinker interferes with X-gal selection.

f Beware of a pUC18 variant with a single bp deletion in codon 2 of *lacZ'* (51).

g pSP68 and pSP69 are the same as pSP18 and pSP19, respectively. pSP6/T7-19 is a 2885 bp derivative of pSP19 with the T7 promoter 'downstream' of the multiple cloning site. pSP6/T3 is a 2853 bp derivative of pSP64 with the T3 promoter inserted in the *Pvu*II site.

Table 2. Important restriction sites in *E. coli* plasmid vectors (for refs see *Table 1*)

Vector	Restriction sites[a]	Map page
c2RB	neo: *Sma*I; intergenic: *Bam*HI, *Cla*I, *Eco*RI, *Hin*dIII	200
pBluescript-II	promoter distal: *Kpn*I, *Apa*I, *Dra*II, *Xho*I, *Sal*I, *Hin*cII, *Acc*I, *Cla*I, *Hin*dIII, *Eco*RV, *Eco*RI, *Pst*I, *Sma*I, *Bam*HI, *Spe*I, *Xba*I, *Not*I, *Xma*III, *Sac*II, *Bst*XI, *Sac*I	200
pBR322	amp: *Pst*I, *Pvu*I, *Sca*I; tet: *Bam*HI, *Eco*RV, *Nru*I, *Sal*I, *Sph*I, *Xma*III	201
pBR325	amp: *Pst*I, *Pvu*I, *Xmn*I; cml: *Eco*RI, *Nco*I; tet: *Bam*HI, *Eco*RV, *Nru*I, *Sal*I, *Sph*I, *Xma*III	201
pBS	promoter distal: M13mp19 polylinker	201
pBS+/−	promoter distal: M13mp19 polylinker	201
pcos1EMBL	kan: *Sma*I, *Xho*I; tet: *Bam*HI, *Sal*I	203
pDR540	promoter distal: *Bam*HI	202
pDR720	promoter distal: *Bam*HI, *Sal*I, *Sma*I	202
pEMBL8	*lacZ'*: M13mp8 polylinker	203
pEMBL9	*lacZ'*: M13mp9 polylinker	203
pFB series	amp: *Pst*I, *Pvu*I, *Sca*I; tet: *Bam*HI, *Eco*RV, *Nru*I, *Sph*I, *Xma*III + *Sal*I (69) or *Eco*RI (I1) or *Sac*I (I2) or *Apa*I (I13) or *Kpn*I (I14) or *Xho*I (I15)	203
pGEM1,2	promoter distal: M13mp10/11 polylinker	204
pGEM3,4	promoter distal: M13mp18/19 polylinker	204, 205
pGEM3Z	promoter distal: M13mp19 polylinker	204
pGEM3Zf(+/−)	promoter distal: M13mp19 polylinker	205
pGEM4Z	promoter distal: M13mp19 polylinker	205
pGEM5Zf(+/−)	promoter distal: *Apa*I, *Aat*II, *Sph*I, *Eag*I, *Nco*I, *Sac*II, *Eco*RV, *Spe*I, *Not*I, *Pst*I, *Sal*I, *Nde*I, *Sac*I, *Bst*XI, *Nsi*I	206
pGEM7Zf(+/−)	promoter distal: *Apa*I, *Aat*II, *Sph*I, *Xba*I, *Xho*I, *Eco*RI, *Kpn*I, *Sma*I, *Csp*45I, *Cla*I, *Hin*dIII, *Bam*HI, *Sac*I, *Bst*XI, *Nsi*I	206
pGEM9Zf(−)	promoter distal: *Nsi*I, *Spe*I, *Hin*dIII, *Xba*I, *Eco*RI, *Sal*I, *Sac*I	207
pGEM11Zf(+/−)	promoter distal: *Sfi*I, *Sac*I, *Eco*RI, *Sal*I, *Xho*I, *Bam*HI, *Apa*I, *Xba*I, *Not*I, *Nsi*I, *Hin*dIII	207
pGEM13Zf(+)	promoter distal: *Not*I, *Sfi*I, *Hin*dIII	207
pGEMEX-1	promoter distal: as pGEM11Zf(+/−)	208
pGEX2T, 3X	glutathione S-transferase fusion: *Bam*HI, *Sma*I, *Eco*RI	208
pJB8	intergenic: *Bam*HI, *Cla*I, *Eco*RI*, *Hin*dIII, *Sal*I	208
pKC30	promoter distal: *Hpa*I	209
pKK175-6	upstream of tet: M13mp8 polylinker	209
pKK223-3	promoter distal: M13mp8 polylinker	209
pKK232-8	upstream of cml: M13mp8 polylinker	210
pKK233-2	promoter distal: *Nco*I, *Pst*I, *Hin*dIII	210
pKT279, 280, 287	promoter distal: *Pst*I	210
pMC1871	*lacZ* fusion: *Sma*I	211
pNH8a, 16a, 18a	8a: *Sma*I, *Bam*HI, *Xba*I, *Sal*I, *Spe*I, *Nru*I, *Xho*I, *Sac*I; 16a and 18a: *Bam*HI, *Xba*I, *Sal*I, *Spe*I, *Nru*I, *Sac*I, *Xho*I	211

Table 2. Continued

Vector	Restriction sites[a]	Map page
pNO1523	str: *Hpa*I, *Sma*I	211
pRIT2T, 5	protein A fusion: *Eco*RI, *Xcy*I, *Sma*I, *Bam*HI, *Sal*I, *Pst*I	212
pSELECT-1	promoter distal: *Eco*RI, *Sac*I, *Kpn*I, *Sma*I, *Bam*HI, *Xba*I, *Sal*I, *Pst*I, *Sph*I, *Hin*dIII	213
pSL1180, 1190	*lacZ'*: polylinker includes all possible hexamers, *Not*I, *Sfi*I	212, 213
pSP18, 19	promoter distal: M13mp18/19 polylinker	214
pSP64, 65	promoter distal: M13mp10/11 polylinker	214, 215
pSP64(polyA)	promoter distal: M13mp10/11 polylinker	214
pSP70, 71	promoter distal: *Xho*I, *Pvu*II, *Hin*dIII, *Eco*RI, *Cla*I, *Eco*RV, *Bgl*II	215
pSPORT1	promoter distal: *Aat*II, *Sph*I, *Mlu*I, *Spl*I, *Sna*BI, *Hin*dIII, *Bam*HI, *Xba*I, *Not*I, *Xma*III, *Spe*I, *Sst*I, *Sal*I, *Sma*I, *Eco*RI, *Bsp*MII, *Rsr*II, *Kpn*I, *Pst*I	216
pSP72, 73	promoter distal: *Xho*I, *Pvu*II, *Hin*dIII, *Sph*I, *Pst*I, *Sal*I, *Acc*I, *Xba*I, *Bam*HI, *Sma*I, *Kpn*I, *Sac*I, *Eco*RI, *Cla*I, *Eco*RV, *Bgl*II	215, 216
pSPT18, 19	promoter distal: M13mp18/19 polylinker	216
pT3/T7-1	promoter distal: *Hin*dIII*, *Sph*I, *Pst*I, *Sal*I*, *Xba*I, *Bam*HI, *Sma*I, *Kpn*I, *Apa*I, *Xho*I, *Cla*I, *Eco*RV, *Eco*RI	217
pT3/T7-2	promoter distal: *Hin*dIII, *Sph*I, *Pst*I, *Sal*I, *Xba*I, *Not*I, *Bst*XI, *Sac*II, *Sac*I, *Eco*RI	217
pT3/T7-3	promoter distal: *Sac*I*, *Bst*XI, *Sac*II, *Not*I, *Xba*I, *Bam*HI, *Sma*I, *Kpn*I*, *Eco*RI, *Eco*RV, *Hin*dIII, *Cla*I, *Xho*I, *Apa*I	217
pT7/T3-18U	promoter distal: M13mp18 polylinker	218
pT7/T3-19U	promoter distal: M13mp19 polylinker	218
pT7-1, 2	promoter distal: M13mp10/11 polylinker	218
pT712, 13	promoter distal: M13mp10/11 polylinker	218
pTZ18R, 18U	promoter distal: M13mp18 polylinker	219
pTZ19R, 19U	promoter distal: M13mp19 polylinker	219
pUC7	*lacZ'*: M13mp7 polylinker	220
pUC8	*lacZ'*: M13mp8 polylinker	220
pUC9	*lacZ'*: M13mp9 polylinker	220
pUC12	*lacZ'*: M13mp10 polylinker	220
pUC13	*lacZ'*: M13mp11 polylinker	220
pUC18	*lacZ'*: M13mp18 polylinker	220
pUC19	*lacZ'*: M13mp19 polylinker	220
pUC118	*lacZ'*: M13mp18 polylinker	219
pUC119	*lacZ'*: M13mp19 polylinker	219
pWE15, 16	promoter distal: *Bam*HI	220
pYEJ001	cml fusion: *Eco*RI; promoter distal: *Hin*dIII	220

[a] Only the restriction sites relevant to the normal usage of the vector are given. Sites marked with an asterisk result in removal of a dispensable segment of the vector. Note that the individual member of a polylinker pair carried by a vector cannot be identified if it does not lie within an ORF. Designations such as M13mp10/11 are used in these cases: check *Figure 1* for the orientation with respect to promoters. For gene abbreviations see *Table 1*, footnote a. For M13mp polylinkers see *Figure 3*.

Figure 1. Maps for *E. coli* plasmid cloning vectors. Plasmids c2RB, pcos1EMBL, pEMBL 8/9 and pJB8 are redrawn from ref. 67; pBluescript II, pBS +/-, pNH8a, pNH16a, pNH18a (and MCS2 and 3), pWE15 and pWE16 are reproduced with permission from Stratagene (4, 36, 48); pBR322, pDR540, pDR720, pFB-, pGEX2T, pGEX3X, pKK175-6, pKK223-3, pKK232-8, pKK233-2, pMC1871, pNO1523, pRIT2T, pRIT5, pSL1180, pSL1190, pSPT18/19, pT7T3 18U/19U, pUC and pYEJ001 are reproduced with permission from Pharmacia LKB (68); pBR325 and pUC118/119 are redrawn from ref. 1; all pGEM vectors shown, pGEMEX-1, pSELECT-1, pSP64, pSP64(poly A), pSP65, pSP70, pSP71, pSP72 and pSP73 are reproduced with permission from Promega Corp. (18); pKC30, pKT280, pT3/T7-1, pT3/T7-2 and pT3/T7-3 are reproduced with permission from Clontech Labs (40); pSP18, pSP19, pSPORT 1 and pT712/13 are redrawn with permission from Bethesda Research Laboratories (69); pT7-1, pT7-2, pTZ18R, pTZ19R, pTZ18U and pTZ19U are reproduced with permission from United States Biochemical Corp. (43).

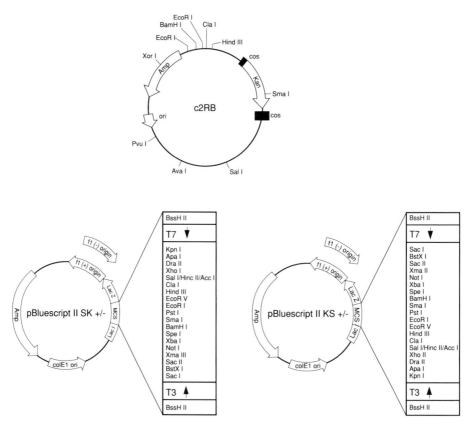

Figure 1. Continued.

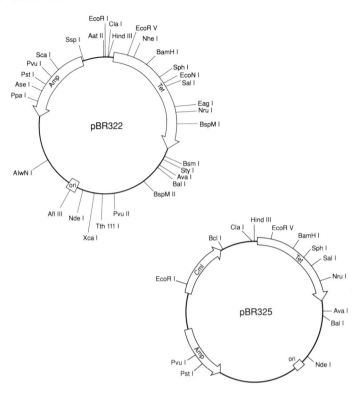

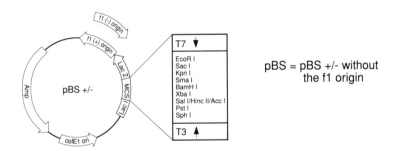

pBS = pBS +/- without
the f1 origin

Figure 1. Continued.

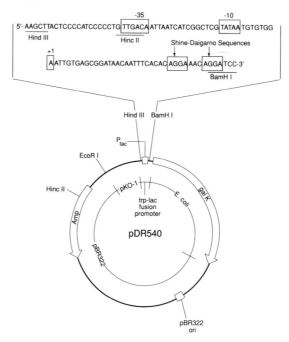

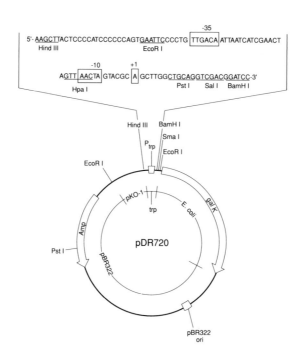

Figure 1. Continued.

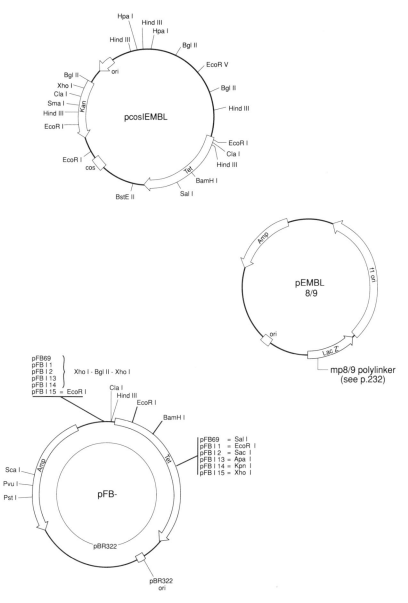

mp8/9 polylinker
(see p.232)

Cloning Vectors

Figure 1. Continued.

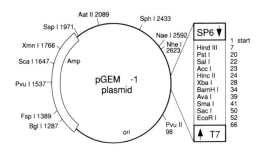

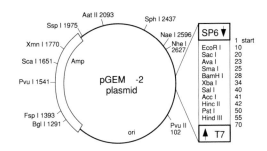

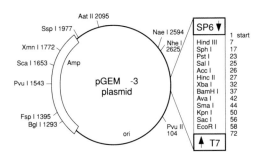

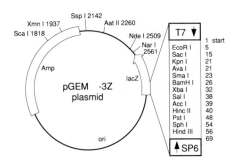

Figure 1. Continued.

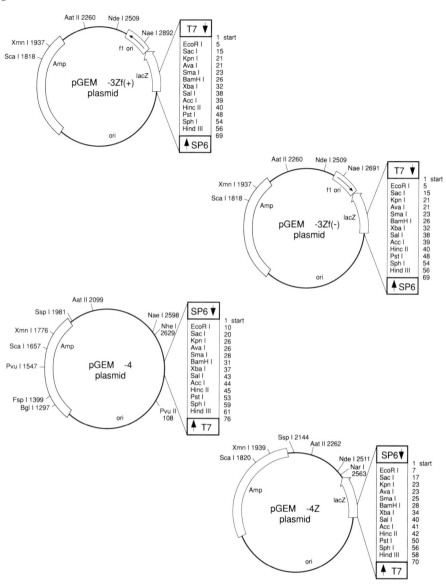

Figure 1. Continued.

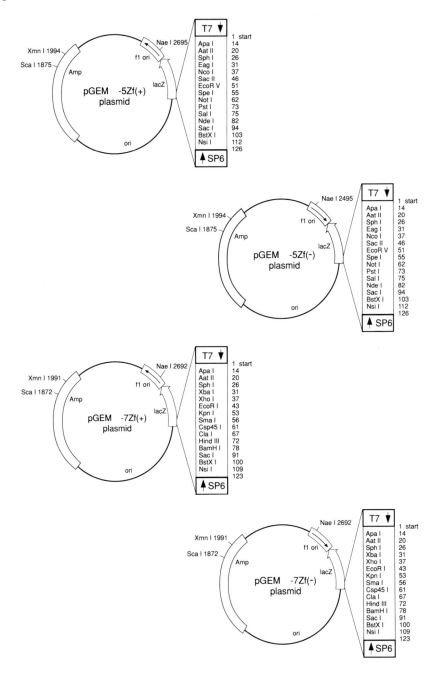

Figure 1. Continued.

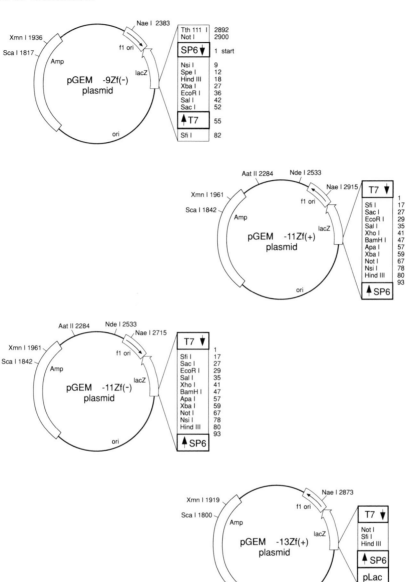

Figure 1. Continued.

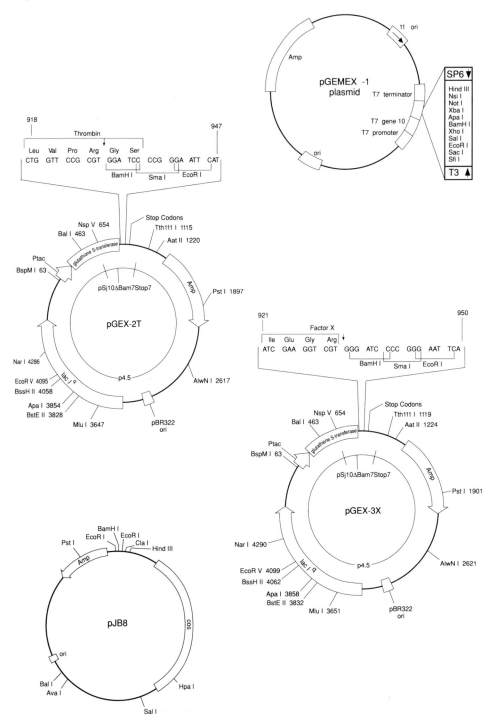

Figure 1. Continued.

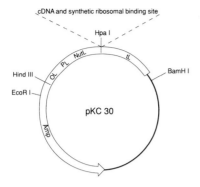

pKC 30

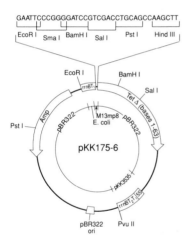

pKK175-6

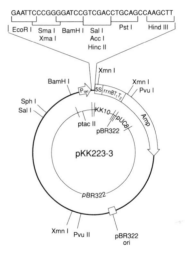

pKK223-3

Figure 1. Continued.

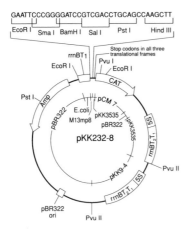

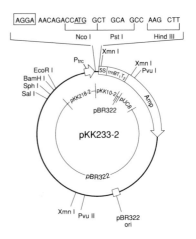

pKT 279
ATG AGT ATT CAA CAT TTC CGT GTC GCC CTT
ATT CCC TTT TTT GCG GCA TTT TGC CTT
CCT GTT TTT GCT CAC CG**C TGC AG**

pKT 280
ATG AGT ATT CAA CAT TTC CGT GTC GCC CTT
ATT CCC TTT TTT GCG GCA TTT TGC CTT
CCT GTT TTT GCT **CAC CCG CTG CAG**

pKT 287
ATG AGT ATT CAA CAT TTC CGT GTC GCC CTT
ATT CCC TTT TTT GCG GCA TTT TGC CTT
CCT GTT TTT GCT **CAC CCA GAA ACG GCT
GCA G**

Figure 1. Continued.

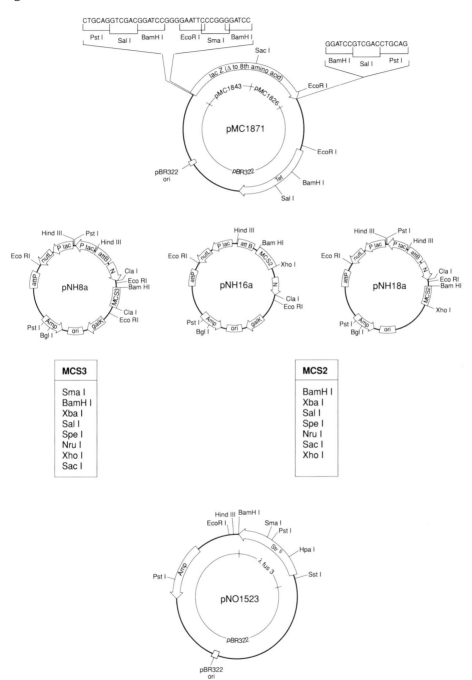

Figure 1. Continued.

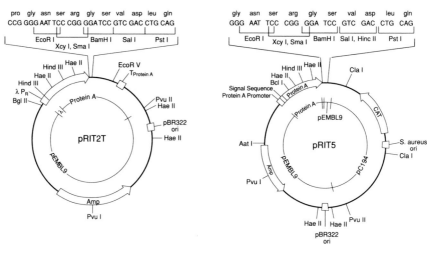

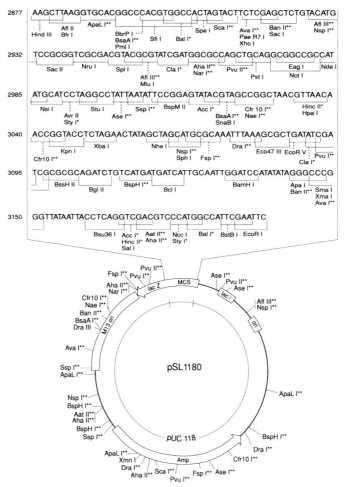

Figure 1. Continued.

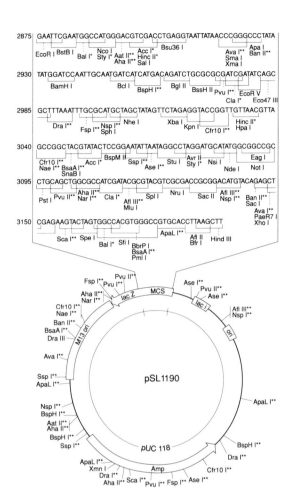

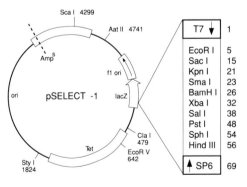

Figure 1. Continued.

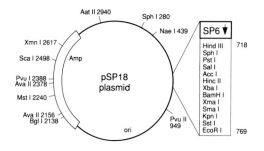

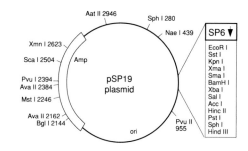

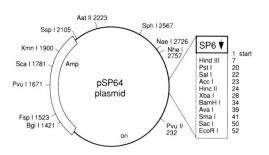

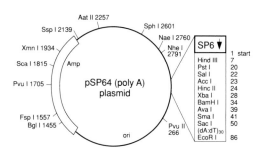

Figure 1. Continued.

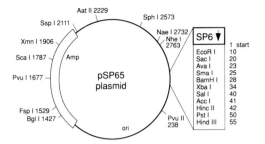

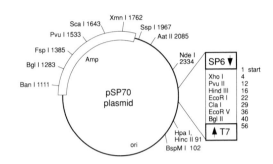

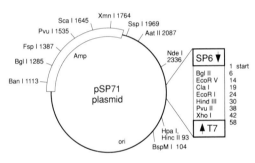

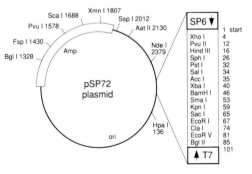

Figure 1. Continued.

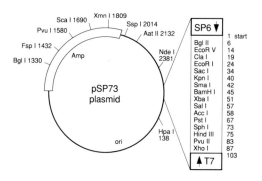

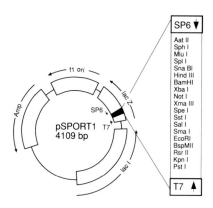

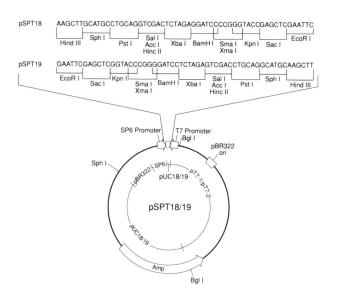

Figure 1. Continued.

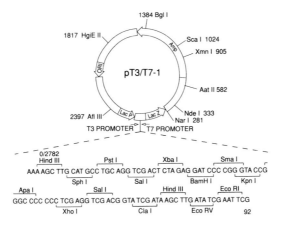

pT3/T7-1

1384 Bgl I
1817 HgiE II
Sca I 1024
Xmn I 905
ORI
Amp
Aat II 582
2397 Afl III
Nde I 333
Nar I 281
Lac P   Lac Z
T3 PROMOTER   T7 PROMOTER

0/2782
Hind III   Pst I   Xba I   Sma I
AAA AGC TTG CAT GCC TGC AGG TCG ACT CTA GAG GAT CCC CGG GTA CCG
       Sph I   Sal I   Sal I   BamH I   Kpn I
Apa I
GGC CCC CCC TCG AGG TCG ACG GTA TCG ATA AGC TTG ATA TCG AAT TCG
       Xho I   Cla I   Eco RV   92
                                          Hind III   Eco RI

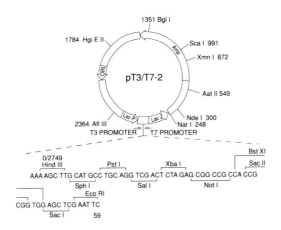

pT3/T7-2

1351 Bgl I
1784 Hgi E II
Sca I 991
Xmn I 872
ORI
Amp
Aat II 549
2364 Afl III
Nde I 300
Nar I 248
Lac P   Lac Z
T3 PROMOTER   T7 PROMOTER

                                              Bst XI
0/2749   Sac II
Hind III   Pst I   Xba I
AAA AGC TTG CAT GCC TGC AGG TCG ACT CTA GAG CGG CCG CCA CCG
       Sph I   Sal I   Not I
              Eco RI
CGG TGG AGC TCG AAT TC
       Sac I   59

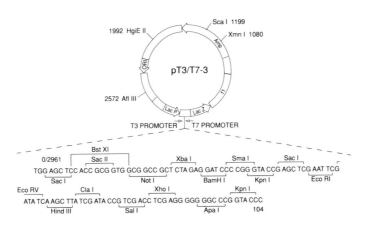

pT3/T7-3

Sca I 1199
Xmn I 1080
1992 HgiE II
ORI
Amp
f1
2572 Afl III
Lac P   Lac Z
T3 PROMOTER   T7 PROMOTER

              Bst XI
0/2961   Sac II   Xba I   Sma I   Sac I
TGG AGC TCC ACC GCG GTG GCG GCC GCT CTA GAG GAT CCC CGG GTA CCG AGC TCG AAT TCG
       Sac I   Not I   BamH I   Kpn I   Eco RI
Eco RV   Cla I   Xho I   Kpn I
ATA TCA AGC TTA TCG ATA CCG TCG ACC TCG AGG GGG GGC CCG GTA CCC
       Hind III   Sal I   Apa I   104

Figure 1. Continued.

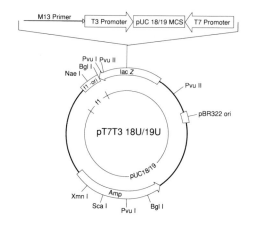

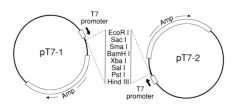

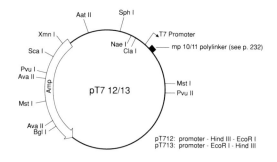

Figure 1. Continued.

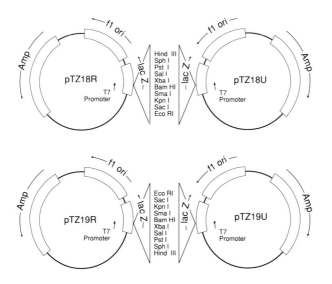

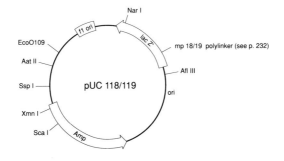

Figure 1. Continued.

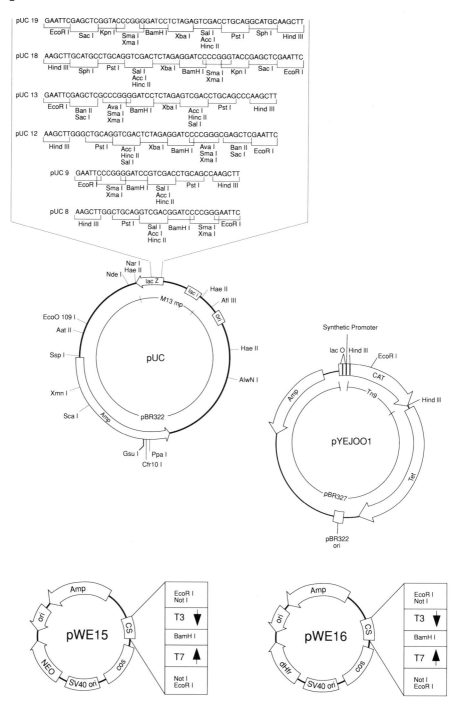

Table 3. Genetic markers for *E. coli* plasmid vectors

Marker	Source	Gene product	Phenotype
amp	Tn3	β-lactamase	ampicillin resistance
cml[a]	Tn9	acetyltransferase	chloramphenicol resistance
dhfr	mouse	dihydrofolate reductase	methotrexate resistance
galK	*E. coli*	galactokinase	Gal$^+$ (radiochemical)
kan	Tn903	phosphotransferase	kanamycin resistance
lacZ'	*E. coli*	β-galactosidase α-peptide	Lac$^+$ (histochemical)
neo	Tn5	phosphotransferase	kanamycin resistance
tet	pSC101	membrane protein, prevents tet uptake	tetracycline resistance

[a] Also sometimes designated as cat.

Table 4. Promoters for *E. coli* plasmid vectors

Promoter	Origin	Induction[a]
For protein expression		
lac	*E. coli lac* operon	IPTG
lac–tac	hybrid	IPTG
P$_L$	phage	*c*Its857 repressor
P$_R$	phage	*c*Its857 repressor
tac	*trp–lac* hybrid	IPTG
trc	*trp–lac* hybrid	IPTG
trp	*E. coli trp* operon	IAA
For RNA expression[b]		
SP6	SP6 phage	specific for SP6 RNA polymerase
T3	T3 phage	specific for T3 RNA polymerase
T7	T7 phage	specific for T7 RNA polymerase

[a] Abbreviations: IAA, 3-β-indoleacrylic acid; IPTG, isopropyl-β-D-thiogalactoside.
[b] For details of these RNA polymerases, see Chapter 5, Section 2.

2. *E.COLI* CLONING VECTORS BASED ON PHAGE LAMBDA

Details concerning phage λ vectors are given in *Table 5*. Genotypes are described in *Table 6* with the genetic markers explained in *Table 7*. Maps for these vectors are given in Figure 2. These maps show the positions and sizes of insertions and deletions, as well as the map locations and coordinates of the restriction sites relevant to cloning. Note that with several of these vectors there is confusion regarding the detailed restriction map and that the exact positions of non-cloning sites are often uncertain. Reliable maps showing a limited number of non-cloning restriction sites have been published (67) but these do not cover the vast majority of commercially-available restriction enzymes. It is recommended that restriction maps for non-cloning enzymes are worked out *de novo* for a particular vector as required, using as a guide the comprehensive listing of restriction sites for wild-type phage λ that is given in Chapter 8, *Table 7*.

Table 5. Details of phage λ cloning vectors

Vector	Type	Cloning sites	Capacity (kb)	Arms (kb) left	Arms (kb) right	Properties of recombinants	Recognition of recombinants	References
λ1059	replacement	BamHI	10–20	20.0	8.8	cI⁻ Red⁻ Gam⁻	Spi	52
λ2001	replacement	XbaI, SacI, XhoI, BamHI, HindIII, EcoRI	9–23	20.0	8.8	cI⁻ Red⁻ Gam⁻	Spi	53
Charon 4, 4A[a]	replacement	EcoRI	7–20	19.6	11.0	cI⁻ Red⁻ Gam⁻	Bio⁻ Lac⁻	54–57
	insertion	XbaI	0–6	24.8	20.5	cI⁻ Red⁻ Gam⁻	Bio⁺ Lac⁺	
Charon 30	replacement	BamHI	7–17	22.4	8.8	cI⁻ Red⁻ Gam⁻	Spi	57
		EcoRI	5–15	21.3	11.7	uncertain	none	
		HindIII	0–9	23.1	15.5	uncertain	none	
		SalI	0–10	28.1	10.1	uncertain	none	
		XhoI	0–9	28.8	9.9	uncertain	rarely Spi	
Charon 34	replacement	BamHI, EcoRI, HindIII, SacI, XbaI SalI + SalI, EcoRI, SacI, BamHI, HindIII or XbaI	9–20	19.5	10.6	cI⁻ Red⁻ Gam⁺	none	58
			10–21	19.5	9.6	cI⁻ Red⁻ Gam⁺	none	
Charon 35	replacement	BamHI, EcoRI, HindIII, SacI, XbaI SalI + SalI, EcoRI, SacI, BamHI, HindIII or XbaI	9–20	19.5	10.6	cI⁻ Red⁻ Gam⁺	none	58
			10–21	19.5	9.6	cI⁻ Red⁻ Gam⁺	none	
Charon 40	replacement	EcoRI, ApaI, XbaI, SfiI, SacI, AvrII, SalI, SpeI, HindIII, XhoI, BamHI, NaeI, KpnI, NotI, SmaI, XmaIII	9–24	19.2	9.6	cI⁻ Red⁻ Gam⁺	Lac⁻	59
λDASH[b]	replacement	XbaI, SalI, EcoRI, HindIII, SacI, XhoI, BamHI	9–22	20.0	8.8	cI⁻ Red⁻ Gam⁻	Spi	4

Table 5. Continued

Vector	Type	Cloning sites	Capacity (kb)	Arms (kb) left	right	Properties of recombinants	Recognition of recombinants	References
λEMBL3 λEMBL3A λEMBL4	replacement	*SalI, BamHI, EcoRI*	7–20	19.9	8.8	cI⁻ Red⁻ Gam⁻	Spi	60
λFIX[b]	replacement	*XbaI, SacI, SalI, XhoI, EcoRI*	9–22	20.0	8.8	cI⁻ Red⁻ Gam⁻	Spi	4
λgt10	insertion	*EcoRI*	0–6	32.7	10.6	cI⁻ Red⁺ Gam⁺	cI⁻	61
λgt11	insertion	*EcoRI*	0–7.2	19.6	24.1	cIts Red⁺ Gam⁺	Lac⁻	62
λgt18 λgt19	insertion	*SalI, EcoRI*	0–7.7	19.6	23.6	cIts Red⁻ Gam⁻	Lac⁻	63
λgt20 λgt21	insertion	*SalI, XbaI, EcoRI*	0–8.2	19.6	23.1	cIts Red⁻ Gam⁻	Lac⁻	64
λgt22[c] λgt23	insertion	*NotI, XbaI, SacI, SalI, EcoRI*	0–8.2	19.6	23.1	cIts Red⁻ Gam⁻	Lac⁻	64,65
λgtWES.λB'	replacement	*EcoRI*	2–13	21.2	13.7	cIts Red⁻ Gam⁺	none	66
		SacI	2–13	21.5	17.2	cIts Red⁻ Gam⁺	none	
λZAP II	insertion	*SacI, NotI, XbaI, SpeI, EcoRI, XhoI*	0–10	22.0	18.8	cIts Red⁺ Gam⁺	Lac⁻	3

[a] Charon 4A is the same as Charon 4 but carries the *A*am32 and Bam1 mutations (see *Table 6*).
[b] λDASH II and λFIX II are updated versions with additional *NotI* cloning sites.
[c] λgt22A is the same as λgt22 but with the *SacI* site replaced by *SpeI*.

Cloning Vectors

Table 6. Genotypes of phage λ vectors (for refs see *Table 5*; also see refs 1, 67)

Vector	Genotype
λ1059	h$^\lambda$*sbh*Iλ1° *b*189 <*int*29 *nin*L44 *c*Its857 p*ac*l29> Δ[*int–c*III] KH54 *sr*Iλ4° *nin*5 *chi*3
λ2001	λ*sbh*Iλ1° *b*189 <polycloning site *sr*Iλ3° *nin*L44 *bio* polycloning site> KH54 *chi*C *sr*Iλ4° *nin*5 *shn*dIIIλ6° *sr*Iλ5°
Charon 4	λ*lac*5 *bio*256 KH54 *sr*Iλ4° *nin*5 *QSR*80
Charon 4A	λ*A*am32 *Bam*1 *lac*5 *bio*256 KH54 *sr*Iλ4° *nin*5 *QSR*80
Charon 30	λ*sbh*Iλ1° [*sbh*Iλ2-(*b*1007)-*sbh*Iλ4]^{dup} KH54 *sr*Iλ4° *nin*5 *shn*dIIIλ6° *sr*Iλ5°
Charon 34	λ*sbh*Iλ1° *lac*5 *sr*I *lacZ* <polycloning site-*E. coli* DNA-polycloning site> *sr*Iλ3 WL113 KH54 *nin*5 *shn*dIIIλ6° *sr*Iλ5°
Charon 35	same as Charon 34 except that the stuffer is a different piece of *E. coli* DNA
Charon 40	λ*sbh*Iλ1° *sap*Iλ1° *sk*Iλ1° *sk*Iλ2° *ssm*Iλ1 <polycloning site-polystuffer-polycloning site> *ss*Iλ2 WL113 KH54 *ssm*Iλ3° *nin*5 *shn*dIIIλ6° *sr*Iλ5°
λDASH	λ*sbh*Iλ1° *b*189 <T3 promoter-polycloning site *sr*Iλ3° *nin*L44 *bio* polycloning site-T7 promoter> KH54 *chi*C *sr*Iλ4° *nin*5 *shn*dIIIλ6° *sr*Iλ5°
λEMBL3 λEMBL3A λEMBL4	λ(EMBL3A only: *A*am32 *Bam*1) *sbh*Iλ1° *b*189 <polycloning site *int*29 *nin*L44 *trpE* polycloning site> KH54 *chi*C *sr*Iλ4° *nin*5 *sr*Iλ5°
λFIX	same as λDASH except the T3 and T7 promoters are inverted
λgt10	λ*sr*Iλ1° *b*527 *sr*Iλ3° *imm*434 (*sr*I434$^+$) *sr*Iλ4° *sr*Iλ5°
λgt11	λ*lac*5 *shn*dIIIλ2–3 *sr*Iλ3° *c*Its857 *sr*Iλ4° *nin*5 *sr*Iλ5° *Sam*100
λgt18, 19	λ*lac*5 *shn*dIIIλ2–3 *sr*Iλ3° *ss*Iλ1–2 *c*Its857 *sr*Iλ4° *nin*5 *sr*Iλ5° *Sam*100
λgt20, 21	λ*lac*5 *sx*Iλ1° *chi* *shn*dIIIλ2–3 *sr*Iλ3° *ss*Iλ1–2 *c*Its857 *sr*Iλ4° *nin*5 *sr*Iλ5° *Sam*100
λgt22, 23	same as λgt20 and 21 except a different polycloning site
λgtWES.λB'	λ*W*am403 *E*am1100 *inv* (*sr*Iλ1–*sr*Iλ2) Δ(*sr*Iλ2–*sr*Iλ3) *c*Its857 *sr*Iλ4° *nin*5 *sr*Iλ5°*Sam*100
λZAP II	λ*sbh*Iλ1° *chi*A131 <T *amp* ColE1 *ori lacZ'* T3 promoter-polycloning site-T7 promoter I> *sr*Iλ3° *c*Its857 *sr*Iλ4° *nin*5 *sr*Iλ5°

Table 7. Genetic markers carried by phage λ vectors (see also ref. 67)

Marker	Description
Aam	Amber mutation in gene A
$b2$, $b189$, $b527$, $b1007$	Deletions within the b region that prevent the phage entering the lysogenic cycle
Bam	Amber mutation in gene B
bio, $bio252$, $bio256$	Substitutions from the bio region of *E. coli*
chiA131, chiC	Directional recombination sites
cIts857	A temperature-sensitive mutation in cI — lysogens carrying this marker can be induced by heat shock
$cos2$	Defective cos sites
Dam	Amber mutation in gene D, resulting in intracellular accumulation of phage λ structural proteins
Eam	Amber mutation in gene E, resulting in intracellular accumulation of phage λ structural proteins
exo bet	Mutations in these genes (the red region) result in the Spi⁻ phenotype, meaning that phage can infect $recA^+$ hosts that are lysogenic for P2
gamam	Amber mutation in gam. A mutation in gam results in the Spi⁻ phenotype (see exo bet)
imm434	Substitution from phage φ434
int29	Substitution from phage φ29
intam	Amber mutation in int
KH54	Deletion that prevents lysogeny
$lac5$, lacUV5, lacZ, lacZ'	Substitutions from the lac region of *E. coli*
$nin5$, ninL44	Deletions that remove t_{R2}, allowing delayed early transcription independent of the N gene product
QSR80	Substitution from phage φ80
$red3$	See exo bet
s....$λ$.	Designates a deletion — the letters immediately following the s indicate the restriction site(s) and the number following the λ denotes the sites involved (e.g ss*II*λ1–2 = deletion between *Sal*I sites 1 and 2). λ1°, λ2° indicate exonuclease digestion at the site denoted
Sam100	Amber mutation in gene S, which results in intracellular accumulation of phage particles
trpE	Substitutions from the trp region of *E. coli*
Wam	Amber mutation in gene W
WL113	Deletion that removes kil, $cIII$, $Ea10$ and ral

Figure 2. Maps for phage λ cloning vectors. A: Genetic map of wild-type phage λ, showing the positions of deletions carried by the cloning vectors. B: Physical maps for the vectors. Figure redrawn, with permission from *Molecular Cloning (2nd edn)* by J. Sambrook, E.F. Fritsch and T. Maniatis. Copyright 1989 Cold Spring Harbor Laboratory Press.

Notes: 1. Charon 4A is the same as Charon 4 but carries the *A*am32 and *B*am1 mutations (see *Table 6*). 2. The insertion in Charon 30 is *Bam*H1 fragment 3/4 (coordinates 22350–34500), minus the *b*1007 deletion from the wild-type genome. This region therefore occurs twice in this vector. 3. The stuffer fragment in Charon 40 is variable in length, so the coordinates of the P2 sites cannot be given. P2 contains the same sites as P1 but in reverse orientation. 4. λEMBL3A is the same as λEMBL3 but carries the *A*am2 and *B*am1 mutations (see *Table 6*). 5. The insertion in λgtWES.λB is *Eco*RI fragment 2 (coordinates 21230–26100) from the wild-type genome in reverse orientation. For references see *Table 5*.

(A)

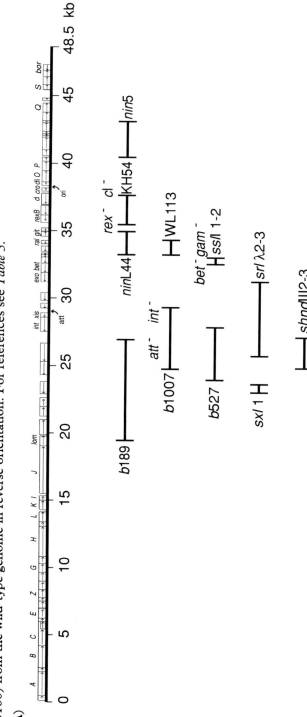

Figure 2. Continued.

(B)

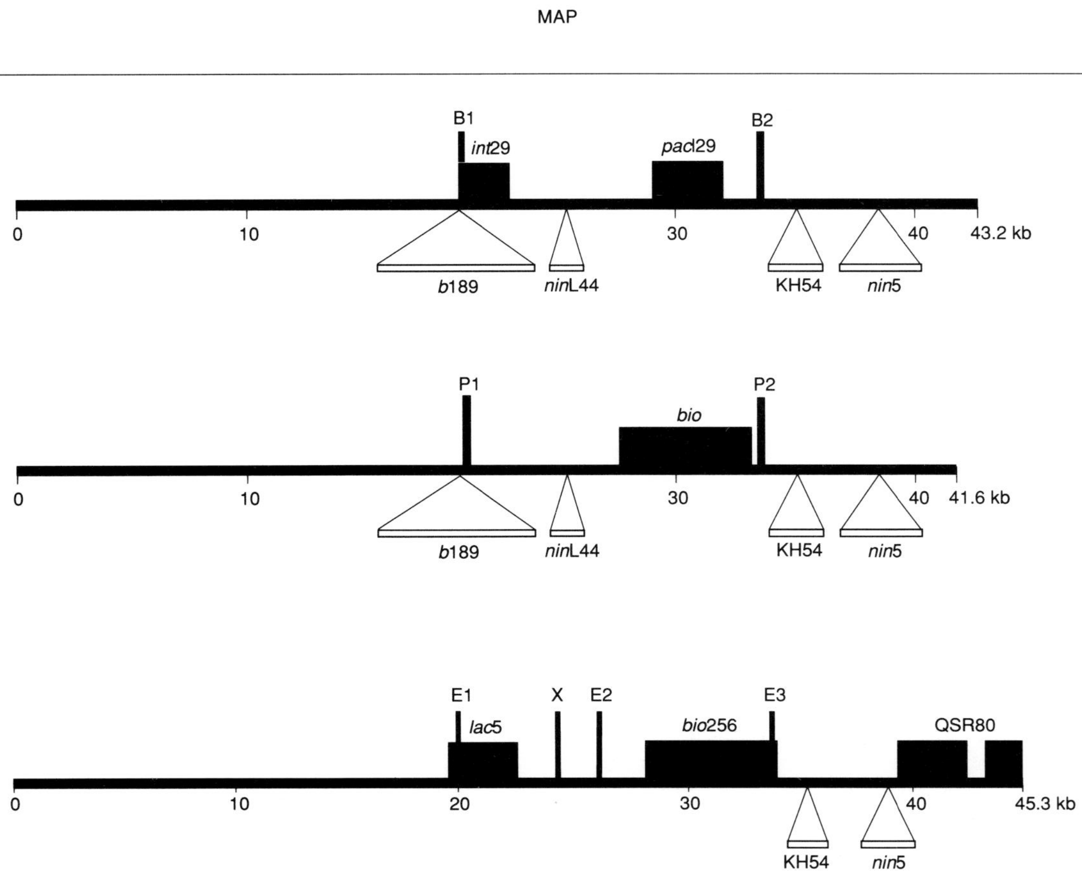

VECTOR

MAP

CLONING SITES

Abbrev.	Sites.	Coords.
B1	*Bam*HI	20020
B2	*Bam*HI	34440
P1	*Xba*I	19990
	*Sac*I	20000
	*Xho*I	20000
	*Bam*HI	20010
	*Eco*RI	20010
	*Xba*I	20020
P2	*Xba*I	32760
	*Eco*RI	32760
	*Hin*dIII	32760
	*Bam*HI	32770
	*Xho*I	32770
	*Sac*I	32770
	*Xba*I	32770
E1	*Eco*RI	19600
X	*Xba*I	24820
E2	*Eco*RI	26420
E3	*Eco*RI	34320

λ 1059

λ 2001

CHARON 4
CHARON 4A
(see note 1)

Cloning Vectors

227

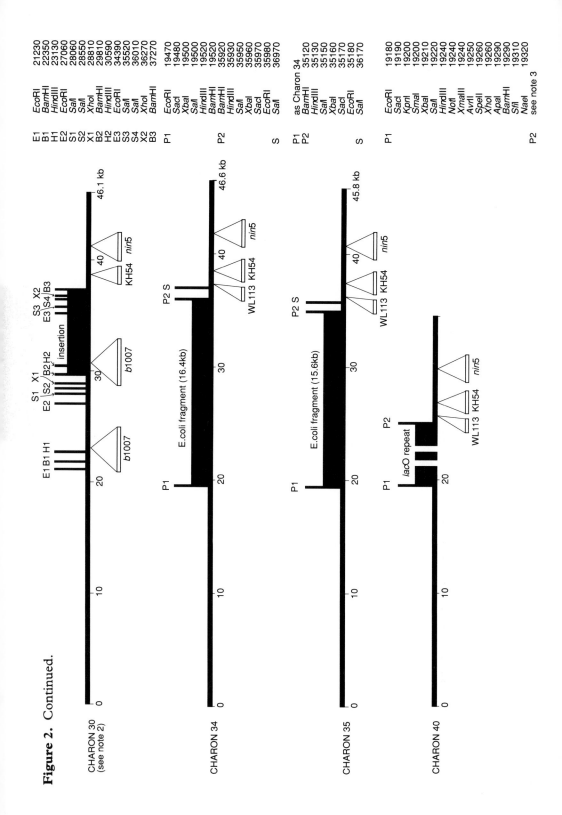

Figure 2. Continued.

Figure 2. Continued.

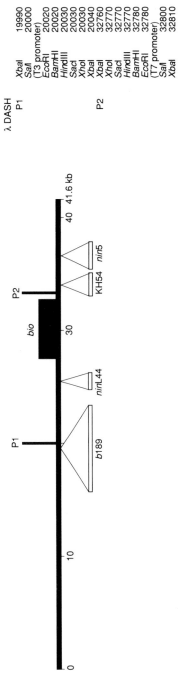

λ DASH
and
λ FIX

	10		20	30	40	41.6 kb

0

b189 · ninL44 · bio · KH54 · nin5

P1 · P2

λ DASH

P1		
	Xbal	19990
	Sall	20000
	(T3 promoter)	
	EcoRI	20020
	BamHI	20020
	HindIII	20030
	Sacl	20030
	Xhol	20030
	Xbal	20040
P2	Xbal	32760
	Xhol	32770
	Sacl	32770
	HindIII	32770
	BamHI	32780
	EcoRI	32780
	(T7 promoter)	
	Sall	32800
	Xbal	32810

λ FIX

P1		
	Xbal	19990
	Sacl	20000
	(T7 promoter)	
	Sall	20020
	Xhol	20020
	EcoRI	20030
	Xbal	20030
P2	Xbal	32760
	EcoRI	32760
	Xhol	32760
	Sall	32770
	(T3 promoter)	
	Sacl	32790
	Xbal	32790

Cloning Vectors

229

Figure 2. Continued.

Figure 2. Continued.

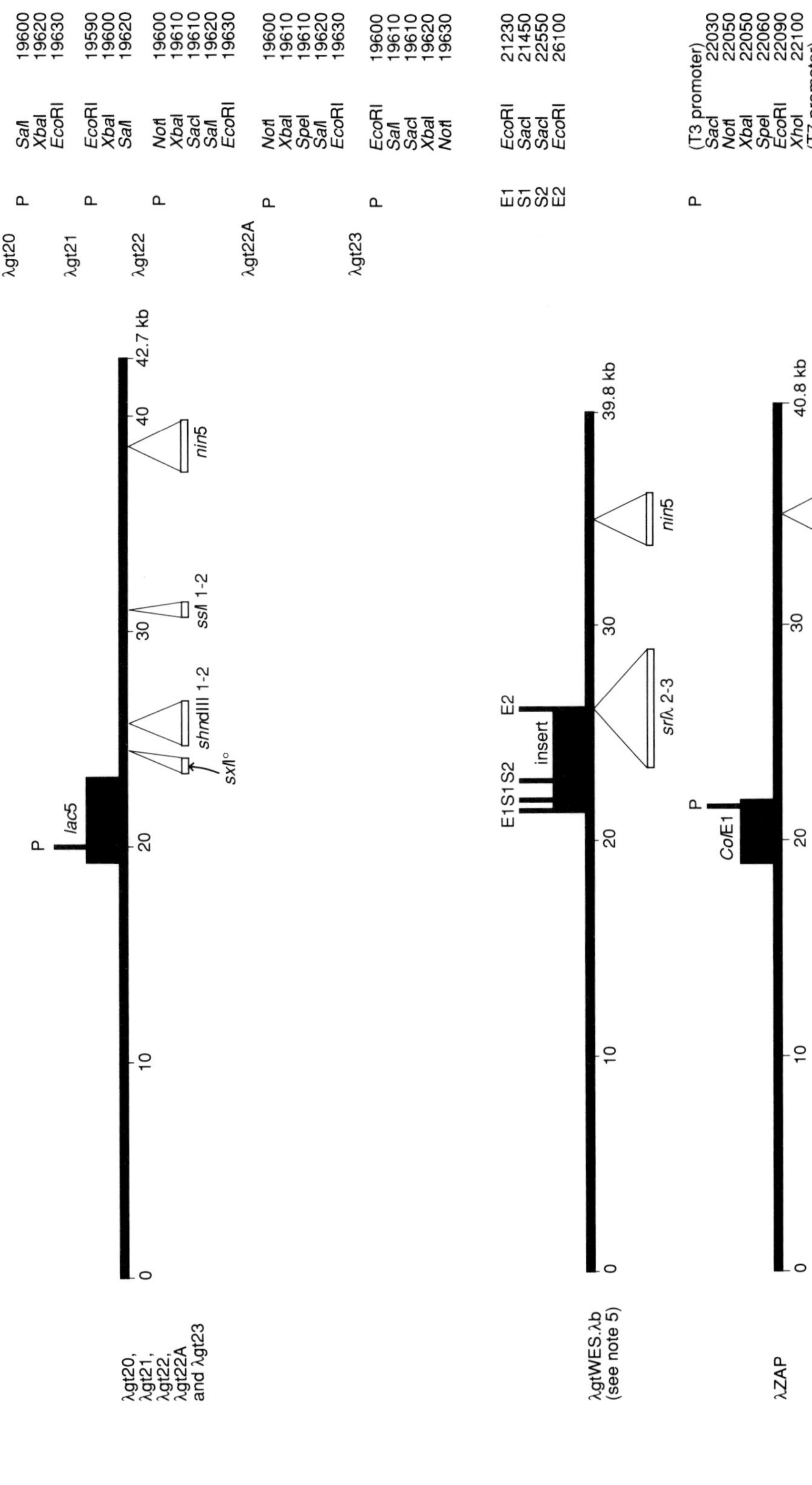
λgt20, λgt21, λgt22, λgt22A and λgt23
0
10
20
30
40
42.7 kb
P
lac5
sxIl°
shndIII 1-2
ssI 1-2
nin5
λgtWES.λb (see note 5)
0
10
20
30
39.8 kb
E1S1S2
insert
E2
srIλ 2-3
nin5
λZAP
0
10
20
30
40.8 kb
P
ColE1
nin5
λgt20 P SalI 19600
XbaI 19620
EcoRI 19630
λgt21 P EcoRI 19590
XbaI 19600
SalI 19620
λgt22 P NotI 19600
XbaI 19610
SacI 19610
SalI 19620
EcoRI 19630
λgt22A P NotI 19600
XbaI 19610
SpeI 19610
SalI 19620
EcoRI 19630
λgt23 P EcoRI 19600
SalI 19610
SacI 19610
XbaI 19620
NotI 19630
E1 EcoRI 21230
S1 SacI 21450
S2 SacI 22550
E2 EcoRI 26100
P (T3 promoter)
SacI 22030
NotI 22050
XbaI 22050
SpeI 22060
EcoRI 22090
XhoI 22100
(T7 promoter)

3. *E.COLI* CLONING VECTORS BASED ON M13 PHAGE

The M13mp series vectors are 7.2 kb in size and carry the *lacZ'* gene. The sequences and restriction sites of the polylinkers carried by the different M13mp vectors are shown in *Figure 3*.

Figure 3. Polylinkers in M13mp vectors.

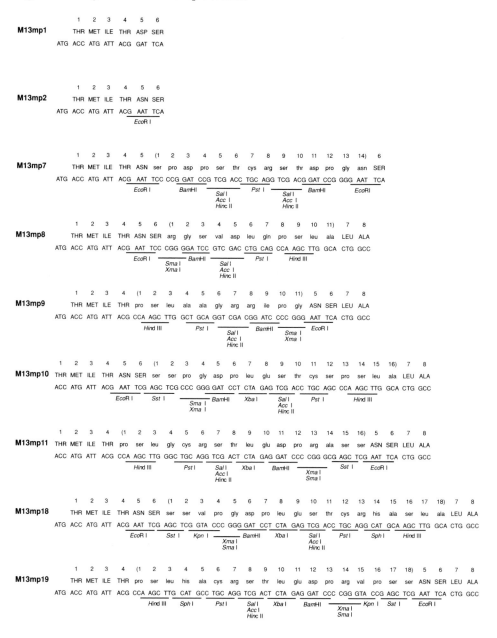

4. REFERENCES

1. Pouwells, P.H., Enger-Valk, B.E. and Brammar, W.J. (1985) *Cloning Vectors.* (Supplements 1986, 1987). Elsevier, Amsterdam.

2. Bates, P.F. and Swift, R.A. (1983) *Gene*, **26**, 137.

3. Short, J.M., Fernandez, J.M., Sorge, J.A. and Huse, W.D. (1988) *Nucleic Acids Res.*, **16**, 7583.

4. Anon (1990) *Product Catalog.* Stratagene, Palo Alto.

5. Bolivar, F., Rodriguez, R.L., Greene, P.J., Betlach, M.C., Heyneker, H.L., Boyer, H.W., Crosa, J.H. and Falkow, S. (1977) *Gene*, **2**, 95.

6. Sutcliffe, J.G. (1978) *Nucleic Acids Res.*, **5**, 2721.

7. Bolivar, F. (1978) *Gene*, **4**, 121.

8. Yanisch-Perron, C., Vieira, J. and Messing, J. (1985) *Gene*, **33**, 103.

9. Melton, D.A., Krieg, P.A., Rebagliati, M.R., Maniatis, T., Zinn, K. and Green, M.R. (1984) *Nucleic Acids Res.*, **12**, 7035.

10. Poustka, A., Rackwitz, H.-R., Frischauf, A.-M., Hohn, B. and Lehrach, H. (1984) *Proc. Natl. Acad. Sci. USA*, **81**, 4129.

11. Babineau, D., Vetter, D., Andrews, B.J., Gronostajski, R.M., Proteau, G.A., Beatty, L.G. and Sadowski, P.D. (1985) *J. Biol. Chem.*, **260**, 12313.

12. Thogersen, H.C., Morris, H.R., Rand, K.N. and Gait, M.J. (1985) *Eur. J. Biochem.*, **147**, 325.

13. de Boer, H.A., Comstock, L.J. and Vasser, M. (1983) *Proc. Natl. Acad. Sci. USA*, **80**, 21.

14. Russell, D.R. and Bennett, G.N. (1982) *Gene*, **20**, 231.

15. Dente, L., Cesareni, G. and Cortese, R. (1983) *Nucleic Acids Res.*, **11**, 1645.

16. Barany, F. (1985) *Proc. Natl. Acad. Sci. USA*, **82**, 4202.

17. Barany, F. (1985) *Gene*, **37**, 111.

18. Anon (1989/1990) *Biological Research Products Catalog.* Promega Corp, Madison.

19. Smith, D.B. and Johnson, K.S. (1989) *Gene*, **67**, 31.

20. Ish-Horowicz, D. and Burke, J.F. (1981) *Nucleic Acids Res.*, **9**, 2989.

21. Rosenberg, M., Ho, Y. and Shatzman, A. (1983) *Methods Enzymol.*, **101**, 123.

22. Shimatake, H. and Rosenberg, M. (1981) *Nature*, **292**, 128.

23. Ho, Y.-S., Lewis, M. and Rosenberg, M. (1982) *J. Biol. Chem.*, **257**, 9128.

24. Brosius, J. (1984) *Gene*, **27**, 151.

25. Brosius, J., Ullrich, A., Raker, M.A., Gray, A., Dull, T.J., Gutell, R.R. and Noller, H.F. (1981) *Plasmid*, **6**, 112.

26. Brosius, J. and Holy, A. (1984) *Proc. Natl. Acad. Sci. USA*, **81**, 6929.

27. Gentz, R., Langner, A., Chang, A.C.Y., Cohen, S.N. and Bujard, H. (1981) *Proc. Natl. Acad. Sci. USA*, **78**, 4936.

28. Brosius, J. (1984) *Gene*, **27**, 161.

29. Amann, E. and Brosius, J. (1985) *Gene*, **40**, 183.

30. Shimizu, Y., Takabayashi, E., Yano, S., Shimizu, N., Yamada, K. and Gushima, H. (1988) *Gene*, **65**, 141.

31. Straus, D. and Gilbert, W. (1985) *Proc. Natl. Acad. Sci. USA*, **82**, 2014.

32. Talmadge, K., Stahl, S. and Gilbert, W. (1980) *Proc. Natl. Acad. Sci. USA*, 77, 3369.

33. Talmadge, H. and Gilbert, W. (1980) *Gene*, **12**, 235.

34. Shapira, S.K., Chou, J., Richard, F.V. and Casadaban, M.J. (1983) *Gene*, **25**, 71.

35. Casadaban, M.J., Arias, A.M., Shapira, S.K. and Chou, J. (1983) *Methods Enzymol.*, **100**, 293.

36. Hasan, N. and Szybalski, W. (1987) *Gene*, **56**, 145.

37. Dean, D. (1981) *Gene*, **15**, 99.

38. Nilsson, B., Abrahmsen, L. and Uhlen, M. (1985) *EMBO J.*, **4**, 1075.

39. Brosius, J. (1989) *DNA*, **8**, 759.

40. Anon (1988) *Products and Protocols Catalog.* Clontech Labs, Palo Alto.

41. Mead, D.A., Szczesna-Skorupa, E. and Kemper, B. (1986) *Protein Eng.*, **1**, 67.

42. Tabor, S. and Richardson, C.C. (1985) *Proc. Natl. Acad. Sci. USA*, **82**, 1074.

43. Anon (1990) *Molecular Biology Reagents Catalog.* United States Biochemical Corp., Cleveland.

44. Vieira, J. and Messing, J. (1982) *Gene*, **19**, 259.

45. Messing, J. (1983) *Methods Enzymol.*, **101**, 20.

46. Norrander, J., Kempe, T. and Messing, J. (1983) *Gene*, **26**, 101.

47. Vieira, J. and Messing, J. (1987) *Methods Enzymol.*, **153**, 3.

48. Wahl, G.M., Lewis, K.A., Ruiz, J.C., Rothenberg, B., Zhao, J. and Evans, G.A. (1987) *Proc. Natl. Acad. Sci. USA*, **84**, 2160.

Cloning Vectors

49. Lau, Y.-F. and Kan, Y.W. (1983) *Proc. Natl. Acad. Sci. USA*, **80**, 5225.

50. Rossi, J.J., Soberon, X., Marumoto, Y., McMahon, J. and Itakura, K. (1983) *Proc. Natl. Acad. Sci. USA*, **80**, 3203.

51. Lobet, Y., Peacock, M. and Cieplak, W. (1989) *Nucleic Acids Res.*, **17**, 4897.

52. Karn, J., Brenner, S., Barnett, L. and Cesareni, G. (1980) *Proc. Natl. Acad. Sci. USA*, 77, 5172.

53. Karn, J., Matthes, H.W.D., Gait, M.J. and Brenner, S. (1984) *Gene*, **32**, 217.

54. Blattner, F.R., Williams, B.G., Blechl, A.E., Thompson, K.D., Faber, H.E., Furlong, L.-A., Grunwald, D.J., Kiefer, D.O., Moore, D.D., Schumm, J.W., Sheldon, E.L. and Smithies, O. (1977) *Science*, **196**, 161.

55. Williams, B.G. and Blattner, F.R. (1979) *J. Virol.*, **29**, 555.

56. de Wet, J.R., Daniels, D.L., Schroeder, J.L., Williams, B.G., Thompson, K.D., Moore, D.D. and Blattner, F.R. (1980) *J. Virol.*, **33**, 401.

57. Nordstrom, K., Molin, S. and Hansen, H.A. (1980) *Plasmid*, **4**, 215.

58. Loenen, W.A.M. and Blattner, F.R. (1983) *Gene*, **26**, 171.

59. Dunn, I.S. and Blattner, F.R. (1987) *Nucleic Acids Res.*, **15**, 2677.

60. Frischauf, A.-M., Lehrach, H., Poustka, A. and Murray, N. (1983) *J. Mol. Biol.*, **170**, 827.

61. Huynh, T.V., Young, R.A. and Davis, R.W. (1985) in *DNA Cloning: A Practical Approach*. (D.M. Glover ed.) IRL Press, Oxford, Vol. 1, p.49.

62. Young, R.A. and Davis, R.W. (1983) *Proc. Natl. Acad. Sci. USA*, **80**, 1194.

63. Han, J.H., Stratowa, C. and Rutter, W.J. (1987) *Biochemistry*, **26**, 1617.

64. Han, J.H. and Rutter, W.J. (1988) in *Genetic Engineering: Principles and Methods*. (J.K. Setlow ed.) Plenum, New York, Vol. 10, p.195.

65. Han, J.H. and Rutter, W.J. (1987) *Nucleic Acids Res.*, **15**, 6304.

66. Leder, P., Tiemeier, D. and Enquist, L. (1977) *Science*, **196**, 175.

67. Sambrook, J., Fritsch, E.F. and Maniatis, T. (1989) *Molecular Cloning: A Laboratory Manual (2nd edn)*. Cold Spring Harbor Laboratory Press, New York.

68. Anon (1986) *Molecular Biologicals Catalog*. Pharmacia LKB, Sweden.

69. Anon (1990) *Catalog and Reference Guide*. GIBCO-BRL Life Technologies, Gaithersburg.

CHAPTER 7
GENOMES AND GENES

This chapter provides information on selected genomes and genes. Although the data are not comprehensive (it would obviously be impossible to include information on all organisms), the examples chosen are representative of different groups.

1. GENOME SIZES AND SEQUENCE COMPLEXITY

Table 1 shows the genome sizes for representative organisms. Indirect methods for estimating genome size appear to be accurate to about ±10%, usually being on the low size; at least this has been the case for those small genomes that have now been completely sequenced. However, the reader is urged to be cautious in using the data provided in this Table.

To the molecular biologist the main practical importance of genome size is its relationship to the required size for a genomic library. The relevant formula is:

$$N = \frac{\ln (1-P)}{\ln (1-a/b)}$$

where N = the number of clones required, P = probability that a given sequence will be present, a = average size of the DNA fragments inserted in the vector, and b = total size of the genome. In *Figure 1* the relationship between N and b is plotted for P = 0.95 (95% probability) and P = 0.99 (99% probability), and for a = 17 kb (fragments suitable for a phage λ replacement vector), a = 35 kb (cosmid vector) and a = 100 kb (YAC vector).

Sequence complexity is also relevant to the isolation of genes from genomic libraries. *Table 2* details the proportion of single copy DNA in different genomes.

Mitochondrial and chloroplast genomes have been characterized from a wide range of organisms. Sizes for representative species are shown in *Table 3*. Recently DNA has also been isolated from basal bodies (1) but as yet the genetic significance of these molecules is unknown.

Table 1. Sizes of genomes of representative organisms

Group	Representative organisms	Estimated genome size (kb)[a]
Microorganisms		
Bacteria		
	Haemophilus influenzae	1200
	Escherichia coli	4000
	Bacillus megaterium	30 000

Table 1. Continued

Group	Representative organisms	Estimated genome size (kb)[a]
Fungi		9 400–175 000
	Hansenula holstii	10 300
	Saccharomyces carlsbergensis	15 000
	Saccharomyces cerevisiae	20 000
	Torulopsis holmii	21 600
	Aspergillus nidulans	25 400
	Phycomyces blakesleeanus	30 100
	Achyla bisexualis	41 400
	Dictyostelium discoideum	47 000
Algae		37 500–190 000 000
Protozoa		37 500–330 000 000
	Tetrahymena pyriformis	190 000
Animals		
Porifera		56 500
Coelenterata		280 000–685 000
	Cassiopeia sp.	310 000
Nematoda		75 000–620 000
Annelida		660 000–6 750 000
	Nereis sp.	1 400 000
Arthropoda		
Crustacea		660 000 – 21 250 000
	Plagusia depressa	1 400 000
	Limulus polyphemus	2 650 000
Insecta		47 000 – 12 000 000
Diptera		125 000 – 835 000
	Drosophila melanogaster	165 000
	Chironomus tetans	197 500
	Culex pipens	820 000
	Musca domestica	840 000
Lepidoptera		470 000 – 565 000
	Bombyx mori	490 000
Orthoptera		3 100 000 – 11 900 000
	Locusta migratoria	5 000 000
Mollusca		375 000 – 5 100 000
	Fissurella bandadensis	470 000
	Tectorius muricatus	630 000
	Aplysia californica	1 700 000
Echinodermata		470 000 – 4 150 000
	Strongylocentrotus purpuratus	845 000
Protochordata		140 000 – 565 000
	Asidea atra	150 000

Table 1. Continued

Group	Representative organisms	Estimated genome size (kb)[a]
Agnatha		1 320 000 – 2 650 000
Pisces		
Chondrichthyes		2 650 000 – 6 950 000
	Carcharias obscurus	2 650 000
Osteichthyes		375 000 – 135 000 000
	Rutilus rutilus	4 500 000
	Protopterus sp.	47 000 000
Amphibia		
Anura		950 000–10 150 000
	Xenopus laevis	2 900 000
	Bufo bufo	6 600 000
Urodela		14 200 000–78 500 000
	Ambystoma mexicanum	35 700 000
	Notophthalamus viridescens	42 500 000
	Amphiuma sp.	78 500 000
Reptilia		1 600 000–5 100 000
	Python reticulatus	1 600 000
	Natrix natrix	2 350 000
	Caiman crocodylus	2 450 000
	Terrapene carolina	3 850 000
Aves		1 125 000–1 975 000
	Gallus domesticus	1 125 000
Mammalia		
Marsupalia		2 800 000–4 420 000
Placentalia		2 350 000–5 550 000
	Homo sapiens	2 800 000
	Microtus agrestis	3 000 000
	Mus musculus	3 300 000
	Peromyscus eremicus	4 420 000
	Dipodomys ordii monoensis	5 200 000
Plants		
Bryophyta		600 000–4 050 000
Pteridophyta		950 000–300 000 000
Gymnospermae		4 900 000–47 000 000
Angiospermae		95 000–120 000 000
	Arabidopsis thaliana	190 000
	Oryza sativa	565 000
	Lycopersicon esculentum	700 000
	Daucus carota ssp *carota*	950 000
	Brassica napus	1 500 000
	Medicago sativa	1 600 000

GENOMES AND GENES

237

Table 1. Continued

Group	Representative organisms	Estimated genome size (kb)[a]
	Solanum tuberosum	2 000 000
	Nicotiana tabacum	3 500 000
	Pisum sativum	4 700 000
	Secale cereale	8 275 000
	Zea mays	15 000 000
	Allium cepa	17 000 000
	Tulipa polychroma	23 000 000
	Lilium davidii	40 000 000
	Fritillaria assyriaca	120 000 000

[a] Per single chromosome set. Data taken from refs 2 and 3 and papers cited therein.

Figure 1. Graph for calculation of the approximate number of clones required for gene libraries constructed with different fragment sizes. *a,* average size of the DNA fragments inserted in the vector; *P,* the probability that a given sequence will be present, e.g. 0.99 = 99% probability of the given sequence being present in the library.

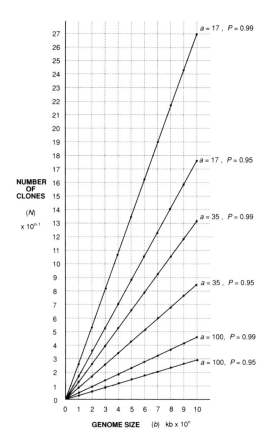

Table 2. Sequence complexity of representative genomes

Group	Organism	Approximate single copy DNA content of the genome (%)[a]
Fungi	*Phycomyces blakesleeanus*	65
	Dictyostelium discoideum	70
	Achyla bisexualis	82
	Torulopsis holmii	84
	Saccharomyces carlbergensis	89
	Hansenula holstii	92
	Aspergillus nidulans	97
Protozoa	*Tetrahymena pyriformis*	90
Coelenterata	*Aurelia aurita*	70
Mollusca	*Aplysia californica*	55
Arthropoda		
Crustacea	*Limulus polyphemus*	70
Insecta	*Drosophila melanogaster*	60
	Musca domestica	90
	Chironomus tetans	95
Echinodermata	*Strongylocentrotus purpuratus*	75
Protochordata	*Ciona intestinalis*	70
Pisces	*Rutilus rutilus*	54
Amphibia	*Bufo bufo*	20
	Xenopus laevis	75
Reptilia	*Natrix natrix*	47
	Terrapene carolina	54
	Caiman crocodylus	66
	Python reticulatus	71
Aves	*Gallus domesticus*	80
Mammalia	*Homo sapiens*	64
	Mus musculus	70
Planta	*Lolium multiflorum*	36

[a] Data taken from ref. 2 and papers cited therein.

Table 3. Sizes of organelle genomes of representative organisms

Group	Representative organisms	Estimated genome size (kb)[a]
MITOCHONDRIAL GENOMES		
Animals		14.5–23.0
	Ascaris suum	14.5
	Paracentrotus lividus	15.697
	mouse	16.295
	cow	16.338
	Homo sapiens	16.569
	Cnemidophorus exsanguis	17.4
	Xenopus laevis	18.4
	Drosophila melanogaster	19.0
Fungi		18.0–175
Myxomycetes	*Physarum polycephalum*	69
Oomycetes	*Phytophora infestans*	36.2
	Saprolegnia ferax	46.5
	Achyla sp.	50.0
	Pythium ultimum	57
Zygomycetes	*Phycomyces blakesleeanus*	25.6
Ascomycetes	*Cephalosporium acremonium*	27
	Aspergillus nidulans	33.3
	Claviceps purpurea	45
	Penicillium chrysogenum	49.2
	Neurospora crassa	60
	Podospora anserina	94
	Brettanomyces custersii	108
	Cochliobolus heterostrophus	115
Basidiomycetes	*Coprinus cinereus*	43.3
	Schizophyllum commune	50.3–52.2
	Ustilago cynodontis	76.5
	Coprinus stercorarius	91
	Suillus grisellus	121
	Agaricus bitorquis	148–174
Yeasts	*Torulopsis glabrata*	18.9
	Schizosaccharomyces pombe	19
	Saccharomyces exiguus	23.7
	Kloeckera africana	27.1
	Kluyveromyces lactis	37
	Hansenula petersonii	42
	Yarrowia lipolytica	44–48
	Hansenula mrakii	55
	Saccharomyces cerevisiae	68–78
Other Lower Eukaryotes		
	Hydra sp.	15
	Chlamydomonas reinhardtii	15.8
	Trypanosoma brucei (maxicircle)	22

Table 3. Continued

Group	Representative organisms	Estimated genome size (kb)[a]
	Tetrahymena sp.	40–50
	Paramecium sp.	40–50
Plants		150–2500
	Brassica oleracea	160
	Brassica hirta	208
	Brassica campestris	218
	Brassica nigra	231
	Raphanus sativa	242
	Chenopodium album	270
	Lycopersicon esculentum	270
	Zea mays	570
	Cucumis melo	2500

CHLOROPLAST GENOMES

Group	Representative organisms	Estimated genome size (kb)[a]
Angiosperms		120–217
	Pisum sativum	120
	Brassica oleracea	150
	Nicotiana tabacum	156
	Lycopersicon esculentum	158
	Spirodela sp.	180
Gymnosperms	*Ginkgo biloba*	158
Ferns	*Osmunda cinnamomea*	144
	Asplenium nidus	150
	Pteris vittata	150
	Adiantum capillus-veneris	153
Bryophytes	*Marchantia polymorpha*	121
	Sphaerocarpus donnellii	125
Green algae	*Codium fragile*	85
	Chlorella ellipsoidea	174
	Chlamydomonas reinhardtii	195
	Chlamydomonas smithii	199
	Chlamydomonas eugametos	243
	Chlamydomonas moewusii	292
	Acetabularia mediterranea	~2000
Other algae	*Dictyota dichotoma*	123
	Vaucheria sessilus	125
	Cyanophora paradoxa	127
	Euglena gracilis	130–152
	Phylaiella littorali	140
	Sphacelaria sp.	150
	Olisthodiscus luteus	154

BASAL BODY DNA

Group	Representative organisms	Estimated genome size (kb)[a]
	Chlamydomonas reinhardtii	6000–9000

[a] Data taken from various sources including refs 1, 4–6 and papers cited therein.

Genomes/Genes

2. PHYSICAL DATA FOR GENOMES

Table 4 gives melting temperatures, buoyant densities and estimated GC contents for a range of genomes.

Table 4. Physical data for representative genomes[a]

Organism	T_m (°C)	Buoyant density (g cm^{-3})	%GC
Bacteriophages			
T1	–	1.705	48.0
T2	83.0	1.700	34.6
T3	90.0	1.712	49.6
T4	84.0	1.698	34.5
T5	–	1.702	39.0
T6	83.0	1.707	34.5
T7	89.5	1.710	47.4
lambda	89.0	1.710	48.6
Viruses			
Adenoviruses	88.8–94.2	1.708–1.720	56.0–57.0
Herpes virus	97.0	1.727	–
Polyoma virus	89.2	1.709	–
Prokaryotes			
Agrobacterium tumefaciens	94.2–94.8	1.718–1.719	58.0–59.7
Anacystis nidulans	–	1.714	54.3
Bacillus amyloliquefaciens	87.1–87.6	1.707–1.708	44.0
Bacillus cereus	83.0	1.696	–
Bacillus licheniformis	88.6	1.705	46.9
Bacillus megaterium	85.0	1.697	–
Bacillus subtilis	86.3–87.1	1.705–1.706	42.6
Bacillus thuringiensis	83.5	1.695	–
Clostridium butyricum	82.1	–	37.4
Clostridium perfringens	80.5	1.691	–
Desulphovibrio desulphuricans	90.5	1.716–1.720	56.3–57.4
Diplococcus pneumoniae	85.5	1.701	–
Erwinia herbicola	92.1–92.2	1.714–1.715	–
Escherichia coli	90.5–91.9	1.710	51.0
Haemophilus aegyptius	86.0	1.698	–
Haemophilus influenzae	85.6	1.696	–
Klebsiella pneumoniae	92.5–92.9	1.712–1.715	–
Lactobacillus brevis	–	1.702–1.706	43.2–43.9
Lactobacillus casei	–	1.705–1.706	43.4–49.4
Lactobacillus lactis	–	1.710	48.2–48.3
Micrococcus luteus	95.1–98.9	–	65.4–74.3
Mycobacterium tuberculosis	–	1.724	65.0
Neisseria gonorrheae	89.5	1.710	–
Neisseria meningitidis	91.0	1.710	51.3
Pasteurella pestis	88.5	1.706	–
Proteus mirabilis	85.3	1.700	–
Pseudomonas aeruginosa	97.5–97.7	1.726–1.727	–
Pseudomonas denitrificans	92.9	1.716	–

Table 4. Continued

Organism	T_m (°C)	Buoyant density (g cm^{-3})	%GC
Pseudomonas fluorescens	94.5–95.3	1.721–1.724	59.5
Pseudomonas putida	93.4	1.719	–
Rhizobium japonicum	95.0	1.722	–
Salmonella typhimurium	91.8	1.712	–
Serratia marcescens	93.5–94.9	1.717–1.718	–
Shigella dysenteriae	90.5	1.710	–
Staphylococcus aureus	82.9–83.5	1.693	32.4–37.7
Streptomyces albus	100.5	1.730	72.3
Streptomyces griseolus	–	1.729	72.4
Streptomyces griseus	–	1.730	72.1
Vibrio cholerae	88.5–88.7	1.706	–
Xanthomonas campestris	97.3	1.727	–
Eukaryotic Microorganisms			
Dictyostelium discoideum	79.5	1.682	23.0
Euglena gracilis	90.0	1.706	49.9
Saccharomyces cerevisiae	85.2	1.699	41.0
Schizosaccharomyces pombe	87.0	–	44.8
Tetrahymena pyriformis	81.2	1.690	–
Invertebrates			
Drosophila melanogaster	86.5	1.702	39.1
Paracentrotus lividus	–	1.697	35.4
Strongylocentrotus purpuratus	–	1.699	36.6
Vertebrates			
Bos taurus	87.0	1.700	43.2
Canis familiaris	–	1.700	44.1
Felis domesticus	–	1.700	42.8
Gallus domesticus	87.5	1.700	45.0
Homo sapiens	86.5	1.699	40.3
Iguana iguana	–	1.702	43.9
Mus musculus	86.5	1.701	40.3
Rattus sp.	87.5	1.701	41.8
Salmo salar	87.5	1.703	43.5
Xenopus laevis	–	1.699	40.9
Plants			
Nicotiana tabacum	85.5	1.700	–
Triticum vulgare	88.5	1.702	45.8

[a] Data taken from refs 7 and 8 and papers cited therein.

3. THE GENETIC CODE

The three-letter and one-letter abbreviations for the 20 common amino acids are given in *Table 5*. The 'universal' genetic code is shown in *Table 6*. The known variations are detailed in *Table 7*, but note that in at least one case (maize CGG) the codon is thought to be corrected by RNA editing prior to translation (9,10).

Genomes/Genes

Table 5. Amino acid abbreviations

Amino acid	Three-letter abbreviation	One-letter abbreviation
Alanine	Ala	A
Arginine	Arg	R
Asparagine	Asn	N
Aspartic acid	Asp	D
Cysteine	Cys	C
Glutamic acid	Glu	E
Glutamine	Gln	Q
Glycine	Gly	G
Histidine	His	H
Isoleucine	Ile	I
Leucine	Leu	L
Lysine	Lys	K
Methionine	Met	M
Phenylalanine	Phe	F
Proline	Pro	P
Serine	Ser	S
Threonine	Thr	T
Tryptophan	Trp	W
Tyrosine	Tyr	Y
Valine	Val	V

Table 6. The genetic code

First position (5' end)	Second position				Third position (3' end)
	U	C	A	G	
U	Phe	Ser	Tyr	Cys	U
	Phe	Ser	Tyr	Cys	C
	Leu	Ser	Stop (Ochre)	Stop (Umber)	A
	Leu	Ser	Stop (Amber)	Trp	G
C	Leu	Pro	His	Arg	U
	Leu	Pro	His	Arg	C
	Leu	Pro	Gln	Arg	A
	Leu	Pro	Gln	Arg	G
A	Ile	Thr	Asn	Ser	U
	Ile	Thr	Asn	Ser	C
	Ile	Thr	Lys	Arg	A
	Met	Thr	Lys	Arg	G
G	Val	Ala	Asp	Gly	U
	Val	Ala	Asp	Gly	C
	Val	Ala	Glu	Gly	A
	Val	Ala	Glu	Gly	G

Table 7. Variations on the genetic code[a]

Organism	Genes	Codon	Universal meaning	Actual meaning	References
Candida cylindracea	all nuclear (?)	CUG	Leu	Ser	15
Drosophila sp.	all mitochondrial	UGA	stop	Trp	20
		AGA	Arg	Ser	
		AUA	Ile	Met (initiation)	
Escherichia coli	formate dehydrogenase	UGA	stop	selenocys	18
Fungi	all mitochondrial (?)	UGA	stop	Trp	22
Maize	all mitochondrial (?)	CGG	Arg	Trp	9, 10, 23
Mammals	all mitochondrial	UGA	stop	Trp	19
		AGA	Arg	stop	
		AGG	Arg	stop	
		AUA	Ile	Met (initiation)	
	glutathione peroxidase	UGA	stop	selenocys	16, 17
Mycoplasma	all (?)	UGA	stop	Trp	26
Protozoans	all nuclear (?)	UAA	stop	Gln	11–14
		UAG	stop	Gln	
Saccharomyces cerevisiae	all mitochondrial	UGA	stop	Trp	21
		CUN	Leu	Thr	
		AUA	Ile	Met (elongation)	

[a] For a recent review see ref. 24.

4. CODON USAGE

(Data kindly supplied by Toshimichi Ikemura.)

The data presented in *Table* 8 were computed from 11 415 genes contained in GenBank Release 62.0 (December 1989). A more comprehensive breakdown of codon usage in nuclear genes has been published (25).

Table 8. Frequency (per 1000) of codon usages for different organisms[a]

		Mammals									Other vertebrates			
		BOV	DOG	HAM	HUM	MUS	PIG	RAB	RAT	SHP	BMO	CHK	FSB	XEL
No. GENES		208	22	41	1490	640	60	91	682	32	23	210	76	111
ARG	CGA	5.5	3.9	5.1	5.4	6.1	5.8	4.4	5.8	5.0	2.8	4.2	3.9	6.8
	CGC	12.2	12.1	14.4	11.3	10.6	13.1	12.1	11.0	11.2	11.6	13.4	14.9	9.7
	CGG	11.2	9.9	10.2	10.4	9.1	10.9	10.4	8.8	8.0	2.4	8.7	4.6	6.0
	CGU	3.7	3.5	7.9	4.7	4.9	4.3	3.9	5.4	4.3	8.3	6.4	9.5	10.0
	AGA	9.9	6.6	10	9.9	11.2	9.8	9.2	10.5	6.7	9.9	9.8	15.8	14.9
	AGG	11.4	6.6	8.0	11.1	11.2	11.4	10.9	10.6	10.2	6.6	10.3	14.9	11.9
LEU	CUA	4.9	3.7	5.8	6.2	6.6	5.1	4.6	6.2	4.7	3.8	4.8	6.1	7.9
	CUC	21.2	33.1	17.3	19.9	20.0	21.7	22.7	19.4	27.3	24.6	17.2	22.1	12.6
	CUG	46.6	51.3	36.1	42.5	40.8	49.6	51.6	40.5	50.9	21.5	42.2	48.3	26.2
	CUU	10.6	7.2	12.3	10.7	11.5	8.5	9.6	11.7	11.9	10.4	10	8.6	15.7
	UUA	4.0	2.9	3.7	5.3	4.3	4.0	4.5	4.3	3.9	5.7	4.3	4.7	8.0
	UUG	9.6	8.8	10.1	11.0	11.4	10.2	9.1	11.5	9.7	13.0	9.6	10.5	13.4
SER	UCA	7.6	5.3	7.5	9.3	10.2	8.0	8.3	5.7		4.3	7.8	9.0	12.4
	UCC	17.6	22.6	18.9	17.7	17.9	21.4	18.5	16.6	23.8	15.4	15.6	22.4	14.3
	UCG	4.5	5.9	3.5	4.2	3.9	4.2	4.3	3.7	3.2	10.6	5.1	3.3	3.2
	UCU	11.2	8.6	14.1	13.2	15.1	11.5	13.4	13.2	8.3	8.3	12.3	21.3	18.0
	AGC	18.7	19.9	12.9	18.7	19.6	18.7	17.4	17.0	23.0	16.1	20.4	18.9	17.1
	AGU	8.6	5.9	8.6	9.4	9.6	8.1	7.7	9.2	5.4	7.8	7.8	8.5	12.5
THR	ACA	11.4	8.8	14.5	14.4	15.6	11.3	11.6	13.9	7.8	6.9	14.6	9.0	18.1
	ACC	21.1	23.2	23.1	23.0	23.0	25.3	23.4	20.7	30.9	21.8	20.7	26.7	17.0
	ACG	7.8	8.1	4.0	6.7	6.2	8.3	8.4	5.4	7.1	5.9	7.6	4.8	3.3
	ACU	9.6	7.9	13.1	12.7	13.5	9.3	10.5	13.1	10.5	11.4	12.7	10.7	19.1
PRO	CCA	12.0	10.7	13.3	14.6	15.8	11.7	13.1	13.6	11.9	4.5	12.8	12.0	16.9
	CCC	19.2	21.7	18.0	20.0	18.3	24.7	22.0	16.5	30.6	16.8	18.2	17.7	12.0
	CCG	7.9	5.9	2.7	6.6	5.8	8.8	8.0	5.2	7.9	3.5	6.2	7.9	3.4
	CCU	14.6	12.9	18.2	15.5	17.6	13.5	14.4	15.0	13.4	9.9	12.1	11.1	16.3
ALA	GCA	13.1	8.3	13.9	14.0	13.4	12.7	11.2	14.0	4.4	6.9	16.3	16.1	21.3
	GCC	35.8	39.7	26.3	29.1	26.5	34.0	31.7	29.2	32.7	29.8	27.5	40.4	20.9
	GCG	9.3	8.6	4.9	7.2	6.2	9.5	8.3	5.7	8.7	11.1	7.5	9.7	3.7
	GCU	19.1	15.1	22.7	19.6	20.2	19.0	15.9	22.3	16.2	26.7	23.3	19.0	26.4
GLY	GGA	16.2	12.0	19.0	17.1	17.9	15.0	16.4	15.5	11.0	33.6	15.4	14.6	22.8
	GGC	28.1	35.3	26.3	25.4	23.6	33.5	27.5	23.4	26.9	47.3	23.5	23.2	15.6
	GGG	19.2	14.0	13.3	17.3	15.8	18.7	17.9	14.9	12.9	9.0	15.3	9.0	11.0
	GGU	11.8	10.8	15.9	11.2	11.8	10.9	8.9	11.7	10.1	54.6	11.1	11.4	15.8
VAL	GUA	5.1	2.8	6.6	5.9	6.3	4.5	4.8	6.3	3.4	9.0	6.0	5.0	9.8
	GUC	18.4	22.6	16.7	16.3	17.1	18.5	18.1	16.8	14.5	26.5	16.2	22.6	13.2
	GUG	32.9	37.3	33.6	30.9	30.7	28.7	35.2	30.8	30.9	21.8	30.5	24.1	21.6
	GUU	9.9	7.7	12.7	10.4	10.2	9.0	10.3	9.8	5.6	11.6	11.4	7.8	17.0
LYS	AAA	21.6	17.3	21.8	22.2	21.6	20.0	19.2	23.0	18.4	9.2	23.5	16.5	35.2
	AAG	37.1	36.6	41.1	34.9	37.9	37.0	37.0	44.2	38.0	27.7	44.9	38.9	37.4
ASN	AAC	22.4	28.5	22.9	22.6	23.3	24.4	24.7	23.8	25.1	30.5	25.6	31.1	18.9
	AAU	12.5	13.4	15.3	16.6	16.2	14.2	14.0	14.8	11.5	10.2	15.3	7.9	19.3
GLN	CAA	9.7	6.8	9.2	11.1	10.9	7.8	8.5	10.4	9.2	8.3	10.0	9.5	15.2
	CAG	34.4	29.2	34.7	33.6	32.3	31.5	29.8	32.5	41.5	18.7	32.0	25.5	25.2
HIS	CAC	14.0	14.7	12.5	14.2	14.4	13.8	15.5	13.6	15.5	10.9	14.5	11.3	13.3
	CAU	7.5	4.8	8.3	9.3	9.3	6.5	7.0	8.1	6.3	2.1	7.9	6.1	12.3
GLU	GAA	24.4	16.7	26.9	26.8	25.5	22.5	22.8	28.9	16.9	17.5	29.3	19.1	39.1
	GAG	45.4	39.2	42.9	41.4	41.2	40.0	45.9	45.1	42.4	26.0	49.2	38.5	33.8
ASP	GAC	31.5	33.8	30.2	29.0	28.8	29.3	33.5	31.3	29.2	30.3	29.7	40.0	22.6
	GAU	19.2	18.2	24.6	21.7	21.3	19.7	20.8	22.5	13.1	14.0	25.4	16.9	29.6
TYR	UAC	20.3	26.7	18.3	18.8	19.0	20.6	19.9	18.5	18.9	41.6	19.1	18.4	15.0
	UAU	10.5	7.7	11.4	12.5	12.2	9.9	10.9	11.5	10.3	8.3	9.4	7.5	14.9
CYS	UGC	13.9	20.2	14.3	14.5	12.6	17.5	10.8	11.2	37.9	39.7	13.7	17.3	10.6
	UGU	9.4	7.7	7.7	9.9	10.7	10.5	6.7	8.8	15.8	24.6	7.0	13.4	10.6
PHE	UUC	26.6	30.7	23.7	22.6	23.5	22.1	32.4	25.5	28.6	29.3	20.8	25.9	17.4
	UUU	15.8	13.6	15.9	15.8	15.4	13.9	16.0	17.0	14.7	5.2	13.8	11.4	18.9
ILE	AUA	5.2	5.5	4.9	5.8	5.6	4.8	5.8	5.6	5.3	5.7	6.4	5.7	9.6
	AUC	25.8	42.1	27.7	24.3	24.6	27.5	30.4	27.6	28.7	32.6	28.3	29.5	18.7
	AUU	13.1	10.1	20.4	14.9	15.0	15.1	14.5	17.6	14.3	13.0	16.9	10.4	20.1
MET	AUG	22.8	23.7	28.4	22.3	22.8	20.2	23.7	25.8	24.6	23.9	24.8	34.5	23.7
TRP	UGG	12.6	17.3	8.9	13.8	13.6	15.3	12.7	12.3	9.4	13.5	11.3	8.5	9.9
TER	UAA	1	0.9	1.3	0.7	0.8	0.9	0.4	1	0.6	2.4	1.0	2.2	1.5
	UAG	0.6	1.3	0.5	0.5	0.5	0.5	0.4	0.6	1.0	1.7	0.4	2.4	0.5
	UGA	1.4	1.8	1.2	1.2	1.5	1.5	1.3	1.3	2.5	1.4	0.8	1.4	0.7
TOTAL		70093	5438	13557	601683	225107	20666	43123	242936	7737	4227	93409	12710	41059

Invertebrates			Eukaryotic microorganisms									Plants		
CEL	DRO	SUS	ASN	DDI	NEU	PFA	SCM	TRB	YSC	YSK	YSP	ATH	BLY	MZE
31	266	24	20	57	31	39	21	39	452	27	36	32	29	62
4.1	6.6	8.4	8.3	0.3	2.2	1.5	9.3	8.0	2.3	1.1	6.7	3.6	0.9	2.1
12.8	17.4	9.6	16.1	0.1	22.3	0.1	2.7	15.4	2.0	0.7	6.4	4.7	14.9	16.2
1.5	6.8	2.4	8.8	0.1	4.8	0.0	2.1	9.5	1.1	1	2.8	2.5	10.3	6.9
23.7	9.7	19.8	11.1	17.1	15.7	3.0	15.5	15.0	7.5	2.4	21.0	14.0	3.6	4.7
16.9	4.6	8.8	5.0	21.7	3.7	14.6	11.0	5.2	24.0	25.0	10.7	14.8	2.1	3.9
1	5.7	9.2	5.4	0.7	5.4	2.0	3.0	6.9	7.5	3.3	4.5	11.6	9.6	14.4
2.0	6.4	5.4	7.7	1.6	2.4	3.5	8.0	8.1	11.8	10.7	5.7	6.9	1.6	9.7
27.2	12.2	20.5	22.6	14.4	43.4	2.1	9.1	19.9	4.1	2.4	7.2	19.9	36.7	26.6
4.4	38.4	16.0	23.2	0.4	13.6	0.2	7.4	18.9	8.3	3.5	6.1	5.9	23.1	28.4
32.2	6.9	15.7	16.3	13.3	15.1	16.7	16.7	23.0	9.6	10.1	26.0	24.4	6.9	15.9
2.3	4.5	3.9	5.6	48.9	0.6	52.1	22.2	6.1	24.5	40.1	21.5	4.4	0.4	4.5
16.5	14.8	8.9	11.3	10.4	9.0	7.5	17.4	13.7	32.1	16.1	23.5	22.0	4.5	15.5
10.7	6.1	9.9	10.6	32.8	2.1	25.0	20.3	11.6	15.6	16.1	16.9	10.8	1.9	8.2
15.7	19.6	14.4	20.1	6.0	23.1	7.2	6.6	12.1	14.4	6.7	14.4	12.2	21.6	15.1
5.9	17.0	4.5	13.3	0.7	11.2	0.9	6.8	7.5	6.5	3.4	7.3	4.9	9.3	10.3
18.6	6.0	11.7	17.1	17.6	11.9	13.1	10.8	11.1	24.6	30.2	32.5	17.4	3.3	8.6
5.4	18.2	14.5	12.9	1.7	13.2	4.4	9.1	13.4	7.1	3.1	9.6	11.0	17.6	16.8
3.7	9.1	9.9	7.7	12.9	3.1	20.5	17.6	13.2	11.7	17.3	12.2	8.7	3.7	4.7
6.9	9.0	12.2	11.6	19.6	5.4	22.8	22.7	17.8	15.6	18.0	10.7	14.8	4.5	7.1
20.2	24.9	28.7	22.0	15.2	34.9	5.0	8.5	13.1	13.9	7.1	13.9	19.4	23.6	21.2
3.3	13.0	5.7	10.0	0.4	7.5	2.0	5.7	16.4	6.7	3.0	4.5	5.0	21.1	10.4
16.4	8.1	10.3	16.1	31.2	12.3	16.8	19.1	11.7	22.0	23.5	25.7	18.4	3.7	8.2
34.0	11.9	19.9	12.9	36.1	4.6	29.7	17.1	13.3	21.4	14.9	10.5	16.6	11.2	17.3
2.4	19.5	12.5	18.2	0.3	25.9	3.0	6.8	12.0	5.9	2.0	11.1	8.2	16.6	16.2
2.8	16.0	4.9	12.9	0.3	8.7	0.5	3.8	9.7	4.1	2.2	4.1	7.2	17.2	19.1
4.1	5.8	13.3	15.0	2.5	11.7	10.6	11.7	10.8	12.8	15.5	23.4	16.5	5.7	13.6
10.3	11.0	20.6	16.5	18.1	4.7	29.9	14.4	23.2	15.3	16.2	12.4	14.8	7.7	16.8
29.4	39.6	38.0	30.9	12.3	52.4	7.4	7.8	17.7	15.5	5.7	13.9	19.7	53.8	35.3
2.7	13.1	6.9	15.6	0.2	10.5	0.6	4.4	18.8	5.1	3.1	4.0	8.0	29.1	27.8
37.6	15.3	29.0	24.8	23.9	27.6	19.2	20.5	21.6	28.3	22.3	36.2	39.8	10.1	28.2
48.6	19.5	37.7	13.7	5.2	6.3	32.0	27.7	15.3	8.9	18.0	15.6	34.1	7.9	9.0
2.6	31.2	18.0	25.3	1.8	36.7	2.0	21.4	18.6	8.9	3.1	8.0	10.8	61.0	30.1
1.5	4.5	7.2	11.4	0.9	3.7	1.8	4.9	10.7	5.1	2.2	3.7	7.5	20.0	14.0
8.3	16.7	23.9	20.1	60.5	39.3	22.3	60.1	30.5	34.9	28.0	31.3	29.3	10.7	14.5
3.5	4.9	10.0	4.8	13.8	2.0	21.3	10.8	12.2	10	18.0	11.8	4.3	2.6	4.8
20.2	15.9	26.0	25.7	10.7	43.0	2.7	11.6	11.2	14.9	6.3	13.7	19.6	31.8	23.8
5.7	27.6	13.8	14.0	1.3	8.0	2.4	10.4	26.3	9.5	5.2	7.0	17.9	32.0	30.5
23.9	11.0	16.1	18.8	36.9	19.1	19.7	23.1	24.2	26.6	23.1	30.9	29.6	7.3	13.2
17.4	14.4	15.2	14.2	55.2	3.4	91.2	47.0	23.1	37.7	66.9	35.2	18.6	6.2	8.4
59.2	42.0	57.9	36.4	13.7	51.3	15.6	18.8	35.4	35.2	21.2	26.8	38.2	47.1	36.1
30.2	27.8	26.2	29.7	13.8	34.3	23.5	12.1	24.6	25.8	19.1	21.4	25.3	37.1	27.2
18.9	21.1	8.4	11.1	41.8	5.1	80.7	33.7	18.0	31.4	52.3	32.9	11.0	6.2	7.4
41.6	13.3	12.3	14.6	44.3	6.1	25.4	19.5	14.7	29.8	19.6	27.2	17.4	11.6	27.0
15.8	37.8	23.6	26.9	0.7	30.5	2.4	7.6	19.3	10.4	5.6	8.4	19.5	25.3	34.0
10.5	17.1	10.8	15.9	6.0	18.1	11.3	8.7	14.5	8.2	4.8	6.2	11.6	17.9	15.1
10.8	10.9	10.6	9.6	13.5	5.2	20.8	22.9	10	12.3	13.4	16.1	7.8	3.5	8.2
44.5	17.3	22.8	21.6	55.2	8.7	75.7	35.8	32.3	48.9	54.4	40.9	29.8	7.0	12.6
49.3	44.8	30.1	37.0	7.8	48.9	7.5	13.6	34.1	16.9	10.4	22.2	35.3	39.1	38.9
23.0	25.6	28.2	28.1	6.8	33.4	7.5	17.1	25.4	22.3	11.9	16.5	25.2	54.6	27.4
36.0	26.8	25.8	25.8	48.9	19.7	58.3	48.1	28.8	37.0	50.8	39.6	27.9	7.5	13.0
13.9	20.2	21.2	20.5	10.6	25.3	5.6	16.9	16.4	16.5	12.5	14.1	20.7	28.7	22.4
7.7	10.3	7.6	9.8	23.1	4.5	33.2	36.8	11.3	16.6	42.7	19.9	8.0	2.9	5.7
9.4	15.6	12.6	9.0	1.9	8.3	2.1	6.4	9.5	3.7	3.6	5.1	9.8	21.4	14.5
9.7	6.0	8.7	4.1	15.4	1.3	12.6	20.3	6.6	7.6	14.7	9.0	7.7	4.8	4.9
23.2	23.5	21.0	22.7	18.8	31.4	9.7	16.9	18.6	20.0	16.3	13.7	29.3	32.0	31.3
8.5	11.5	9.6	9.8	21.4	6.1	23.1	20.5	18.3	23.2	31.6	27.0	13.2	6.2	9.8
1.6	6.9	4.3	4.4	5.6	1.4	22.6	17.1	8.9	12.8	44.7	8.3	6.2	4.0	5.5
28.7	26.6	37.6	31.5	20.9	34.9	3.5	11.4	16.1	18.4	10	13.8	29.8	38.0	26.3
21.6	15.6	11.5	16.3	47.8	16.7	26.4	28.0	21.6	31.1	26.3	34.5	23.3	6.4	11.8
21.6	24.3	29.4	23.2	21.9	22.7	16.0	23.1	30.1	21.5	21.8	23.1	29.3	26.8	25.0
6.7	10.3	13.0	12.7	10.6	13.3	3.6	5.7	7.2	10.5	12.7	8.4	11.3	21.6	11.0
1.1	1.1	2.0	0.8	2.6	1.7	1.2	2.3	1.1	1.1	1.4	1.6	1.4	1.5	0.4
0.3	0.6	0.9	0.8	0.2	0.5	0.3	0.2	0.3	0.4	0.0	0.4	0.4	0.4	1.4
0.1	0.4	0.3	0.4	0.1	0.4	0.2	1.5	0.1	0.6	0.3	0.2	1	1.5	1.5
20742	129868	7375	10455	19742	11754	23756	5277	24830	223055	15260	15865	11521	8534	19217

Table 8. Continued

No. GENES		Plants					Prokaryotes								
		PEA	POT	SOY	TOM	WHT	ANA	ATU	AVI	BME	BST	BSU	BTH	CHT	CLO
		25	24	44	30	34	25	32	37	21	23	150	40	23	24
ARG	CGA	3.0	5.6	5.6	2.2	2.3	2.4	11.8	0.3	3.4	4.6	3.7	8.6	6.4	0.5
	CGC	3.6	2.1	8.1	3.0	7.5	13.2	23.6	38.7	12.2	21.9	8.9	2.8	7.0	0.7
	CGG	0.7	0.4	2.1	1.0	4.6	8.7	15.4	7.5	0.4	12.0	5.5	2.0	2.9	0.7
	CGU	9.5	10.1	7.4	8.8	1.1	19.7	15.0	12.5	21.1	6.0	8.9	11.7	9.0	2.6
	AGA	18.0	7.0	13.9	13.9	4.1	6.8	8.2	0.3	3.9	3.2	12.4	21.9	17.3	24.5
	AGG	10.2	9.8	12.5	7.8	7.1	1.0	7.9	1.5	0.2	2.9	3.8	4.8	5.7	5.8
LEU	CUA	5.4	7.8	6.9	5.0	12.1	16.0	8.8	0.8	10.8	4.6	5.0	10.9	12.7	9.4
	CUC	11.7	10.7	16.7	10.9	18.6	11.0	21.3	20.7	1.9	14.7	9.6	3.6	6.4	1.5
	CUG	5.6	4.3	10.1	2.2	15.5	16.4	28.1	67.4	5.3	14.4	21.1	5.0	5.9	3.6
	CUU	21.2	25.0	24.4	23.7	6.5	5.7	20.6	1.4	24.8	12.2	23.5	16.8	17.1	17.0
	UUA	9.3	17.8	8.5	7.9	1.8	23.7	4.7	0.2	38.6	12.5	19.1	40.9	31.7	39.4
	UUG	20.5	21.0	21.2	26.3	15.3	28.9	16.5	5.1	6.4	17.4	12.7	9.5	23.5	10.6
SER	UCA	15.9	17.0	13.4	18.7	14.6	8.2	10.7	0.1	18.1	6.0	14.5	15.9	12.0	17.7
	UCC	13.1	8.7	9.2	12.1	10.1	9.6	13.7	19.1	1.6	7.6	8.5	7.8	13.4	4.4
	UCG	2.3	3.4	3.8	2.9	9.8	2.1	15.1	10.7	3.2	12.6	6.3	4.9	6.6	2.6
	UCU	19.8	14.8	17.5	20.0	14.8	16.7	8.2	1	14.2	5.2	15.3	20.5	34.7	14.9
	AGC	11.7	9.2	15.6	14.2	12.8	11.7	17.0	19.6	13.8	13.5	14.4	6.1	9.4	8.6
	AGU	13.5	15.0	13.6	10.6	2.9	7.3	8.0	3.6	7.8	5.5	6.1	18.6	11.3	13.4
THR	ACA	17.5	20.4	14.2	15.1	4.6	18.8	8.7	0.3	29.2	12.2	23.9	28.4	22.0	26.2
	ACC	12.1	8.4	13.6	11.5	15.9	25.6	17.9	37.9	2.3	15.2	7.6	8.9	8.3	7.6
	ACG	3.3	2.8	2.9	2.4	4.5	2.4	16.3	4.4	14.3	24.9	13.6	12.5	5.6	2.2
	ACU	23.9	33.5	17.5	18.5	11.8	12.0	7.6	2.9	9.9	6.6	9.2	19.8	24.7	20.9
PRO	CCA	24.7	20.6	30.5	28.7	71.2	9.6	10.9	1	18.2	8.2	6.7	18.3	9.3	20.0
	CCC	5.2	8.0	9.1	9.5	11.1	10.8	10.7	17.0	1.1	3.6	2.7	1.9	3.6	2.5
	CCG	4.9	2.7	4.2	1.3	19.4	1.6	18.9	27.1	6.9	24.2	15.0	6.7	2.7	4.3
	CCU	20.9	16.1	22.3	17.9	10.3	15.5	10.5	1.6	13.3	6.0	10.3	13.3	16.4	13.0
ALA	GCA	20.5	26.8	20.2	23.7	11.2	28.4	20.2	3.5	39.7	9.9	22.6	20.5	22.7	27.3
	GCC	11.5	9.3	15.9	18.7	19.5	17.4	29.6	65.0	6.9	29.5	12.7	7.4	8.9	5.2
	GCG	4.7	4.1	4.5	2.4	13.8	11.8	30.1	22.2	15.8	29.2	19.3	8.4	8.7	4.4
	GCU	39.3	40.1	23.9	42.8	9.6	46.0	19.3	7.3	31.7	13.8	19.4	18.5	38.7	25.9
GLY	GGA	33.0	35.2	23.3	36.2	25.9	8.5	11.7	3.1	22.7	17.8	21.8	25.8	25.9	36.7
	GGC	10.6	15.5	12.0	10.1	28.0	15.1	25.4	57.1	18.2	36.1	23.5	7.0	7.0	9.2
	GGG	7.1	6.5	11.3	7.9	28.5	4.4	12.5	6.0	3.4	19.7	9.4	11.7	10.4	3.2
	GGU	27.6	25.3	21.7	24.3	9.6	42.8	13.9	15.2	32.6	10.5	14.1	15.8	14.2	24.9
VAL	GUA	10.2	10.7	8.1	9.8	4.4	24.0	8.6	2.7	33.5	9.1	14.6	27.1	16.9	25.2
	GUC	8.7	8.3	8.8	10.8	14.8	9.9	20.2	30.7	6.9	28.5	17.1	7.1	7.8	2.6
	GUG	16.9	14.8	25.8	17.5	12.9	8.2	18.6	35.2	10.6	23.8	16.5	10.8	9.7	6.4
	GUU	35.5	33.8	25.2	32.7	11.8	24.5	17.3	5.7	25.8	14.5	19.2	19.9	22.1	26.4
LYS	AAA	30.2	32.3	23.1	24.4	4.5	39.9	16.9	6.8	52.9	42.7	52.7	35.7	53.4	56.0
	AAG	38.3	29.6	35.7	35.8	17.4	14.3	23.8	41.1	12.2	15.7	20.1	8.0	19.9	14.9
ASN	AAC	24.3	13.9	29.1	21.7	14.2	27.2	17.6	26.2	23.5	26.6	20.9	18.3	18.3	20.2
	AAU	24.5	34.1	20.9	28.4	6.7	11.1	16.7	5.4	18.9	15.7	22.8	54.5	31.1	48.3
GLN	CAA	20.9	22.4	27.4	24.7	171.8	41.6	15.2	4.8	38.6	24.4	21.8	37.6	28.0	21.5
	CAG	9.2	7.7	20.3	10.2	79.4	13.8	20.7	28.7	11.7	10.9	18.9	7.6	11.2	6.6
HIS	CAC	8.2	6.2	9.3	4.9	8.2	8.2	9.7	17.6	7.3	8.6	8.0	2.7	4.6	3.7
	CAU	10.7	14.1	12.2	6.7	7.1	4.4	12.0	5.7	11.9	14.5	14.5	16.4	8.6	11.2
GLU	GAA	38.6	37.0	35.9	30.0	7.8	56.0	29.4	34.3	58.8	41.9	53.1	48.5	41.2	51.9
	GAG	31.6	21.3	35.5	28.0	19.7	14.1	34.3	44.8	17.4	17.8	23.7	17.0	17.2	11.0
ASP	GAC	17.5	14.2	17.2	13.0	13.0	25.1	26.1	43.5	18.2	28.2	22.0	9.9	17.3	15.4
	GAU	30.9	34.7	30.6	25.0	4.0	25.9	27.8	15.6	33.1	31.8	34.3	44.0	34.3	48.0
TYR	UAC	15.8	19.7	15.0	23.0	24.5	21.9	8.9	19.3	12.4	20.0	11.7	9.9	13.2	10.4
	UAU	18.4	19.1	15.1	12.8	12.5	12.7	12.3	6.2	17.0	22.7	21.1	38.5	15.3	39.7
CYS	UGC	5.2	11.1	10.3	12.1	14.8	5.2	6.9	15.4	1.2	5.1	4.0	2.9	5.5	3.0
	UGU	8.4	15.0	5.6	7.8	4.9	4.4	2.8	2.1	0.9	1.9	3.4	7.0	7.1	9.6
PHE	UUC	21.3	14.1	24.7	21.0	14.1	17.1	20.6	29.2	12.9	15.7	14.0	9.2	19.1	10.1
	UUU	21.4	29.5	19.7	25.1	15.0	18.3	19.3	2.1	21.4	24.2	25.5	34.4	28.4	26.5
ILE	AUA	11.3	9.9	12.2	8.2	5.4	5.4	8.9	0.6	1.8	3.6	8.0	19.7	19.1	44.7
	AUC	20.6	12.6	14.1	16.9	19.7	29.1	28.0	49.5	21.4	28.6	27.1	11.5	20.3	8.0
	AUU	24.2	29.5	23.0	30.6	10.7	27.0	19.4	4.2	39.1	33.5	34.7	34.2	33.5	28.7
MET	AUG	18.3	25.8	20.0	29.0	13.8	17.9	25.7	28.5	26.0	24.2	26.5	15.0	20.9	26.0
TRP	UGG	13.7	10.1	12.0	19.5	7.9	8.9	10.0	9.4	7.1	18.8	9.0	14.1	10.5	10.9
TER	UAA	2.2	2.2	1.3	2.7	0.4	2.6	0.7	0.7	3.0	1.3	2.1	0.9	1.6	1.7
	UAG	0.4	0.3	1.1	0.3	0.4	1.2	0.7	0.1	0.5	0.5	0.5	0.2	0.5	0.2
	UGA	0.8	1.0	0.8	1.3	2.2	0.5	1.8	2.1	0.2	0.9	0.8	0.2	1	0.2
	TOTAL	7277	6753	14032	6964	11404	5745	9971	13040	5648	8497	44057	30410	7322	12030

| ECO | FDI | HAL | KPN | PSE | RCA | RHL | RHM | SSP | STA | STM | STR | STY | SYN | VIB |
941	26	20	62	105	21	22	33	31	46	53	48	99	23	24
3.0	2.9	5.7	2.4	4.4	1.6	8.6	4.7	2.9	3.6	4.5	3.0	3.4	3.2	5.1
21.7	18.4	20.9	39.0	29.7	36.9	30.2	33.0	3.1	3.3	34.7	6.1	24.8	19.2	10.3
4.4	7.0	11.9	12.5	9.6	18.7	14.5	13.6	1.6	0.8	29.9	1.7	5.1	10.9	2.6
25.5	22.6	7.2	8.8	9.6	7.6	8.3	10.1	4.9	8.9	7.2	16.2	22.8	15.2	15.3
2.0	6.6	2.3	1.5	2.6	0.7	4.0	2.7	13.5	14.0	1.6	7.3	2.2	0.7	5.5
1.2	1.2	1.0	1.7	3.3	1.0	5.0	6.1	13.3	1.9	6.0	1.7	1.2	0.3	1.5
2.9	14.1	3.1	3.3	3.1	0.0	8.0	4.1	21.7	7.4	0.5	8.4	3.9	4.6	11.6
9.6	7.2	35.2	17.6	18.4	19.4	26.9	33.1	7.8	1.9	34.2	6.8	10.4	25.5	8.9
54.8	9.3	28.1	66.5	46.8	59.4	31.1	34.0	11.1	2.6	51.9	6.7	55.8	39.0	17.5
9.9	7.0	5.7	6.8	7.4	12.3	18.9	17.4	15.4	10.4	2.2	19.2	10.1	7.8	12.9
10.2	26.3	1.8	4.2	3.1	0.0	4.6	2.2	35.8	40.9	0.3	23.6	12.3	4.6	17.5
11.2	26.9	9.4	4.5	13.5	3.6	18.2	10.5	14.5	10.9	3.2	19.3	12.1	30.9	11.3
6.4	6.2	3.1	2.7	5.0	0.5	5.9	2.7	16.4	19.3	1.6	16.6	5.9	3.3	11.7
9.6	15.3	20.1	10.7	12.0	7.3	10.2	15.1	5.9	2.1	22.5	3.3	11.5	9.7	8.1
7.8	1.4	23.1	11.3	14.5	23.6	14.9	20.9	5.7	3.0	17.7	2.9	8.7	14.4	5.9
10.5	22.8	1.8	3.9	5.0	1.0	5.6	4.2	8.8	14.0	0.8	16.1	8.5	10.7	17.5
14.9	21.9	18.0	23.6	24.6	10.7	13.8	15.4	12.9	8.9	16.0	8.1	17.2	16.1	14.7
7.2	9.3	4.3	2.2	9.6	0.7	3.6	2.7	15.6	15.3	1.7	13.7	6.5	6.2	11.8
6.4	12.4	6.1	2.4	5.5	0.9	7.3	6.2	17.4	33.6	1.7	25.1	4.7	3.2	13.3
24.2	26.7	34.4	31.6	29.4	31.4	19.8	24.9	8.2	3.2	43.2	12.8	25.6	28.3	16.8
12.2	1.4	29.3	14.7	13.6	14.2	18.3	18.1	17.4	7.9	17.2	7.5	17.1	9.6	9.7
10.6	16.6	6.6	2.8	8.0	0.9	9.2	6.8	19.2	18.8	1.3	26.3	7.6	11.6	13.8
8.1	5.6	4.9	3.6	5.2	0.7	10.1	7.1	7.4	18.3	1	13.2	6.0	5.4	14.2
4.0	10.3	9.4	11.4	12.4	20.1	10.8	9.4	7.6	1.5	26.1	2.0	5.4	14.8	5.8
24.1	1.7	17.0	32.5	21.6	34.0	20.9	22.1	12.3	4.8	28.8	3.8	25.6	13.8	8.4
6.4	14.9	3.9	3.8	6.1	1.9	7.1	4.4	8.6	13.6	2.7	11.8	7.8	8.9	12.4
20.4	39.7	12.9	5.1	14.2	1.4	19.1	18.1	19.2	25.8	4.9	27.4	11.4	21.2	19.7
23.0	9.3	45.7	46.4	48.8	56.8	39.2	48.5	7.8	4.5	69.4	11.9	27.7	29.7	18.2
32.9	8.5	40.4	48.3	30.8	47.3	30.0	33.7	15.8	5.8	43.7	8.8	39.5	26.6	16.6
17.8	69.9	10.0	7.3	15.0	6.6	17.6	15.2	14.5	19.4	3.4	39.4	13.8	32.2	17.4
6.6	9.5	7.4	6.8	7.5	1.6	15.2	10.3	16.8	19.1	8.7	14.5	6.3	3.9	6.2
30.5	17.0	43.0	46.2	56.2	57.3	40.2	42.0	12.5	11.3	59.8	10.7	34.6	36.9	23.3
9.4	2.9	17.4	12.6	9.9	14.4	9.9	9.9	15.6	5.6	17.4	7.5	10.2	9.3	9.9
28.6	43.2	13.9	9.1	18.2	8.8	12.6	14.2	16.4	28.7	8.4	29.1	19.1	34.7	24.1
12.2	19.2	3.9	5.3	7.0	0.0	7.0	4.7	26.0	21.0	1.7	14.1	11.6	6.1	12.7
14.0	5.6	46.7	20.3	23.1	34.8	30.0	39.9	13.1	5.2	43.3	12.0	19.0	21.2	13.3
24.7	7.4	29.1	26.1	26.8	33.6	20.9	21.2	14.3	7.7	31.4	10.7	26.4	22.8	17.7
20.8	35.4	8.6	8.7	9.4	6.6	14.6	9.5	22.5	23.2	2.0	29.4	14.5	19.6	19.5
37.4	22.1	9.2	19.2	11.4	11.6	12.4	8.0	34.0	75.5	1.4	56.4	36.0	17.3	38.8
12.1	14.3	13.5	10.2	24.1	27.9	24.9	28.1	39.5	15.9	20.7	26.0	11.7	15.5	19.3
24.6	25.2	34.2	22.5	24.2	24.3	16.9	20.7	27.2	23.0	26.1	20.8	22.5	24.4	29.1
15.9	14.9	2.7	12.0	11.1	3.3	13.6	10.9	23.5	49.3	1.4	38.2	17.9	8.0	33.1
12.9	36.6	4.9	8.2	10.5	5.0	10.2	6.2	22.1	34.0	2.5	29.9	12.3	27.7	24.3
30.2	12.0	26.6	44.1	27.3	23.6	19.1	26.4	10.4	5.7	22.7	10	32.9	22.1	17.1
10.8	3.9	12.7	13.3	14.8	12.1	12.1	12.1	3.7	5.6	21.2	5.5	9.2	15.5	11.2
11.3	5.4	2.5	12.1	9.5	8.1	11.5	8.8	7.6	18.3	2.8	9.9	10.9	5.1	16.8
44.3	46.6	24.2	27.5	26.6	32.6	25.7	29.7	34.4	61.2	10.4	55.0	41.4	30.5	38.8
19.5	7.9	48.3	34.3	29.9	31.0	30.0	35.9	25.0	14.6	47.8	15.6	21.7	19.9	19.2
22.5	19.0	72.7	25.8	35.0	33.1	29.7	33.1	16.0	13.2	58.6	19.9	22.4	24.5	24.9
31.8	33.1	11.3	28.5	20.3	20.6	24.4	22.5	18.6	47.4	3.4	44.6	33.5	21.9	39.6
13.5	29.4	27.2	14.3	20.1	17.7	12.1	7.6	23.1	9.2	24.2	15.1	11.9	21.5	16.8
15.0	12.8	2.9	10.9	11.5	9.9	9.9	9.1	35.0	36.7	1.4	28.5	15.0	7.8	23.9
6.1	6.2	0.8	12.6	9.5	10.7	9.9	8.0	3.5	1.6	7.4	0.9	5.3	3.5	6.2
4.7	3.5	2.7	3.2	2.5	1.0	3.3	2.2	3.9	3.8	0.9	2.3	3.9	2.2	9.5
18.3	14.5	28.1	15.5	24.1	34.1	24.6	26.8	23.8	11.8	27.9	13.3	15.3	30.3	18.9
18.6	16.3	1.8	19.9	8.4	5.5	11.4	7.9	20.1	28.7	0.5	24.4	20.2	20.5	26.0
3.8	3.1	0.4	3.4	3.7	0.0	6.5	7.0	44.2	19.3	1.0	6.3	4.2	0.0	10.5
27.6	28.6	44.9	30.2	31.3	53.7	39.7	41.7	19.0	10.7	27.7	17.2	27.3	35.0	20.7
27.0	27.7	1.8	17.8	11.9	4.3	15.7	10.9	24.4	34.7	0.8	31.2	27.7	22.1	26.0
26.6	20.7	16.0	28.7	23.5	30.0	27.8	25.4	18.4	19.6	15.4	18.1	26.3	19.0	25.9
12.7	6.8	13.1	15.4	14.7	18.0	13.0	9.4	12.7	9.2	17.9	10.6	11.2	20.3	17.9
2.0	2.1	1.6	1.2	0.7	1.0	0.9	0.6	1.8	1.9	0.1	1.4	1.9	0.9	2.5
0.2	2.9	0.4	0.6	0.5	0.0	0.3	0.2	1.0	0.4	0.8	0.3	0.1	1.9	0.1
0.8	0.4	2.0	1	1.5	2.6	2.1	1.9	3.5	0.4	2.3	0.2	0.8	0.6	0.4
316835	4833	4882	22062	38679	5773	6758	12068	4884	16465	16582	24218	34520	6891	8116

Genomes/Genes

GENOMES AND GENES

Table 8. Continued

		Plasmids/Transposons			Viruses									
No. GENES		FPL 26	TIP 40	TRN 47	ADB 28	FLA 127	FLB 47	HIV 156	HPB 38	HS1 121	HS4 81	HS5 53	MCV 32	NDV 45
ARG	CGA	5.3	11.9	6.3	6.7	6.2	4.1	6.0	5.2	6.3	6.0	9.2	7.7	5.3
	CGC	14.3	18.7	24.4	32.4	3.1	1	2.4	10.2	39.2	13.4	22.1	11.1	6.5
	CGG	13.2	13.6	12.4	10.9	5.7	1.9	2.8	9.0	25.1	13.3	13.1	6.1	7.5
	CGU	19.7	12.1	12.4	7.9	2.2	2.8	0.8	8.5	5.3	6.1	14.5	18.4	3.2
	AGA	9.9	8.7	8.0	7.8	32.8	30.6	40.0	19.3	1.5	9.0	6.3	10.9	11.5
	AGG	6.7	8.4	6.9	9.7	18.5	13.6	17.0	11.7	4.5	15.2	5.0	7.5	11.9
LEU	CUA	3.2	9.2	6.8	10.7	11.9	17.8	15.0	21.1	5.4	9.9	8.8	9.0	13.1
	CUC	12.0	20.2	13.8	12.5	13.2	9.2	11.5	25.8	22.5	25.0	22.2	17.9	16.1
	CUG	48.8	23.4	43.8	34.1	16.4	12.7	15.5	20.8	52.6	40.5	37.6	12.5	15.7
	CUU	15.7	21.7	13.9	16.8	15.4	14.8	11.1	21.8	6.7	11.2	8.2	17.7	18.5
	UUA	12.0	7.2	15.7	2.3	7.0	17.7	21.1	15.6	1.6	4.6	5.8	18.2	19.8
	UUG	7.2	20.2	16.4	14.9	12.4	17.6	13.9	24.3	8.8	10.5	15.6	22.3	16.4
SER	UCA	14.1	11.0	9.3	5.9	18.7	18.5	12.8	21.4	1.9	10.7	6.3	16.3	21.9
	UCC	10.2	12.9	7.8	16.6	10.2	8.5	7.9	24.3	18.4	19.0	18.0	20.9	10.5
	UCG	5.1	12.9	9.1	8.2	4.4	2.1	2.3	10.8	15.5	8.5	17.2	11.3	7.1
	UCU	13.6	10.6	8.4	10.4	14.6	13.2	7.6	29.7	4.0	12.2	13.6	26.5	26.7
	AGC	10.9	15.8	13.9	21.3	15.7	10.3	15.1	6.0	17.2	18.2	18.3	8.1	15.6
	AGU	14.1	9.0	11.4	6.6	15.1	12.2	15.0	10.8	2.5	8.3	8.5	14.6	13.5
THR	ACA	12.0	11.7	9.6	10.2	25.0	27.6	28.7	15.2	3.8	13.6	7.3	13.1	35.1
	ACC	14.3	15.7	23.8	30.2	14.0	15.3	14.5	21.3	30.7	30.0	26.9	14.8	22.3
	ACG	15.0	13.9	13.3	8.3	4.5	2.6	4.4	6.6	23.0	13.4	23.0	9.6	4.6
	ACU	10.9	10.6	10.9	10.3	19.5	16.0	15.3	22.0	2.6	10	9.6	23.6	25.8
PRO	CCA	6.0	13.3	7.8	13.0	13.5	19.0	24.1	26.4	6.8	22.0	7.5	8.8	10.8
	CCC	7.4	10.6	7.5	27.6	6.6	7.6	11.2	22.1	44.3	32.2	21.1	14.2	11.1
	CCG	18.7	15.8	14.9	17.9	4.1	2.9	4.1	10.1	27.5	13.6	19.1	11.9	7.1
	CCU	9.0	11.6	7.2	14.7	12.1	12.9	15.5	31.6	4.9	10	11.1	16.7	15.8
ALA	GCA	22.4	23.5	15.8	13.9	25.0	27.8	28.2	13.0	6.3	16.0	7.0	11.3	32.0
	GCC	28.7	27.5	33.3	32.9	11.1	11.9	11.5	13.8	67.3	45.4	32.0	15.2	17.0
	GCG	17.8	22.0	23.2	22.2	4.9	3.1	4.3	5.5	43.5	13.5	19.0	12.5	7.8
	GCU	18.7	21.5	15.6	13.9	16.0	18.7	16.0	16.6	5.8	12.8	13.2	33.2	17.0
GLY	GGA	9.3	18.2	10.1	15.0	31.6	37.6	32.1	23.8	7.5	18.6	9.3	9.8	13.9
	GGC	15.3	21.4	25.6	22.7	9.6	6.4	12.0	12.5	32.6	24.9	23.8	9.2	12.0
	GGG	12.7	11.6	10.8	12.0	20.2	15.7	17.8	16.1	33.4	22.6	10.3	10.6	24.9
	GGU	16.0	17.1	16.0	11.8	12.0	10.3	7.2	8.5	5.6	12.4	14.3	26.5	16.3
VAL	GUA	10.9	8.1	9.3	9.0	12.8	18.4	25.7	8.5	3.7	6.1	11.1	8.1	23.5
	GUC	13.9	13.8	16.6	11.4	10.2	6.6	7.6	10.9	25.8	19.4	20.7	18.1	17.9
	GUG	20.6	16.2	20.0	31.0	18.4	15.9	15.0	15.3	34.0	29.0	35.4	13.6	16.6
	GUU	25.0	16.7	15.5	11.0	13.5	12.9	7.5	18.4	6.4	9.0	9.6	27.5	13.8
LYS	AAA	32.6	19.9	33.7	15.8	30.9	48.4	31.5	14.1	5.1	8.1	14.7	24.2	18.7
	AAG	15.5	19.2	25.0	21.0	20.1	27.9	24.1	6.9	14.7	20.7	19.7	26.7	23.4
ASN	AAC	15.5	17.6	17.3	32.6	23.7	26.7	17.7	12.5	21.9	19.1	24.4	15.6	20.4
	AAU	20.8	17.0	18.1	12.0	35.2	31.5	33.6	18.8	2.6	12.5	10.0	19.2	29.8
GLN	CAA	7.6	20.7	17.6	13.2	21.8	23.6	27.4	21.3	5.1	11.8	12.4	15.4	19.7
	CAG	36.8	19.3	25.4	29.7	18.2	8.1	26.5	12.7	24.9	27.7	25.0	14.2	18.2
HIS	CAC	9.0	9.2	10.5	16.4	6.2	8.1	8.6	10.5	21.1	16.2	18.7	9.8	6.7
	CAU	12.7	14.8	13.2	5.8	10.8	9.4	15.5	12.7	4.5	9.1	8.4	10.6	8.7
GLU	GAA	35.8	32.0	33.6	25.0	38.1	50.9	41.0	13.4	9.5	17.2	22.2	22.9	20.7
	GAG	22.4	30.7	27.3	42.4	29.5	22.6	24.4	14.0	39.5	35.3	31.7	27.8	19.1
ASP	GAC	18.7	25.6	22.3	35.4	22.9	22.2	17.3	17.9	44.3	28.5	31.6	25.5	21.0
	GAU	32.1	31.4	27.4	18.4	25.3	30.6	19.1	7.2	9.3	16.7	16.7	35.9	26.5
TYR	UAC	12.5	9.9	13.0	21.7	12.9	14.6	8.2	9.8	22.3	18.5	26.0	17.5	18.4
	UAU	19.0	15.8	15.4	8.0	15.1	13.2	17.7	17.1	5.3	9.1	10.4	20.5	19.6
CYS	UGC	3.2	9.7	5.6	11.4	11.9	10.3	9.0	14.9	12.0	12.6	12.9	7.5	10.6
	UGU	6.2	4.8	4.2	5.1	9.9	8.6	15.7	25.0	5.7	6.3	10.8	13.8	11.1
PHE	UUC	17.3	18.5	19.7	14.4	19.8	15.5	10.4	26.7	17.9	16.2	18.4	21.5	16.2
	UUU	29.4	20.5	23.7	20.7	16.5	18.9	15.7	28.3	18.5	20.1	21.0	23.8	13.1
ILE	AUA	11.1	10.6	9.7	8.3	26.0	32.2	30.8	7.5	3.4	6.7	5.1	7.1	24.7
	AUC	22.9	26.6	25.1	13.9	19.7	9.6	13.2	17.6	20.9	18.2	25.7	10.6	23.6
	AUU	24.7	22.3	23.9	14.4	23.4	19.4	17.6	20.3	5.6	11.8	10.5	21.3	16.0
MET	AUG	32.4	22.6	21.8	24.2	34.7	37.2	22.0	18.8	18.0	19.7	24.3	26.7	20.7
TRP	UGG	14.8	10.1	15.5	12.0	17.1	10.1	28.8	33.8	11.4	11.9	15.2	10.4	5.4
TER	UAA	3.5	0.7	1.7	1.5	1.3	0.9	1.2	1.5	0.5	0.9	1.3	0.4	0.7
	UAG	0.9	1.1	0.6	0.6	0.5	0.9	1.7	1.2	0.4	0.6	0.2	1	0.3
	UGA	1.6	1.5	1.3	0.7	0.7	0.6	0.5	0.8	0.9	0.5	1.0	4.8	0.6
	TOTAL	4324	12126	13279	10286	50849	19500	46019	10768	70451	40277	20697	5207	26312

							Phages						
PIF	PLY	PPH	REO	SIV	VAC	VAZ	VSV	F1C	LAM	P22	PT4	PT7	PZA
26	29	24	21	37	30	71	57	22	64	20	85	61	23
4.4	3.4	8.0	8.8	3.7	5.3	11.9	3.7	3.5	7.1	9.9	5.9	8.1	4.2
0.4	4.1	6.0	6.9	4.0	1.5	11.7	1	8.5	16.1	15.3	6.1	14.6	4.0
1.8	4.0	4.8	5.8	2.7	1.0	11.0	3.2	0.9	10.1	7.8	1.2	3.0	4.0
1.0	4.4	9.3	12.3	1.9	6.1	14.7	3.1	17.2	16.8	13.8	20.5	23.0	11.3
29.2	28.1	17.6	13.5	32.6	24.4	9.9	19.8	5.2	9.4	12.9	8.4	5.5	11.6
10.5	16.2	7.6	9.2	16.4	4.1	5.8	9.6	1.7	4.8	7.8	1.7	3.4	5.7
16.8	23.9	12.7	11.7	16.7	19.6	11.1	11.0	4.4	3.1	7.5	6.9	10.3	14.6
8.3	12.6	2.6	9.1	12.7	6.7	7.1	8.7	10.9	9.1	9.9	4.2	10.7	7.6
11.8	20.4	10.5	19.3	14.8	6.8	12.3	11.6	15.3	36.3	27.0	7.0	21.8	5.9
11.6	19.5	11.4	16.0	13.4	12.5	17.3	11.7	23.7	13.6	18.9	21.3	16.2	13.4
28.0	16.3	35.7	16.2	20.7	20.7	30.8	16.8	31.4	8.8	12.9	25.7	12.3	22.6
12.9	13.1	17.8	26.8	14.9	19.4	17.4	19.1	15.0	5.7	9.6	10.8	10.1	12.2
37.5	14.1	13.0	26.9	13.0	13.4	13.1	16.8	16.3	13.3	13.5	17.0	9.1	17.5
6.3	9.6	7.9	12.0	9.9	9.7	14.4	15.5	16.1	11.3	6.6	3.4	12.6	4.5
3.9	2.1	3.6	13.0	2.9	7.5	12.2	4.4	5.7	8.8	8.7	4.0	4.3	4.2
17.4	11.6	19.8	17.3	12.2	25.6	14.8	17.7	50.1	6.8	12.0	24.7	18.0	14.4
10.4	12.6	8.5	9.3	10.5	6.2	9.0	8.5	5.7	17.0	14.7	5.4	7.2	9.7
13.6	15.1	17.9	13.5	9.4	16.4	9.8	14.4	6.5	11.5	10.8	11.5	9.2	19.6
40.9	22.0	41.6	16.7	21.9	20.2	23.8	28.7	5.9	13.0	17.1	16.6	10.1	31.8
8.1	18.1	13.1	9.8	9.9	9.4	19.7	13.3	11.5	21.2	14.1	6.5	18.5	6.8
5.6	3.0	6.5	16.8	4.1	8.4	17.4	5.1	5.7	19.4	9.0	4.9	7.7	8.7
19.0	17.3	22.9	23.4	20.5	23.1	12.1	19.6	32.0	9.2	10.8	27.6	19.8	17.5
21.5	20.3	21.8	19.6	30.5	12.5	17.3	22.4	7.4	8.8	12.6	11.9	9.1	7.5
6.7	14.6	9.0	6.9	16.4	3.7	17.5	10.8	4.6	4.8	1.5	1.2	1.8	2.3
3.4	4.2	4.5	9.0	5.2	7.0	14.2	5.2	8.5	17.2	9.6	5.7	7.6	4.7
9.4	19.2	23.7	13.4	15.3	13.8	11.4	20.6	23.3	7.3	13.2	14.2	14.8	14.1
20.6	19.5	26.9	17.1	25.9	12.2	20.6	14.7	16.3	28.6	33.9	20.6	16.9	15.8
5.1	14.7	9.3	13.4	14.8	9.5	21.0	8.2	9.2	27.9	16.5	5.7	13.9	8.9
2.5	3.9	2.6	15.8	6.2	7.8	18.6	2.0	7.8	25.7	18.0	6.8	14.5	9.9
10.7	25.6	17.6	29.9	17.6	16.4	15.2	14.4	30.5	17.0	24.9	34.8	43.5	22.2
26.4	22.1	14.8	22.7	31.6	23.0	22.7	39.5	2.2	13.0	15.3	19.2	11.8	17.2
6.7	13.5	12.5	8.2	13.8	3.6	9.1	6.7	25.1	21.4	15.6	9.4	14.5	9.6
12.0	15.6	9.4	11.1	13.8	3.4	15.4	17.5	5.2	14.5	9.9	3.9	8.9	9.9
9.8	15.0	16.5	17.1	10.2	14.2	14.4	10.1	44.9	19.9	15.9	32.0	38.8	32.5
16.8	18.9	24.4	13.6	18.6	23.6	19.2	11.6	15.0	19.9	12.0	18.4	17.2	13.7
9.0	6.1	4.0	13.6	10.6	8.5	9.1	14.0	8.5	11.5	10.2	5.7	11.8	7.3
10.6	17.1	18.2	29.6	13.1	8.9	16.6	14.9	3.9	24.1	12.9	5.1	15.0	17.4
15.8	17.3	16.5	21.2	8.6	23.9	23.6	20.9	49.5	18.9	20.7	32.2	19.6	31.6
46.1	33.9	33.1	15.4	35.9	48.9	24.4	39.2	37.3	36.8	43.8	65.4	22.2	42.7
18.3	20.6	13.9	21.3	26.4	26.6	10.7	23.1	17.9	19.9	35.4	16.9	44.6	30.9
24.1	13.4	13.6	18.3	17.7	21.0	19.1	9.9	10.2	21.4	22.8	15.6	33.1	37.7
42.9	26.0	28.9	26.8	27.3	47.5	21.6	21.6	44.4	18.5	21.0	41.0	12.3	30.7
27.9	23.4	25.4	18.1	33.9	15.3	21.6	17.9	17.9	9.6	11.7	22.7	16.8	20.1
15.3	22.9	19.0	25.0	24.2	6.3	13.5	10.9	21.6	33.1	27.0	11.8	19.6	12.7
5.2	8.6	9.7	6.5	9.0	4.4	10.8	11.3	2.8	6.8	7.2	4.7	13.7	5.6
14.3	15.4	18.9	8.2	15.9	12.7	15.6	20.5	7.0	10.8	10.8	13.2	6.3	9.4
37.1	36.9	34.2	17.1	36.8	43.1	31.9	37.9	19.4	37.2	39.0	57.0	28.8	43.8
15.3	26.9	17.2	25.8	28.2	14.9	19.8	24.8	17.0	27.6	33.0	10.7	41.9	22.4
16.7	21.3	23.2	24.4	15.1	14.8	21.6	23.9	19.8	24.9	24.0	14.9	37.9	30.4
40.2	32.1	35.4	41.0	26.8	48.3	32.0	34.9	35.7	32.5	30.3	48.6	24.7	33.7
11.3	13.0	9.0	10.4	13.1	12.7	13.1	15.3	5.9	12.4	17.7	9.5	20.0	17.5
22.7	20.4	32.0	17.7	20.9	34.4	20.9	23.1	34.2	17.7	12.6	29.7	12.4	27.3
5.0	9.0	9.1	5.0	13.9	3.1	6.3	8.4	4.4	8.5	7.5	3.3	5.9	1.2
11.3	11.9	19.2	7.4	14.8	11.5	14.9	11.9	8.1	3.7	4.8	6.4	6.4	5.2
11.9	11.7	3.9	14.1	9.1	14.8	6.5	22.1	18.3	14.9	15.6	11.3	24.5	20.7
14.0	27.2	37.3	16.5	17.7	34.0	33.8	22.8	36.8	19.4	10.5	30.4	10.3	23.8
46.9	17.2	23.4	16.1	18.0	26.7	19.0	19.0	9.8	8.6	12.0	10.1	4.7	16.3
21.5	5.4	2.8	20.7	11.0	19.8	10.4	25.3	9.4	20.8	22.8	12.2	22.8	22.2
28.4	17.6	25.2	21.1	15.9	43.7	25.9	32.6	35.3	22.8	26.4	51.0	24.8	27.8
22.7	27.0	18.1	31.4	18.2	26.6	20.7	23.4	17.4	27.8	30.3	26.9	30.7	25.2
12.3	16.1	13.0	13.4	25.6	6.6	10.8	21.7	9.6	16.7	15.6	14.0	15.8	12.5
1.1	1.7	2.3	1.2	0.9	1.9	1.3	1.4	2.8	1.8	2.7	2.8	2.5	1.9
0.3	1	0.9	0.3	1.4	0.3	0.3	0.1	0.4	0.3	0.0	0.3	0.4	0.7
0.4	0.2	0.5	0.5	1.1	0.4	0.3	0.3	1.5	2.5	3.3	1.1	1.5	1.4
14387	10056	6466	10846	10935	11778	37973	30799	4590	14041	3337	20311	13700	5758

Genomes/Genes

Table 8. Continued

Organelle genomes

		MPO CP	MZE CP	PEA CP	SPI CP	TOB CP	YSC MT
No. GENES		54	33	21	31	67	20
ARG	CGA	11.6	12.0	10.7	11.1	14.7	0.2
	CGC	2.0	7.3	4.8	5.1	5.1	0.4
	CGG	1.1	3.1	3.9	4.3	3.7	1.1
	CGU	16.1	14.2	14.3	16.2	19.0	3.4
	AGA	16.0	16.4	12.9	15.8	16.8	19.7
	AGG	0.9	4.0	4.4	6.6	5.5	3.0
LEU	CUA	5.5	13.9	14.6	16.4	13.6	8.1
	CUC	0.9	9.7	5.4	5.2	4.3	1.3
	CUG	0.8	7.8	6.5	6.5	6.8	6.0
	CUU	21.6	18.4	19.7	21.8	19.4	5.4
	UUA	72.7	33.2	37.5	34.1	31.9	96.5
	UUG	6.2	19.8	20.2	21.7	20.0	12.7
SER	UCA	13.9	13.6	11.0	11.9	9.8	31.5
	UCC	3.0	14.2	11.3	10.5	10.3	4.5
	UCG	1.8	5.7	3.5	7.1	4.4	2.8
	UCU	25.4	13.8	16.7	19.9	18.3	17.1
	AGC	2.0	5.6	4.4	4.1	5.9	3.2
	AGU	16.5	16.2	13.1	15.0	13.5	10.5
THR	ACA	21.4	17.2	15.5	14.6	14.4	21.2
	ACC	2.7	11.1	11.6	11.6	13.2	4.1
	ACG	1.8	5.6	5.9	4.9	4.9	2.6
	ACU	26.5	22.0	25.3	22.4	22.6	16.5
PRO	CCA	16.3	14.0	13.1	12.1	11.9	16.1
	CCC	1.8	9.3	6.0	8.2	7.2	3.2
	CCG	1.9	5.9	4.2	5.8	5.7	1.9
	CCU	21.6	17.0	21.1	19.6	20.4	19.9
ALA	GCA	22.3	22.3	25.6	18.6	21.7	21.9
	GCC	3.3	13.8	12.8	11.0	13.2	5.1
	GCG	2.5	7.6	8.9	8.4	7.6	2.8
	GGU	37.3	36.1	42.6	31.3	36.9	31.9
GLY	GGA	32.7	28.5	28.2	28.1	29.8	13.5
	GGC	4.1	11.8	9.5	9.8	9.6	2.1
	GGG	4.4	14.9	13.3	13.3	13.6	3.9
	GGU	31.6	27.3	37.0	28.5	34.6	41.8
VAL	GUA	21.0	23.4	27.7	24.0	26.8	29.0
	GUC	2.3	9.7	7.8	6.1	7.6	5.1
	GUG	2.2	14.1	8.6	7.1	8.3	4.1
	GUU	29.0	21.8	21.4	20.5	22.9	22.9
LYS	AAA	61.5	34.5	29.7	35.4	34.0	41.2
	AAG	2.6	13.0	8.0	10.5	11.3	12.9
ASN	AAC	6.6	10	9.0	13.4	13.8	10.7
	AAU	43.3	22.6	28.5	31.6	29.0	62.4
GLN	CAA	35.9	28.5	29.8	25.6	26.0	22.1
	CAG	2.1	8.6	8.7	7.1	8.6	4.1
HIS	CAC	3.2	5.9	5.9	7.2	7.0	3.6
	CAU	18.2	16.8	19.7	21.5	17.8	18.0
GLU	GAA	45.7	40.1	36.1	38.3	41.0	29.8
	GAG	3.6	12.9	12.3	11.5	13.2	6.7
ASP	GAC	3.4	8.9	8.4	9.2	8.6	10.3
	GAU	29.9	28.1	30.3	30.3	28.7	27.2
TYR	UAC	4.4	6.9	5.1	6.7	8.4	7.7
	UAU	30.8	24.2	24.4	23.5	22.8	35.4
CYS	UGC	1.9	2.9	1.8	2.3	2.2	0.9
	UGU	8.9	6.6	8.0	6.9	7.1	6.7
PHE	UUC	4.9	17.5	19.0	22.5	18.3	25.9
	UUU	53.9	30.7	33.7	37.7	28.0	33.0
ILE	AUA	27.3	20.9	17.6	22.5	18.9	14.8
	AUC	4.7	15.3	14.8	15.1	17.6	12.0
	AUU	59.9	40.0	43.5	33.8	38.5	69.7
MET	AUG	22.2	24.7	21.5	23.2	23.4	28.1
TRP	UGG	16.6	14.3	20.0	22.1	16.3	6.4
TER	UAA	3.2	1.7	2.0	1.9	2.5	3.4
	UAG	0.3	1.1	0.3	0.6	0.8	0.2
	UGA	0.0	0.8	0.9	0.5	0.6	10.1
	TOTAL	15798	9220	6641	10488	16950	5337

[a] Abbreviations: **Mammals:** BOV, cow; DOG, dog; HAM, hamster; HUM, human; MUS, mouse; PIG, pig; RAB, rabbit; RAT, rat; SHP, sheep. **Other vertebrates:** BMO, *Bombyx mori*; CHK, chicken; FSB, fish; XEL, *Xenopus laevis*. **Invertebrates:** CEL, *Caenorhabditis elegans*; DRO, *Drosophila melanogaster*; SUS, sea urchin; **Eukaryotic microorganisms:** ASN, *Aspergillus nidulans*; DDI, *Dictyostelium discoideum*; NEU, *Neurospora crassa*; PFA, *Plasmodium falciparum*; SCM, *Schistosoma mansoni*; TRB, *Trypanosoma brucei*; YSC, *Saccharomyces cerevisiae;* YSK, *Kluveromyces lactis*; YSP, *Schizosaccharomyces pombe*. **Plants:** ATH, *Arabidopsis thaliana*; BLY, barley; MZE, maize; PEA, pea; POT, potato; SOY, soybean; TOM, tomato; WHT, wheat. **Prokaryotes:** ANA, *Anabaena*; ATU, *Agrobacterium tumefaciens*; AVI, *Azotobacter vinelandii*; BME, *Bacillus megaterium*; BST, *Bacillus stearothermophilus*; BSU, *Bacillus subtilis*; BTH, *Bacillus thuringiensis*; CHT, *Chlamydia trachomatis*; CLO, *Clostridium*; ECO, *Escherichia coli*; FDI, *Fremyella diplosiphon*; HAL, *Halobacterium halobium*; KPN, *Klebsiella pneumoniae*; PSE, *Pseudomonas*; RCA, *Rhodobacter capsulatus*; RHL, *Rhizobium leguminosarum*; RHM, *Rhizobium meliloti*; SSP, *Sulfolobus* virus-like particle; STA, *Staphylococcus aureus*; STM, *Streptomyces*; STR, *Streptococcus*; STY, *Salmonella typhimurium*; SYN, *Synechococcus*; VIB, *Vibrio*. **Plasmids/ Transposons:** FPL, F plasmid; TIP, Ti plasmid; TRN, Tn transposon; **Viruses:** ADB, adenovirus type 2; FLA, influenza virus A; FLB, influenza virus B; HIV, human immunodeficiency virus; HPB, hepatitis B virus; HS1, herpes simplex virus type 1; HS4, Epstein–Barr virus; HS5, human cytomegalovirus; MCV, cucumber mosaic virus; NDV, Newcastle disease virus; PIF, human parainfluenza virus; PLY, polyoma virus; PPH, human papillomavirus; REO, reovirus; SIV, simian immunodeficiency virus; VAC, vaccinia virus; VAZ, varicella-zoster virus; VSV, vesicular stomatitis virus. **Phages:** F1C, phage f1; LAM, phage λ; P22, phage P22; PT4, phage T4; PT7, phage T7; PZA, phage PZA. **Organelle genomes:** MPO CP, *Marchantia polymorpha* chloroplast; MZE CP, maize chloroplast; PEA CP, pea chloroplast; SPI CP; spinach chloroplast; TOB CP, tobacco chloroplast; YSC MT, *Saccharomyces cerevisiae* mitochondrion.

Data reproduced from ref. 25 by permission of Oxford University Press.

5. REFERENCES

1. Hall, J.L., Ramanis, Z. and Luck, D.J.L. (1989) *Cell*, 59, 121.

2. John, B. and Miklos, G. (1988) *The Eukaryote Genome in Development and Evolution.* Unwin Hyman, London.

3. Bennett, M.D. and Smith, J.B. (1976) *Phil. Trans. R. Soc. Ser. B*, 274, 227.

4. Brown, T.A. (1987) in *Gene Structure in Eukaryotic Microbes.* (J.R. Kinghorn ed.) IRL Press, Oxford, pp. 141.

5. Grossman, L.I. and Hudspeth, M.E.S. (1985) in *Gene Manipulation in Fungi.* (J.W. Bennett and L.L. Lasure eds) Academic Press, pp. 65.

6. Palmer, J.D. (1985) *Annu. Rev. Genet.*, 19, 325.

7. Thiery, J.P., Macaya, G. and Bernardi, G. (1976) *J. Mol. Biol.*, 108, 219.

8. de Ley, J. (1970) *J. Bacteriol.*, 101, 738.

9. Gualberto, J.M., Lamattina, L., Bonnard, G., Weil, J.-H. and Grienenberger, J.-M. (1989) *Nature*, 341, 660.

10. Covello, P.S. and Gray, M.W. (1989) *Nature*, 341, 662.

11. Caron, F. and Meyer, E. (1985) *Nature*, 314, 185.

12. Preer, J., Preer, L.B., Rudman, B.M. and Barnett, A.J. (1985) *Nature*, 314, 188.

13. Horowitz, S. and Gorovsky, M.A. (1985) *Proc. Natl. Acad. Sci. USA*, 82, 2452.

14. Kuchino, Y., Hanyu, N., Tashiro, F. and Nishimura, S. (1985) *Proc. Natl. Acad. Sci. USA*, 82, 4758.

15. Kawaguchi, Y., Honda, H., Taniguchi-Morimura, J. and Iwasaki, S. (1989) *Nature*, 341, 164.

16. Chambers, I., Frampton, J., Goldfarb, P., Affara, N., McBain, W. and Harrison, P.R. (1986) *EMBO J.*, 5, 1221.

17. Sukenaga, Y., Ishida, K., Takeda, T. and Takagi, K. (1987) *Nucleic Acids Res.*, 15, 7178.

18. Leinfelder, W., Zehelein, E., Mandrand-Berthelot, M.-A. and Bock, A. (1988) *Nature*, 331, 723.

19. Anderson, S., Bankier, A.T., Barrell, B.G., de Bruijn, M.H.L., Coulson, A.R., Drouin, J., Eperon, I.C., Nierlich, D.P., Roe, B.A., Sanger, F., Schreier, P.H., Smith, A.J.H., Staden, R. and Young, I.G.. (1981) *Nature*, 290, 457.

20. Clary, D.O., Wahleithner, J.A. and Wolstenholme, D.R. (1984) *Nucleic Acids Res.*, 12, 3747.

21. Sibler, A.-P., Dirheimer, G. and Martin, R.P. (1981) *FEBS Lett.*, 132, 344.

22. Waring, R.B., Davies, R.W., Lee, S., Grisi, E., Berks, M.M. and Scazzocchio, C. (1981) *Cell*, 27, 4.

23. Fox, T.D. and Leaver, C.J. (1981) *Cell*, 26, 315.

24. Osawa, S., Muto, A., Jukes, T.H. and Ohama, T. (1990) *Proc. R. Soc. Lond. B* 241, 19–28.

25. Wada, K., Aota, S., Tsuchiya, R., Ishibashi, F., Gojobori, T. and Ikemura, T. (1990) *Nucleic Acids Res.*, 18, 2367.

26. Yamao, F., Muto, A., Kawauchi, Y., Iwami, M., Iwagami, S., Azumi, Y. and Osawa, S. (1985) *Proc. Natl Acad. Sci. USA*, 82, 2306.

CHAPTER 8
ELECTROPHORESIS

1. AGAROSE GEL ELECTROPHORESIS

The basic repeating unit of agarose is:

Different grades of agarose are available from a number of suppliers. These differ not only in their gelling/melting temperatures but also in their sieving characteristics for DNA fragments. *Figure 1* provides a guide to the selection of agarose type and concentration for optimal resolution of different sizes of DNA molecule, using the range of agaroses available from FMC BioProducts. The separation of DNA molecules in these agarose systems is illustrated in *Figures 2–4*. The effects of agarose concentration on gelling and melting temperatures and on gel strength are shown in *Figures 5* and *6*.

Figure 1. Optimal range of resolution for different grades of agarose. The charts were developed with data obtained from the electrophoresis of linear, dsDNA in a traditional submarine gel at 5 V cm^{-1} in either TAE or TBE buffer. (**A**) The overall fractionation range of each agarose type; (**B**) the effects of electrophoresis buffer (TAE or TBE) and gel concentration on the resolution of DNA fragments in each type of agarose. The boxed areas on each chart give the size range of DNA best resolved under the conditions that are indicated. Lower or higher molecular weight DNA will migrate in the gel, but may blur or fail to resolve. Agarose concentrations above or below those shown for each type of agarose are difficult to handle. (Figure courtesy of FMC BioProducts, Rockland, ME, USA, © 1988. NuSieve, SeaKem and SeaPlaque are trademarks of FMC Corporation.)

(**A**)

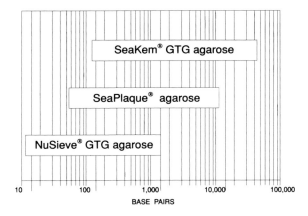

Figure 1. Continued.

(B)

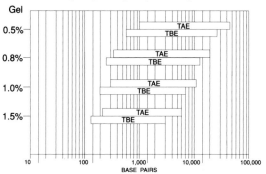

SeaKem® GTG agarose
standard gelling/melting temperature agarose

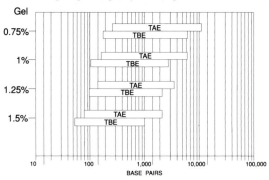

SeaPlaque® agarose
low gelling/melting temperature agarose

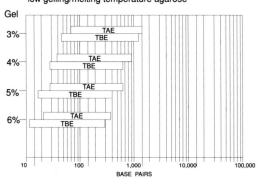

NuSieve® GTG agarose
low gelling/melting temperature agarose

Figure 2. Comparison of the separation of small DNA fragments in NuSieve GTG agarose concentrations from 1 to 9% (w/v) in TAE (**A**) and TBE (**B**) buffers. The sizes (bp) of the fragments (*Hae*III–φX174) are (1) 1353, (2) 1078, (3) 872, (4) 603, (5) 310, (6) 281/271, (7) 234, (8) 194, (9) 118, (10) 72. TD stands for bromophenol blue tracking dye. Insert gels were cast in a Serwer multilane chamber framed with 1% SeaKem LE agarose gel in the same buffer as the inserts. The 22 cm gel was electrophoresed submerged at 5 V cm^{-1} for 3 h. (Figure courtesy of FMC BioProducts.)

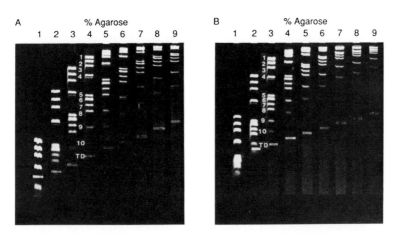

Figure 3. Comparison of the separation of DNA fragments in SeaPlaque GTG agarose at concentrations from 0.75 to 2.0% (w/v) in TAE (**A**) and TBE (**B**) buffers. The sizes (bp) of the fragments (*Hin*dIII–digest of phage λ DNA) are (1) 23 130, (2) 9416, (3) 6557, (4) 4361, (5) 2322, (6) 2027, (7) 564. The 125 bp fragment is too faint to see. TD stands for bromophenol blue tracking dye. Electrophoresis conditions were as described in *Figure 2*, except that the gel was framed with 1% SeaKem GTG agarose gel. (Figure courtesy of FMC BioProducts.)

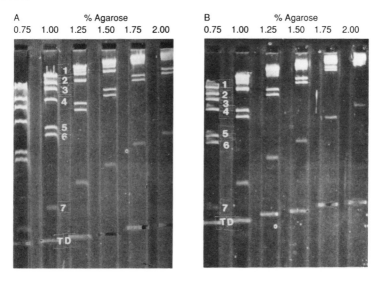

Figure 4. Comparison of the separation of DNA fragments in SeaKem GTG agarose at concentrations from 0.4 to 1.75% (w/v) in TAE (**A**) and TBE (**B**) buffers. The sizes (bp) of the fragments (*Hind*III–digest of phage λ DNA) are (1) 23 130, (2) 9416, (3) 6557, (4) 4361, (5) 2322, (6) 2027, (7) 564. The 125 bp fragment is too faint to see. TD stands for bromophenol blue tracking dye. Electrophoresis conditions were as described in *Figure 3*. (Figure courtesy of FMC BioProducts.)

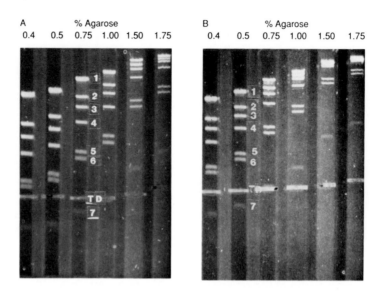

Figure 5. Effect of agarose concentration on gelling (open symbols) and melting (solid symbols) temperatures. (Courtesy of FMC BioProducts.)

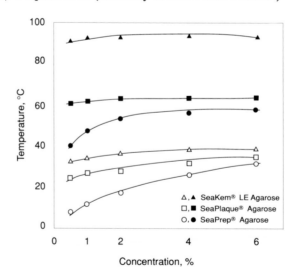

Figure 6. Effect of concentration on gel strength for some typical agars and agaroses. (Courtesy of FMC BioProducts.)

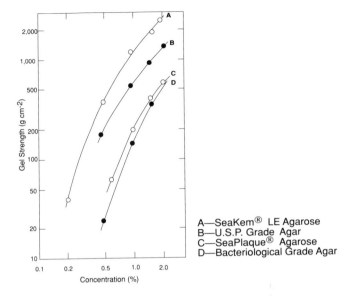

A—SeaKem® LE Agarose
B—U.S.P. Grade Agar
C—SeaPlaque® Agarose
D—Bacteriological Grade Agar

Table 1 gives recipes for three alternative buffers for non-denaturing agarose gel electrophoresis. The buffer used to make the gel is the same as that used in the electrophoresis chambers. Since samples are loaded into wells already covered and filled with running buffer, the loading buffer must have a higher density to allow the sample to fall to the well bottom. One or more dyes are present in the loading buffer, the migration of which during electrophoresis indicates the extent of the separation achieved. Three alternative loading buffers are given in *Table 2*; the choice of buffer is a matter of personal preference. A number of buffer systems exist for denaturing agarose gel electrophoresis. Details of these are given in *Table 3*.

Table 1. Running buffers for non-denaturing gel electrophoresis

Buffer	Components (1× buffer)	Recipe (for 1 liter of 10× buffer)
Tris–borate (TBE)[a, b]	89 mM Tris–borate (pH 8.3) 2 mM Na$_2$.EDTA	108.0 g Tris-base 55.0 g boric acid 9.3 g Na$_2$.EDTA.2H$_2$O
Tris–acetate (TAE)	40 mM Tris–acetate (pH 7.6) 1 mM Na$_2$.EDTA	48.4 g Tris-base 11.4 ml glacial acetic acid 20.0 ml 0.5 M Na$_2$.EDTA (pH 8.0)
Tris–phosphate (TPE)	89 mM Tris–phosphate 2 mM Na$_2$.EDTA	108.0 g Tris-base 15.5 ml 85% phosphoric acid 40.0 ml 0.5 M Na$_2$.EDTA (pH 8.0)

[a] TBE can be used at half strength for agarose gels. 1× TBE is the standard running buffer for DNA separations in polyacrylamide gels.
[b] Concentrated stock solutions of TBE tend to collect a precipitate when stored for long periods of time. Store at room temperature and discard any solutions that develop a precipitate.

Electrophoresis

Table 2. Loading buffers for non-denaturing agarose gel electrophoresis

Loading buffer	Recipe for 6× buffer[a]
SBX	40% (w/v) sucrose 0.25% (w/v) bromophenol blue 0.25% (w/v) xylene cyanol
FBX	15% (w/v) Ficoll 400 0.25% (w/v) bromophenol blue 0.25% (w/v) xylene cyanol
GBX	30% (v/v) glycerol 0.25% (w/v) bromophenol blue 0.25% (w/v) xylene cyanol

[a] One dye can be omitted if desired. Store SBX and GBX at 4°C.

Table 3. Denaturing agarose gel electrophoresis

System	Conditions
Alkaline gels	
Suitable for:	DNA only.
Basis:	Gels are run in an alkaline pH buffer which denatures the DNA.
Gel buffer:	50 mM NaCl, 1 mM Na_2.EDTA. Neutral pH buffer is used since alkaline pH at high temperature hydrolyzes agarose.
Running buffer:	30 mM NaOH, 1 mM Na_2.EDTA. The gel is soaked in running buffer for 30 min before loading the sample.
Sample loading buffer:	DNA is resuspended in 50 mM NaOH, 1 mM Na_2.EDTA, 3% (w/v) Ficoll 400, 0.025% (w/v) bromocresol green, 0.025% (w/v) xylene cyanol.
Visualization:	Autoradiography or ethidium bromide staining.
Reference:	1.
Glyoxal gels	
Suitable for:	DNA, RNA.
Basis:	Nucleic acids are denatured with glyoxal prior to electrophoresis.
Gel buffer:	10 mM sodium phosphate, pH 7.0. Neutral buffer is used since glyoxal dissociates from nucleic acids at pH 8.0 and above.
Running buffer:	Same as gel buffer.
Sample loading buffer:	Nucleic acid (up to 10 µg) is resuspended in 8 µl 1 M glyoxal, 50% (v/v) DMSO, 10 mM sodium phosphate pH 7.0. After incubation at 50°C for 1 h, the sample is cooled and 2 µl of 10 mM sodium phosphate pH 7.0, 50% (v/v) glycerol, 0.4% (w/v) bromophenol blue are added.
Visualization:	Autoradiography, ethidium bromide staining or acridine orange staining.
Reference:	2.

Table 3. Continued

System	Conditions
Methyl-mercuric hydroxide gels[a]	
Suitable for:	DNA, RNA.
Basis:	Gels are run in the presence of methylmercuric hydroxide which denatures nucleic acids.
Gel buffer:	50 mM boric acid, 5 mM sodium borate, 10 mM sodium sulfate, pH 8.2. Agarose is dissolved in this buffer then cooled to 50°C and methylmercuric hydroxide added to 5 mM.
Running buffer:	50 mM boric acid, 5 mM sodium borate, 10 mM sodium sulfate, pH 8.2.
Sample loading buffer:	2 × buffer = 500 μl of 4 × running buffer, 200 μl 30% (w/v) Ficoll 400, 25 μl 1 M methylmercuric hydroxide, plus water to 1.0 ml then bromophenol blue and xylene cyanol each added to 0.1% (w/v) final concentration.
Visualization:	Autoradiography or ethidium bromide staining.
Reference:	3.
Formaldehyde gels	
Suitable for:	RNA better than DNA.
Basis:	Gels are run in the presence of formaldehyde which denatures nucleic acids.
Gel buffer:	40 mM MOPS pH 7.0, 10 mM sodium acetate, 1 mM Na_2.EDTA, 2.2 M formaldehyde.
Running buffer:	40 mM MOPS pH 7.0, 10 mM sodium acetate, 1 mM Na_2.EDTA.
Sample loading buffer:	RNA samples are incubated at 55°C for 15 min in 2 μl of 5 × running buffer, 10 μl formamide, 3.5 μl concentrated formaldehyde (12.3 M) and then 2 μl of 30% (w/v) Ficoll 400, 1 mM Na_2.EDTA, 0.25% (w/v) bromophenol blue, 0.25% (w/v) xylene cyanol are added.
Visualization:	Ethidium bromide staining (but not efficiently) prior to Northern transfer.
Reference:	4.

[a] Methylmercuric hydroxide: Mol. wt, 232.6; Risk, 26/27/28-48; Safety, 28-36-45.

2. POLYACRYLAMIDE GEL ELECTROPHORESIS

Physical and chemical data on reagents required for polyacrylamide gel electrophoresis are given in *Table 4* and their structures are shown in *Figure 7*. The chemistry of acrylamide polymerization is illustrated in *Figure 8*. *Table 5* provides a guide to the selection of polyacrylamide gel concentration for the resolution of DNA molecules of known sizes. *Table 6* gives recipes for polyacrylamide gel preparation. The buffer used is 1× TBE (see *Table 1* for recipe) which is also used in the electrophoresis chamber. Sample loading buffers are as used for agarose gel electrophoresis, i.e. one of those given in *Table 2*. For denaturing polyacrylamide gel electrophoresis include 8 M urea in the gel and use a loading buffer consisting of 10 mM Na_2.EDTA (pH 8.0), 0.1% (w/v) xylene cyanol and 0.1% (w/v) bromophenol blue, made in deionised formamide rather than water. Heat the samples at 80°C for 15 min immediately before loading. In a denaturing gel the dye mobilities are approximately half the values for non-denaturing gels (see *Table 5*).

Table 4. Polyacrylamide gel electrophoresis reagents: physical and chemical data

Compound	Mol. wt	Risk	Safety	Refs
Acrylamide	71.1	23/24/25-33	27-44	5
Ammonium persulfate (AMPS, ammonium peroxydisulfate)	228.2	8-22	24/25	–
N,N'-diallyltartardiamide	228.3	–	–	6
N,N'-methylenebisacrylamide	154.2	20/21/22	24/25	–
Riboflavin	376.4	–	–	–
Riboflavin 5'-phosphate	514.4	–	–	–
TEMED (N,N,N'N',-tetramethylethylenediamine	116.2	11-36/38	–	–

Table 5. Optimal range of resolution for different polyacrylamide gel concentrations

| Acrylamide (%)[a] | Range of resolution | | Dye mobility (bp dsDNA)[b] | |
	dsDNA (bp)	ssDNA (nt)	XC[c]	BPB[c]
3.5	100–1000	750–2000	450	100
5.0	75–500	200–1000	250	65
8.0	50–400	50–400	150	45
12.0	35–250		70	20
15.0	20–150		60	15
20.0	5–100		45	12

[a] 29:1 acrylamide:N,N'-methylenebisacrylamide.
[b] The size of the dsDNA molecules comigrating with the dye in TBE.
[c] Abbreviations: XC, xylene cyanol; BPB, bromophenol blue.

Table 6. Recipes for polyacrylamide gels of different concentrations

| Reagent | Volume required for different gel concentrations (ml) | | | | | |
	3.5%	5.0%	8.0%	12.0%	15.0%	20.0%
30% acrylamide	11.6	16.6	26.6	40.0	50.0	66.6
10 × TBE	10.0	10.0	10.0	10.0	10.0	10.0
water	77.7	72.7	62.7	49.3	39.3	22.7
10% ammonium persulfate	0.7	0.7	0.7	0.7	0.7	0.7
TEMED	0.005	0.005	0.005	0.005	0.005	0.005

Figure 7. Structures of reagents used in polyacrylamide gel electrophoresis.

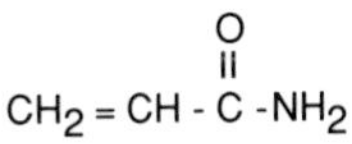

$$CH_2 = CH - \overset{\overset{\displaystyle O}{\|}}{C} - NH_2$$

ACRYLAMIDE

$$CH_2 = CH - CH_2 - NH - \overset{\overset{\displaystyle O}{\|}}{C} - \overset{\overset{\displaystyle OH}{|}}{CH} - \overset{\overset{\displaystyle OH}{|}}{CH} - \overset{\overset{\displaystyle O}{\|}}{C} - NH - CH_2 - CH = CH_2$$

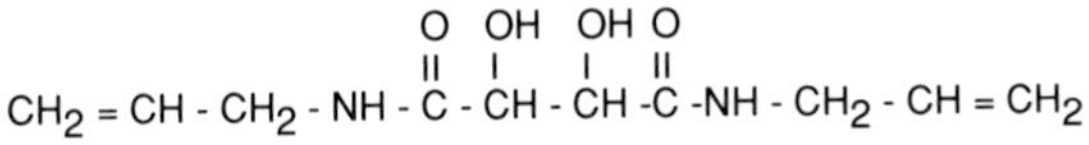

N,N' - DIALLYLTARTARDIAMIDE (DATD)

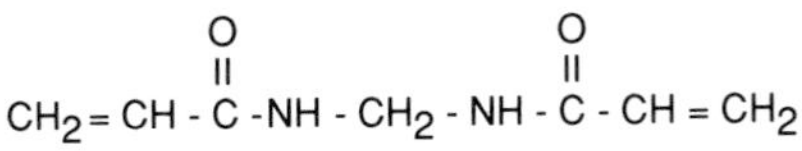

$$CH_2 = CH - \overset{\overset{\displaystyle O}{\|}}{C} - NH - CH_2 - NH - \overset{\overset{\displaystyle O}{\|}}{C} - CH = CH_2$$

N,N' - METHYLENEBISACRYLAMIDE

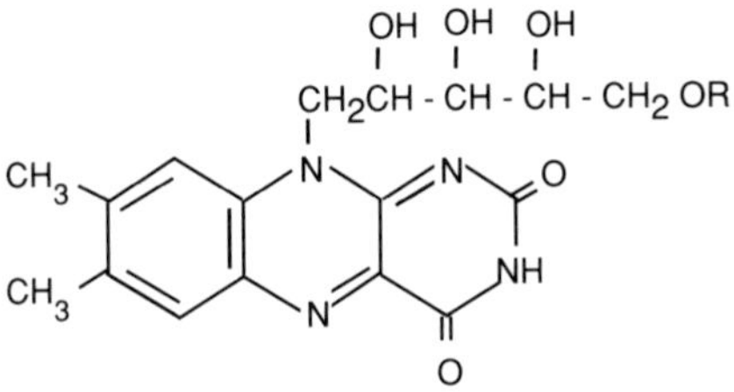

RIBOFLAVIN   R = H

RIBOFLAVIN 5' - PHOSPHATE   R = $-\overset{\overset{\displaystyle OH}{|}}{\underset{\underset{\displaystyle OH}{|}}{P}} = O$

$$\overset{\displaystyle H_3C}{\underset{\displaystyle H_3C}{}} \!\!\! > N - CH_2 - CH_2 - N < \!\!\! \overset{\displaystyle CH_3}{\underset{\displaystyle CH_3}{}}$$

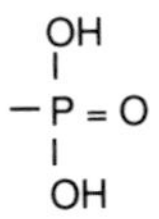

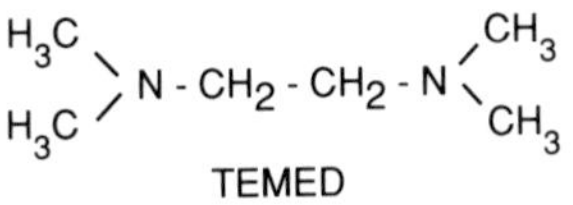

TEMED

Figure 8. Chemistry of acrylamide polymerization.

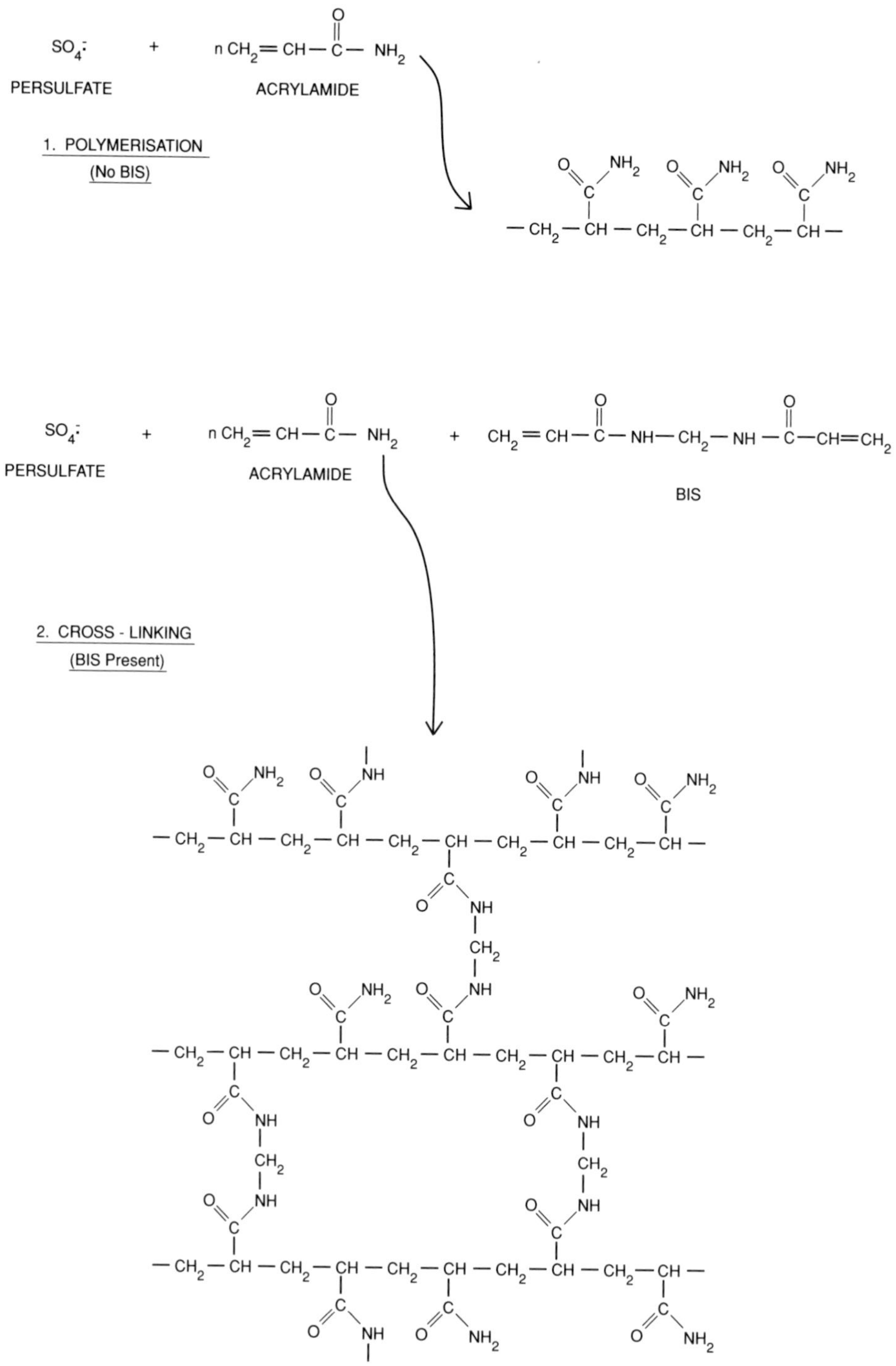

3. DNA FRAGMENT SIZE MARKERS

Tables 7–10 give the sizes of DNA fragments obtained by digestion of phage λ, M13mp18, pBR322 and φX174 DNA with a variety of different restriction enzymes. The format for the data provided is as follows:

Restriction enzyme number of sites: positions of the cut sites
              fragments: sizes of the fragments in bp

See Chapter 4, Section 1 for the recognition site of each restriction enzyme.

Table 7. Restriction fragment lengths for phage λ DNA (48502 bp, linear).

No sites for *Not*I, *Sfi*I, *Spe*I

*Aat*I	6 sites:	12347, 31481, 33000, 39995, 40599, 40617
	fragments:	19044, 12436, 7886, 6995, 1519, 604, 18
*Aat*II	10 sites:	5110, 9399, 11248, 14979, 29041, 40811, 41118, 42252, 45568, 45597
	fragments:	14062, 11770, 5109, 4289, 3731, 3316, 2906, 1849, 1134, 307, 29
*Acc*I	9 sites:	2192, 15262, 18836, 19475, 31303, 32747, 33246, 40203, 42923
	fragments:	13070, 11828, 6957, 5580, 3574, 2720, 2191, 1444, 639, 499
*Acc*III	24 sites:	611, 1827, 2620, 3330, 3342, 5099, 6895, 7812, 9697, 10319, 11987, 13087, 13807, 16041, 17631, 19776, 22344, 24843, 25614, 30843, 31815, 31889, 35450, 40002
	fragments:	8501, 5229, 4552, 3561, 2568, 2499, 2234, 2145, 1885, 1796, 1757, 1668, 1590, 1216, 1100, 972, 917, 793, 771, 720, 710, 622, 610, 74, 12
*Afl*II	3 sites:	6541, 12619, 42631
	fragments:	30012, 6540, 6078, 5872
*Afl*III	20 sites:	459, 629, 5549, 11282, 15373, 17792, 18285, 19997, 20953, 22221, 24134, 24169, 26529, 32765, 39396, 42087, 42364, 43763, 44502, 46983
	fragments:	6631, 6236, 5733, 4920, 4091, 2691, 2481, 2419, 2360, 1913, 1712, 1520, 1399, 1268, 956, 739, 493, 458, 277, 170, 35
*Apa*I	1 site :	10091
	fragments:	38412, 10090
*Apa*LI	4 sites:	5620, 21799, 27174, 40217
	fragments:	16179, 13043, 8286, 5619, 5375
*Ase*I	17 sites:	3418, 19481, 23083, 23692, 24986, 26762, 27202, 30619, 33221, 33735, 33800, 34686, 36274, 43409, 46407, 46852, 48377
	fragments:	16063, 7135, 3602, 3418, 3417, 2998, 2602, 1776, 1588, 1525, 886, 609, 514, 445, 440, 125, 65
*Asu*II	7 sites:	18050, 25886, 27982, 29152, 30398, 34333, 42639
	fragments:	18049, 8306, 7836, 5864, 3935, 2096, 1246, 1170
*Ava*I	8 sites:	4721, 19398, 21000, 27888, 31618, 33499, 38215, 39889
	fragments:	14677, 8614, 6888, 4720, 4716, 3730, 1881, 1674, 1602
*Ava*II	35 sites:	1612, 1922, 2816, 3801, 4314, 4622, 6042, 6440, 8995, 11000, 11045, 12996, 13147, 13737, 13952, 13984, 14329, 15613, 16587, 16610, 16683, 19289, 19356, 19867, 22001, 22243, 28798, 32474, 32562, 39004, 39437, 39479, 47605, 48202, 48474
	fragments:	8126, 6555, 6442, 3676, 2606, 2555, 2134, 2005, 1951, 1612, 1420, 1284, 985, 974, 894, 597, 590, 513, 511, 433, 398, 345, 310, 308, 272, 242, 215, 151, 88, 73, 67, 45, 42, 32, 28, 23

Electrophoresis

Table 7. Continued

*Avr*II	2 sites:	24323, 24397
	fragments:	24322, 24106, 74
*Bal*I	18 sites:	1329, 2209, 3263, 4196, 6499, 6880, 7587, 7981, 8059, 8862, 10612, 10780, 13937, 14906, 21263, 26626, 28621, 36043
	fragments:	12460, 7422, 6357, 5363, 3157, 2303, 1995, 1750, 1328, 1054, 969, 933, 880, 803, 707, 394, 381, 168, 78
*Bam*HI	5 sites:	5506, 22347, 27973, 34500, 41733
	fragments:	16841, 7233, 6770, 6527, 5626, 5505
*Ban*I	25 sites:	1181, 1366, 2332, 5408, 5666, 5672, 5901, 8037, 8044, 8442, 8765, 8989, 10222, 13039, 13643, 14642, 15200, 15238, 16237, 17054, 18557, 21546, 39908, 42798, 45679
	fragments:	18362, 3076, 2989, 2890, 2882, 2823, 2817, 2136, 1503, 1233, 1180, 999, 981, 966, 817, 604, 576, 398, 323, 258, 229, 224, 185, 38, 7, 6
*Ban*II	7 sites:	586, 10091, 19768, 21575, 24777, 25882, 39458
	fragments:	13576, 9677, 9505, 9045, 3202, 1807, 1105, 585
*Bbv*II	24 sites:	8494, 9858, 9920, 11943, 12405, 12810, 13102, 14285, 22177, 23891, 24785, 27553, 28529, 28596, 29485, 29792, 30034, 30072, 30475, 38440, 39262, 39300, 40410, 42448
	fragments:	8494, 7965, 7892, 3354, 2768, 2038, 2023, 1714, 1364, 1183, 1110, 976, 894, 889, 822, 462, 405, 403, 307, 292, 242, 67, 62, 38, 38
*Bcl*I	8 sites:	8845, 9362, 13821, 32730, 37353, 43683, 46367, 47943
	fragments:	18909, 8844, 6330, 4623, 4459, 2684, 1576, 560, 517
*Bgl*I	29 sites:	411, 2667, 3805, 4367, 4458, 4584, 5253, 5439, 6060, 6111, 7557, 8056, 11065, 12715, 12724, 12839, 13205, 14408, 14897, 15164, 17645, 18092, 19341, 20131, 20257, 20467, 21240, 30889, 32330
	fragments:	16173, 9649, 3009, 2481, 2256, 1650, 1446, 1441, 1249, 1203, 1138, 790, 773, 669, 621, 562, 499, 489, 447, 410, 366, 267, 210, 186, 126, 126, 115, 91, 51, 9
*Bgl*II	6 sites:	416, 22426, 35712, 38104, 38755, 38815
	fragments:	22010, 13286, 9688, 2392, 651, 415, 60
*Bsa*I	2 sites:	11424, 42715
	fragments:	31291, 11423, 5788
*Bsp*HI	8 sites:	889, 4650, 4989, 10249, 18275, 28860, 31608, 40642
	fragments:	10585, 9034, 8026, 7860, 5260, 3761, 2748, 889, 339
*Bss*HII	6 sites:	3523, 4127, 5628, 14816, 16650, 28009
	fragments:	20494, 11359, 9188, 3522, 1834, 1501, 604
*Bst*EII	13 sites:	5688, 7059, 8323, 9025, 13349, 13573, 13690, 16013, 17942, 25184, 30006, 36375, 40050
	fragments:	8453, 7242, 6369, 5687, 4822, 4324, 3675, 2323, 1929, 1371, 1264, 702, 224, 117
*Bst*XI	13 sites:	2863, 6714, 8421, 8858, 10923, 13271, 14346, 18037, 19749, 21630, 34604, 38300, 46442
	fragments:	12974, 8142, 3851, 3696, 3691, 2862, 2348, 2065, 2061, 1881, 1712, 1707, 1075, 437
*Cla*I	15 sites:	4200, 15585, 16122, 26618, 30291, 31992, 32965, 33586, 34698, 35052, 36967, 41365, 42022, 43826, 46440
	fragments:	11385, 10496, 4398, 4199, 3673, 2614, 2063, 1915, 1804, 1701, 1112, 973, 657, 621, 537, 354

Table 7. Continued

*Dra*I	13 sites:	93, 8463, 16297, 23113, 23287, 25439, 26135, 26668, 32706, 36305, 36533, 38836, 47432
	fragments:	8596, 8370, 7834, 6816, 6038, 3599, 2303, 2152, 1071, 696, 533, 228, 174, 92
*Dra*III	10 sites:	2960, 5619, 6641, 9005, 14483, 30371, 31915, 41485, 47318, 48440
	fragments:	15888, 9570, 5833, 5478, 2959, 2364, 1544, 1122, 1022, 63
*Eco*81I	2 sites:	26719, 34320
	fragments:	26718, 14183, 7601
*Eco*47III	2 sites:	20998, 37060
	fragments:	20997, 16062, 11442
*Eco*52I	2 sites:	19945, 36655
	fragments:	19944, 16710, 11848
*Eco*NI	9 sites:	13509, 21292, 22377, 25174, 25223, 35521, 38268, 41842, 47213
	fragments:	13509, 10298, 7783, 7747, 5371, 3574, 2797, 1289, 1085, 49
*Eco*O109I	3 sites:	2817, 28799, 48475
	fragments:	25982, 19676, 2816, 28
*Eco*RI	5 sites:	21227, 26105, 31769, 44973
	fragments:	21226, 7421, 5804, 5643, 4878, 3530
*Eco*RV	21 sites:	653, 2087, 6684, 8087, 8825, 13438, 14026, 17770, 18388, 21272, 22951, 26824, 28201, 28214, 33590, 39355, 41276, 41544, 41579, 42234, 45829
	fragments:	5765, 5376, 4613, 4597, 3873, 3744, 3595, 2884, 2674, 1921, 1679, 1434, 1403, 1377, 738, 655, 652, 618, 588, 268, 35, 13
*Esp*I	6 sites:	10299, 10684, 11663, 16520, 20746, 39452
	fragments:	18706, 10298, 9051, 4857, 4226, 979, 385
*Fsp*I	15 sites:	466, 2506, 4273, 5158, 6982, 11566, 11693, 13358, 16049, 21808, 21829, 27952, 32686, 34824, 42383
	fragments:	7559, 6123, 6120, 5759, 4734, 4584, 2691, 2138, 2040, 1824, 1767, 1665, 885, 465, 127, 21
*Gdi*II	21 sites:	2739, 5601, 6008, 8366, 10588, 13481, 14575, 16416, 18547, 19284, 19332, 19944, 20239, 20323, 20928, 20988, 22025, 35465, 36654, 39458, 45214
	fragments:	13440, 5756, 3288, 2893, 2862, 2804, 2739, 2358, 2222, 2131, 1841, 1189, 1094, 1037, 737, 612, 605, 407, 295, 84, 60, 48
*Gsu*I	25 sites:	419, 1511, 5021, 5872, 6816, 7791, 8591, 11421, 11775, 11994, 12075, 12147, 12449, 13271, 14600, 18037, 21655, 24795, 27514, 33942, 34542, 40672, 41046, 42517, 47620
	fragments:	6428, 6130, 5103, 3618, 3510, 3437, 3140, 2830, 2719, 1471, 1329, 1092, 975, 944, 882, 851, 822, 800, 600, 419, 374, 354, 302, 219, 81, 72
*Hgi*AI	28 sites:	5624, 6007, 9490, 10300, 11955, 13294, 13497, 14479, 15216, 16521, 21617, 21803, 21857, 24777, 25882, 26474, 27178, 33472, 35588, 37938, 40221, 40494, 42376, 42517, 44182, 44851, 46703, 47665
	fragments:	6294, 5623, 5096, 3483, 2920, 2350, 2283, 2116, 1882, 1852, 1665, 1655, 1339, 1305, 1105, 982, 962, 838, 810, 737, 704, 669, 592, 383, 273, 203, 186, 141, 54
*Hgi*EII	14 sites:	1785, 2250, 5903, 6555, 12513, 13954, 15877, 17433, 20244, 26435, 35595, 35639, 37999, 42048
	fragments:	9160, 6454, 6191, 5958, 4049, 3653, 2811, 2360, 1923, 1785, 1556, 1441, 652, 465, 44

Electrophoresis

Table 7. Continued

*Hinc*II	35 sites:	200, 735, 5270, 5711, 7951, 8202, 9057, 9627, 11586, 13786, 14994, 17077, 18757, 19842, 20570, 21095, 23148, 26745, 27319, 28929, 31810, 32220, 32748, 33247, 35262, 35616, 37434, 37990, 38549, 39609, 39837, 40943, 43184, 47939, 48299
	fragments:	4755, 4535, 3597, 2881, 2241, 2240, 2200, 2083, 2015, 1959, 1818, 1680, 1610, 1335, 1243, 1208, 1106, 1085, 1060, 855, 728, 574, 570, 559, 556, 535, 528, 499, 441, 410, 360, 354, 251, 228, 204, 199
*Hind*III	7 sites:	23131, 25158, 27480, 36896, 37460, 37585, 44142
	fragments:	23130, 9416, 6557, 4361, 2322, 2027, 564, 125
*Hpa*I	14 sites:	735, 5270, 5711, 7951, 8202, 11586, 14994, 21905, 27319, 31810, 32220, 35262, 39609, 39837
	fragments:	8666, 6911, 5414, 4535, 4491, 4347, 3408, 3384, 3042, 2240, 734, 441, 410, 251, 228
*Kpn*I	2 sites:	17058, 18561
	fragments:	29942, 17057, 1503
*Mae*I	13 sites:	23948, 24324, 24398, 24510, 24542, 25120, 26125, 34681, 35038, 37256, 44655, 47336, 47707
	fragments:	23947,8556,7399,2681,2218,1005,796,578,376,371,357,112,74, 32
*Mlu*I	7 sites:	459, 5549, 15373, 17792, 19997, 20953, 22221
	fragments:	26282, 9824, 5090, 2419, 2205, 1268, 956, 458
*Mme*I	18 sites:	1286, 15900, 16596, 26889, 26966, 27876, 29922, 30451, 34025, 34575, 35854, 39631, 41996, 42146, 44686, 45543, 48162, 48489
	fragments:	14614, 10293, 3777, 3574, 2619, 2540, 2365, 2046, 1286, 1279, 910, 857, 696, 550, 529, 327, 150, 77, 13
*Nae*I	1 site:	20043
	fragments:	28460, 20042
*Nar*I	1 site:	45681
	fragments:	45680, 2822
*Nco*I	4 sites:	19330, 23902, 27869, 44269
	fragments:	19329, 16380, 4572, 4254, 3967
*Nde*I	7 sites:	27632, 29885, 33681, 36114, 36670, 38359, 40133
	fragments:	27631, 8370, 3796, 2433, 2253, 1774, 1689, 556
*Nhe*I	1 site:	34680
	fragments:	34679, 13823
*Nru*I	5 sites:	4593, 28053, 31706, 32410, 41811
	fragments:	23460, 9401, 6692, 4592, 3653, 704
*Nsi*I	14 sites:	10330, 27211, 27377, 28437, 30347, 30994, 32972, 33687, 34213, 35873, 33670, 36676, 37774, 38312
	fragments:	16881, 10329, 10191, 1978, 1910, 1660, 1098, 1060, 797, 715, 647, 538, 526, 166, 6
*Pfl*MI	14 sites:	5072, 6162, 9005, 11166, 18190, 18607, 24573, 25072, 30309, 33334, 33622, 39723, 41532, 42526
	fragments:	7024, 6101, 5977, 5966, 5237, 5071, 3025, 2843, 2161, 1809, 1090, 994, 499, 417, 288
*Ppu*MI	3 sites:	2817, 28799, 48475
	fragments:	25982, 19676, 2816, 28

Table 7. Continued

*Pst*I	28 sites:	2561, 2825, 3630, 3645, 3861, 4375, 4714, 4914, 5125, 5219, 5687, 8525, 9618, 9782, 11768, 11840, 14299, 14386, 16086, 16236, 17395, 19838, 20286, 22426, 26933, 32010, 32257, 37006
	fragments:	11497, 5077, 4749, 4507, 2838, 2560, 2459, 2443, 2140, 1986, 1700, 1159, 1093, 805, 514, 468, 448, 339, 264, 247, 216, 211, 200, 164, 150, 94, 87, 72, 15
*Pvu*I	3 sites:	11937, 26258, 35791
	fragments:	14321, 12712, 11936, 9533
*Pvu*II	15 sites:	212, 1920, 2329, 2388, 3061, 3640, 7834, 12102, 12165, 16081, 19719, 20062, 20698, 22994, 27415
	fragments:	21088, 4421, 4268, 4194, 3916, 3638, 2296, 1708, 636, 579, 532, 468, 343, 211, 141, 63
*Rsr*II	5 sites:	3802, 6043, 13985, 19290, 22244
	fragments:	26259, 7942, 5305, 3801, 2954, 2241
*Sac*I	2 sites:	24777, 25882
	fragments:	24776, 22621, 1105
*Sac*II	4 sites:	20324, 20534, 21610, 40390
	fragments:	20323, 18780, 8113, 1076, 210
*Sal*I	2 sites:	32746, 33245
	fragments:	32745, 15258, 499
*Sca*I	5 sites:	16424, 18687, 25688, 27266, 32805
	fragments:	16423, 15698, 7001, 5539, 2263, 1578
*Sma*I	3 sites:	19400, 31620, 39891
	fragments:	19399, 12220, 8612, 8271
*Sna*I	3 sites:	15260, 18834, 19473
	fragments:	29030, 15259, 3574, 639
*Sna*BI	1 site:	12191
	fragments:	36312, 12190
*Sph*I	6 sites:	2217, 12007, 23947, 24376, 27379, 39423
	fragments:	12044, 11940, 9790, 9080, 3003, 2216, 429
*Spl*I	1 site:	19323
	fragments:	29179, 19323
*Ssp*I	20 sites:	8472, 24078, 24749, 25260, 25627, 26316, 29157, 29367, 29517, 33573, 35004, 38977, 41908, 42828, 43266, 46848, 47324, 47424, 47559, 47696
	fragments:	15606, 8471, 4056, 3973, 3582, 2931, 2841, 1431, 920, 807, 689, 671, 511, 476, 438, 367, 210, 150, 137, 135, 100
*Sty*I	10 sites:	19330, 21212, 23902, 24323, 24397, 27869, 28794, 35017, 36506, 44249
	fragments:	19329, 7743, 6223, 4254, 3472, 2690, 1882, 1489, 925, 421, 74
*Tth*111I	2 sites:	11202, 36120
	fragments:	24918, 12382, 11202
*Xba*I	1 site:	24509
	fragments:	24508, 23994
*Xho*I	1 site:	33499
	fragments:	33498, 15004

Electrophoresis

Table 7. Continued

*Xho*II	21 sites:	416, 1607, 2532, 5506, 6423, 22347, 22426, 24512, 27028, 27973, 29594, 30427, 34500, 35712, 38104, 38665, 38755, 38815, 39577, 41733, 47774
	fragments:	15924, 6041, 4073, 2974, 2516, 2392, 2156, 2086, 1621, 1212, 1191, 945, 925, 917, 833, 762, 729, 561, 415, 90, 79, 60
*Xmn*I	24 sites:	38, 1156, 2324, 8495, 10116, 13107, 16914, 22857, 22876, 23813, 23833, 24233, 24583, 25490, 27257, 29020, 29998, 31090, 33816, 34190, 42482, 44732, 45746, 47569
	fragments:	8292, 6171, 5943, 3807, 2991, 2726, 2250, 1823, 1767, 1763, 1621, 1168, 1118, 1092, 1014, 978, 937, 934, 907, 400, 374, 350, 37, 20, 19

Table 8. Restriction fragment lengths for M13mp18 DNA (7249 bp, circular)

No sites for

*Aat*I	*Aat*II	*Acc*III	*Afl*II	*Apa*I	*Asu*II	*Ava*II
*Avr*II	*Bbv*II	*Bcl*I	*Bsa*I	*Bsp*MII	*Bss*HII	*Bst*EII
*Bst*XI	*Eco*52I	*Eco*57I	*Eco*NI	*Eco*O109I	*Eco*RV	*Esp*I
*Hpa*I	*Mlu*I	*Nco*I	*Nhe*I	*Not*I	*Nru*I	*Nsi*I
*Pfl*MI	*Pma*CI	*Ppu*MI	*Rsr*II	*Sac*II	*Sca*I	*Sfi*I
*Sna*I	*Spe*I	*Sph*I	*Sty*I	*Tth*111I	*Xho*I	

One site for (coordinates in brackets)

*Acc*I (6265)	*Acy*I (6002)	*Alw*NI (2187)	*Apa*LI (4743)
*Ava*II (5914)	*Bal*I (5082)	*Bam*HI (6252)	*Bgl*I (6437)
*Bgl*II (6935)	*Bsm*I (1746)	*Bsp*HI (1299)	*Cfr*10I (5613)
*Dra*III (5721)	*Eco*81I (6508)	*Eco*RI (6231)	*Fsp*I (6425)
*Hgi*EII (6468)	*Kpn*I (6243)	*Nae*I (5615)	*Nar*I (6002)
*Pst*I (6274)	*Pvu*I (6408)	*Sac*I (6237)	*Sma*I (6247)
*Sna*BI (1270)	*Sph*I (6276)	*Xba*I (6258)	

*Acc*II	18 sites:	45, 349, 1120, 1177, 2467, 3356, 3410, 3600, 3953, 4314, 4995, 5490, 5514, 5534, 6030, 6032, 6690, 6753
	fragments:	1290, 889, 771, 681, 658, 541, 496, 495, 361, 353, 304, 190, 63, 57, 54, 24, 20, 2
*Afl*III	3 sites:	196, 3618, 3717
	fragments:	3728, 3422, 99
*Alu*I	27 sites:	41, 65, 205, 231, 335, 935, 1489, 1518, 2964, 3277, 3613, 4097, 5427, 5631, 5962, 6055, 6119, 6214, 6239, 6284, 6377, 6488, 6668, 6731, 6951, 6978, 7179
	fragments:	1446, 1330, 600, 554, 484, 336, 331, 313, 220, 204, 201, 180, 140, 111, 111, 104, 95, 93, 93, 64, 63, 45, 29, 27, 26, 25, 24
*Alw*I	4 sites:	1382, 2221, 6252, 6253
	fragments:	4031, 2378, 839, 1
*Ase*I	6 sites:	4131, 4135, 4238, 4628, 6046, 6105
	fragments:	5275, 1418, 390, 103, 59, 4

Table 8. Continued

*Ava*I	2 sites:	5825, 6247
	fragments:	6827, 422
*Ban*I	7 sites:	1249, 5677, 6001, 6131, 6243, 6465, 6477
	fragments:	4428, 2021, 324, 222, 130, 112, 12
*Ban*II	2 sites:	5647, 6241
	fragments:	6655, 594
*Bbv*I	10 sites:	931, 1367, 2521, 3132, 4871, 5536, 5923, 6052, 6355, 6428
	fragments:	1752, 1739, 1154, 665, 611, 436, 387, 303, 129, 73
*Bsp*MI	3 sites:	1110, 2261, 6271
	fragments:	4010, 2088, 1151
*Cla*I	2 sites:	2528, 6883
	fragments:	4355, 2894
*Dde*I	29 sites:	234, 1099, 1371, 1417, 1784, 1847, 1862, 1877, 1901, 1973, 2015, 2318, 2333, 2348, 2363, 3361, 4013, 4040, 4079, 4093, 4121, 4281, 4881, 5262, 5934, 6509, 6908, 7061, 7189
	fragments:	998, 865, 672, 652, 600, 575, 399, 381, 367, 303, 294, 272, 160, 153, 128, 72, 63, 46, 42, 39, 28, 27, 24, 15, 15, 15, 15, 15, 14
*Dra*I	5 sites:	192, 475, 4624, 6786, 7076
	fragments:	4149, 2162, 365, 290, 283
*Dsa*I	2 sites:	2763, 6617
	fragments:	3853, 3396
*Eae*I	3 sites:	5080, 6036, 6293
	fragments:	6036, 956, 257
*Eco*RII	7 sites:	1015, 1967, 5942, 5999, 6138, 6329, 6456
	fragments:	3975, 1808, 952, 191, 139, 127, 57
*Fin*I	2 sites:	5090, 6529
	fragments:	5810, 1439
*Fnu*4HI	16 sites:	933, 1368, 1395, 2286, 2289, 2313, 2358, 2522, 3133, 4872, 5501, 5515, 5537, 5924, 6356, 6429
	fragments:	1753, 1739, 891, 629, 611, 435, 432, 387, 164, 73, 45, 27, 24, 22, 14, 3
*Fok*I	4 sites:	239, 3547, 6361, 7244
	fragments:	3308, 2814, 883, 244
*Gdi*II	2 sites:	6036, 6293
	fragments:	6992, 257
*Gsu*I	2 sites:	6493, 6899
	fragments:	6843, 406
*Hae*I	4 sites:	2246, 5082, 5346, 6683
	fragments:	2836, 2812, 1337, 264
*Hae*II	6 sites:	2714, 3043, 5563, 5571, 6005, 6450
	fragments:	3513, 2520, 445, 434, 329, 8
*Hae*III	15 sites:	1397, 2246, 2555, 5082, 5240, 5346, 5415, 5726, 5940, 6038, 6295, 6397, 6514, 6683, 7024
	fragments:	2527, 1622, 849, 341, 311, 309, 257, 214, 169, 158, 117, 106, 102, 98, 69
*Hga*I	7 sites:	526, 2164, 2479, 3237, 4083, 5158, 6687
	fragments:	1638, 1529, 1088, 1075, 846, 758, 315

Electrophoresis

Table 8. Continued

*Hgi*AI	3 sites:	4747, 5469, 6241
	fragments:	5755, 772, 722
*Hha*I	26 sites:	47, 1013, 1087, 1179, 1472, 2197, 2469, 2713, 3042, 3098, 3410, 3600, 4314, 4997, 5492, 5505, 5514, 5536, 5562, 5570, 6004, 6032, 6097, 6428, 6449, 6564
	fragments:	966, 732, 725, 714, 683, 495, 434, 331, 329, 312, 293, 272, 244, 190, 115, 92, 74, 65, 56, 28, 26, 22, 21, 13, 9, 8
*Hin*fI	27 sites:	137, 217, 491, 512, 724, 2011, 2497, 2845, 3258, 3418, 3742, 3838, 4072, 4117, 4349, 5120, 5329, 5375, 5438, 5766, 5788, 6041, 6262, 6624, 6885, 6904, 7041
	fragments:	1287, 771, 486, 413, 362, 348, 345, 328, 324, 274, 261, 253, 234, 232, 221, 212, 209, 160, 137, 96, 80, 63, 46, 45, 22, 21, 19
*Hpa*II	18 sites:	315, 966, 1095, 1924, 2378, 2396, 2552, 3370, 3842, 4018, 5614, 6159, 6248, 6463, 6481, 6838, 6961, 7021
	fragments:	1596, 829, 818, 651, 545, 543, 472, 454, 357, 215, 176, 156, 129, 123, 89, 60, 18, 18
*Hph*I	18 sites:	1376, 1503, 1774, 1909, 2398, 2542, 2571, 2620, 2626, 2635, 4847, 4923, 5117, 5706, 5947, 5980, 7005, 7031
	fragments:	2212, 1594, 1025, 589, 489, 271, 241, 194, 144, 135, 127, 76, 39, 39, 33, 26, 9, 6
*Mae*I	5 sites:	3827, 5565, 6259, 6861, 6979
	fragments:	4097, 1738, 694, 602, 118
*Mae*II	22 sites:	771, 790, 943, 1269, 1577, 1688, 2546, 2818, 3151, 4404, 4447, 4635, 5038, 5456, 5463, 5607, 5717, 5760, 5772, 6309, 6578, 6774
	fragments:	1253, 1246, 585, 537, 418, 403, 333, 326, 308, 272, 269, 196, 188, 153, 144, 111, 110, 43, 43, 19, 12, 7
*Mae*III	24 sites:	145, 939, 1102, 1376, 1717, 1774, 1789, 2094, 2542, 2620, 2743, 2909, 3210, 4321, 4384, 4609, 5110, 5528, 5540, 6313, 6333, 6553, 6579, 6637
	fragments:	1111, 794, 773, 757, 501, 448, 418, 341, 305, 301, 274, 225, 220, 166, 163, 123, 78, 63, 58, 57, 26, 20, 15, 12
*Mbo*II	12 sites:	781, 977, 2218, 3912, 4075, 4271, 4937, 5255, 5587, 6391, 6504, 6805
	fragments:	1694, 1241, 1225, 804, 666, 332, 318, 301, 196, 196, 163, 113
*Nci*I	4 sites:	1925, 6248, 6249, 6839
	fragments:	4323, 2335, 590, 1
*Nde*I	3 sites:	2724, 3804, 6847
	fragments:	3126, 3043, 1080
*Nla*III	16 sites:	153, 200, 1110, 1303, 1802, 2037, 2160, 2858, 3535, 3622, 3721, 5182, 6225, 6280, 6860, 6965
	fragments:	1461, 1043, 910, 698, 677, 580, 437, 499, 235, 193, 123, 105, 99, 87, 55, 47
*Nla*IV	18 sites:	1064, 1251, 1543, 1553, 1805, 2053, 2376, 2394, 5646, 5658, 5679, 5869, 6003, 6133, 6245, 6254, 6467, 6479
	fragments:	3252, 1834, 323, 292, 252, 248, 213, 190, 187, 134, 130, 112, 21, 18, 12, 12, 10, 9
*Nsp*BII	4 sites:	1631, 5960, 6053, 6375
	fragments:	4329, 2505, 322, 93
*Ple*I	8 sites:	2011, 2845, 4072, 5329, 5766, 5788, 6262, 6904
	fragments:	2356, 1257, 1227, 834, 642, 474, 437, 22

Table 8. Continued

*Pvu*II	3 sites:	5962, 6055, 6377
	fragments:	6834, 322, 93
*Rsa*I	19 sites:	175, 282, 1023, 1166, 1770, 1797, 1890, 1906, 1971, 2134, 3468, 3669, 4191, 4381, 5385, 5487, 6245, 6844, 7166
	fragments:	1334, 1004, 758, 741, 604, 599, 522, 322, 258, 201, 190, 163, 143, 107, 102, 93, 65, 27, 16
*Sau*3AI	7 sites:	1381, 1713, 2220, 6252, 6405, 6501, 6935
	fragments:	4032, 1695, 507, 434, 332, 153, 96
*Sau*96I	4 sites:	5724, 5914, 5938, 6396
	fragments:	6577, 458, 190, 24
*Scr*FI	11 sites:	1015, 1925, 1967, 5942, 5999, 6138, 6248, 6249, 6329, 6456, 6839
	fragments:	3975, 1425, 910, 383, 139, 127, 110, 80, 57, 42, 1
*Sdu*I	5 sites:	2092, 4747, 5469, 5647, 6241
	fragments:	3100, 2655, 722, 594, 178
*Sec*I	9 sites:	2763, 2894, 5997, 6136, 6247, 6248, 6327, 6617, 6838
	fragments:	3174, 3103, 290, 221, 139, 131, 111, 79, 1
*Sfa*NI	7 sites:	25, 388, 1354, 3979, 4850, 6546, 6559
	fragments:	2625, 1696, 966, 871, 715, 363, 13
*Ssp*I	6 sites:	502, 2662, 5025, 5215, 6769, 6790
	fragments:	2363, 2160, 1554, 961, 190, 21
*Taq*I	12 sites:	337, 1127, 1508, 1949, 2528, 3455, 3694, 4665, 5683, 6235, 6265, 6883
	fragments:	1018, 971, 927, 790, 703, 618, 579, 552, 441, 381, 239, 30
*Taq*II	4 sites:	1423, 2445, 3279, 5704
	fragments:	2968, 2425, 1022, 834
*Tth*111II	7 sites:	3395, 3446, 4257, 4506, 4671, 5878, 6893
	fragments:	3751, 1207, 1015, 811, 249, 165, 51
*Xho*II	3 sites:	2220, 6252, 6935
	fragments:	4032, 2534, 683
*Xmn*I	2 sites:	362, 2650
	fragments:	4961, 2288

Table 9. Restriction fragment lengths for pBR322 DNA (4363 bp, circular)

No sites for

*Aat*I	*Afl*II	*Apa*I	*Asu*II	*Avr*II	*Bcl*I	*Bgl*I
*Bss*HII	*Bst*EII	*Bst*XI	*Dra*III	*Eco*81I	*Esp*I	*Hpa*I
*Kpn*I	*Mlu*I	*Nco*I	*Not*I	*Nsi*I	*Pma*CI	*Sac*I
*Sac*II	*Sfi*I	*Sma*I	*Sna*BI	*Spe*I	*Sph*I	*Xba*I
*Xho*I						

Table 9. Continued

One site for (coordinates in brackets)

*Aat*II (4291)	*Acc*III (1665)	*Afl*III (2476)	*Alw*NI (2886)
*Ase*I (3539)	*Ava*I (1426)	*Bal*I (1447)	*Bam*HI (376)
*Bsa*I (3435)	*Bsm*I (1354)	*Bsp*MI (1061)	*Bsp*MII (1665)
*Cla*I (25)	*Eco*NI (622)	*Eco*RI (4362)	*Eco*RV (188)
*Hin*dIII (30)	*Nde*I (2299)	*Nhe*I (230)	*Nru*I (975)
*Pst*I (3614)	*Pvu*I (3739)	*Pvu*II (2069)	*Sal*I (652)
*Sca*I (3849)	*Sna*I (2246)	*Sph*I (567)	*Ssp*I (4173)
*Sty*I (1370)	*Tth*111I (2223)		

*Acc*I	2 sites:	653, 2248
	fragments:	2768, 1595
*Acc*II	23 sites:	348, 704, 819, 948, 975, 980, 1041, 1107, 1236, 1246, 1391, 1417, 1539, 1636, 2008, 2077, 2079, 2182, 2523, 3104, 3434, 3927, 4259
	fragments:	581, 493, 452, 372, 356, 341, 332, 330, 145, 129, 129, 122, 115, 103, 97, 69, 66, 61, 27, 26, 10, 5, 2
*Acy*I	6 sites:	415, 436, 550, 1207, 3906, 4288
	fragments:	2699, 657, 490, 382, 114, 21
*Alu*I	16 sites:	17, 32, 688, 1091, 2001, 2058, 2069, 2118, 2137, 2418, 2644, 2780, 3037, 3558, 3658, 3721
	fragments:	910, 659, 656, 521, 403, 281, 257, 226, 136, 100, 63, 57, 49, 19, 15, 11
*Alw*I	12 sites:	375, 376, 1097, 1667, 3042, 3116, 3128, 3213, 3226, 3690, 3993, 4011
	fragments:	1375, 727, 721, 570, 464, 303, 85, 74, 18, 13, 12, 1
*Apa*LI	3 sites:	2292, 2790, 4036
	fragments:	2619, 1246, 498
*Ava*II	8 sites:	798, 886, 1135, 1438, 1480, 1759, 3505, 3727
	fragments:	1544, 1411, 101, 279, 249, 222, 88, 42
*Ban*I	9 sites:	77, 120, 414, 435, 549, 767, 1206, 1290, 3317
	fragments:	2027, 1123, 439, 294, 218, 114, 84, 43, 21
*Ban*II	2 sites:	476, 490
	fragments:	4349, 14
*Bbv*I	21 sites:	226, 615, 773, 1406, 1430, 1559, 1562, 1685, 2065, 2068, 2114, 2211, 2380, 2398, 2817, 2882, 2885, 3091, 3419, 3608, 3785
	fragments:	804, 633, 419, 389, 380, 328, 206, 189, 177, 169, 158, 129, 123, 97, 65, 46, 24, 18, 3, 3, 3
*Bbv*II	3 sites:	737, 1600, 4353
	fragments:	2753, 863, 747
*Bgl*I	3 sites:	936, 1170, 3489
	fragments:	2319, 1810, 234
*Bsp*HI	4 sites:	489, 3195, 4203, 4308
	fragments:	2706, 1008, 544, 105
*Cfr*10I	7 sites:	160, 401, 410, 769, 929, 1283, 3448
	fragments:	2165, 1075, 359, 354, 241, 160, 9
*Dde*I	8 sites:	1582, 1744, 2286, 2751, 3160, 3326, 3866, 4292
	fragments:	1653, 542, 540, 465, 426, 409, 166, 162
*Dra*I	3 sites:	3231, 3250, 3942
	fragments:	3652, 692, 19
*Dsa*I	2 sites:	528, 1447
	fragments:	3444, 919

Table 9. Continued

*Eae*I	6 sites:	296, 400, 532, 940, 1445, 3757
	fragments:	2312, 902, 505, 408, 132, 104
*Eco*47III	4 sites:	232, 494, 775, 1727
	fragments:	2868, 952, 281, 262
*Eco*57I	2 sites:	3002, 4050
	fragments:	3315, 1048
*Eco*O109I	4 sites:	525, 1440, 1482, 4345
	fragments:	2863, 915, 543, 42
*Eco*RII	6 sites:	130, 1059, 1442, 2502, 2623, 2636
	fragments:	1857, 1060, 929, 383, 121, 13
*Fin*I	4 sites:	538, 888, 1084, 1761
	fragments:	3140, 677, 350, 196
*Fok*I	12 sites:	99, 147, 1001, 1046, 1695, 1757, 1835, 2023, 2164, 3335, 3516, 3803
	fragments:	1171, 854, 659, 287, 188, 181, 141, 78, 62, 48, 45
*Gdi*II	5 sites:	295, 399, 531, 939, 3756
	fragments:	2817, 902, 408, 132, 104
*Gsu*I	4 sites:	811, 1401, 1983, 3453
	fragments:	1721, 1470, 590, 582
*Hae*II	11 sites:	237, 418, 439, 499, 553, 780, 1210, 1649, 1732, 2354, 2724
	fragments:	1876, 622, 439, 430, 370, 227, 181, 83, 60, 54, 21
*Hae*III	22 sites:	175, 298, 402, 526, 534, 598, 832, 921, 942, 993, 1050, 1063, 1447, 1951, 2491, 2502, 2520, 2954, 3412, 3492, 3759, 4346
	fragments:	587, 540, 504, 458, 434, 267, 234, 213, 192, 184, 124, 123, 104, 89, 80, 64, 57, 51, 21, 18, 11, 8
*Hga*I	11 sites:	380, 659, 934, 966, 1250, 1400, 1994, 2191, 2567, 3145, 3915
	fragments:	828, 770, 594, 578, 376, 284, 279, 275, 197, 150, 32
*Hgi*AI	8 sites:	281, 592, 1179, 1470, 2296, 2794, 3955, 4040
	fragments:	1161, 826, 604, 587, 498, 311, 291, 85
*Hgi*EII	2 sites:	2295, 3056
	fragments:	3602, 761
*Hha*I	31 sites:	104, 236, 264, 417, 438, 498, 552, 704, 779, 819, 950, 1209, 1360, 1422, 1458, 1648, 1731, 2079, 2182, 2212, 2353, 2386, 2656, 2723, 2823, 2997, 3106, 3499, 3592, 3929, 4261
	fragments:	393, 348, 337, 332, 270, 259, 206, 190, 174, 153, 152, 151, 141, 132, 131, 109, 103, 100, 93, 83, 75, 67, 62, 60, 54, 40, 36, 33, 30, 28, 21
*Hinc*II	2 sites:	654, 3910
	fragments:	3256, 1107
*Hinf*I	10 sites:	633, 853, 1007, 1305, 1526, 2032, 2376, 2451, 2847, 3364
	fragments:	1632, 517, 506, 396, 344, 298, 221, 220, 154, 75
*Hpa*II	26 sites:	162, 171, 388, 403, 412, 535, 695, 771, 931, 1021, 1259, 1285, 1486, 1666, 1813, 2122, 2156, 2683, 2830, 2856, 3046, 3450, 3484, 3551, 3661, 3903
	fragments:	622, 527, 404, 309, 242, 238, 217, 201, 190, 180, 160, 160, 147, 147, 123, 110, 90, 76, 67, 34, 34, 26, 26, 15, 9, 9
*Hph*I	12 sites:	126, 408, 453, 1307, 1528, 2085, 2094, 3219, 3446, 3842, 4068, 4083
	fragments:	854, 557, 406, 396, 282, 227, 226, 221, 125, 45, 15, 9
*Mae*I	5 sites:	231, 1490, 2971, 3224, 3559
	fragments:	1481, 1259, 1035, 335, 253

Electrophoresis

Table 9. Continued

*Mae*II	10 sites:	902, 958, 1547, 1571, 1801, 2229, 3179, 3595, 3968, 4288
	fragments:	977, 950, 589, 428, 416, 373, 320, 230, 56, 24
*Mae*III	17 sites:	125, 213, 881, 1148, 1808, 1831, 1917, 2130, 2225, 2832, 2895, 3011, 3294, 3625, 3683, 3836, 4024
	fragments:	668, 660, 607, 464, 331, 283, 267, 213, 188, 153, 116, 95, 88, 86, 63, 58, 23
*Mbo*II	11 sites:	464, 737, 1008, 1600, 2353, 3124, 3215, 3970, 4048, 4157, 4353
	fragments:	771, 755, 753, 592, 473, 273, 271, 196, 109, 91, 78
*Mme*I	4 sites:	197, 284, 2665, 2849
	fragments:	2381, 1711, 184, 87
*Nae*I	4 sites:	404, 772, 932, 1286
	fragments:	3481, 368, 354, 160
*Nar*I	4 sites:	415, 436, 550, 1207
	fragments:	3571, 657, 114, 21
*Nci*I	10 sites:	172, 536, 1260, 1486, 1814, 2122, 2157, 2856, 3552, 3903
	fragments:	724, 699, 696, 632, 364, 351, 328, 308, 226, 35
*Nsp*BII	6 sites:	1139, 2066, 2185, 2815, 3060, 4001
	fragments:	1507, 941, 927, 630, 245, 119
*Pfl*MI	2 sites:	1322, 1371
	fragments:	4314, 49
*Ppu*MI	2 sites:	1440, 1482
	fragments:	4321, 42
*Rsa*I	3 sites:	166, 2284, 3849
	fragments:	2118, 1565, 680
*Sau*3AI	22 sites:	349, 376, 467, 826, 1098, 1129, 1144, 1461, 1668, 3042, 3117, 3128, 3136, 3214, 3226, 3331, 3672, 3690, 3736, 3994, 4011, 4047
	fragments:	1374, 665, 359, 341, 317, 272, 258, 207, 105, 91, 78, 75, 46, 36, 31, 27, 18, 17, 15, 12, 11, 8
*Sau*96I	15 sites:	173, 525, 800, 888, 1137, 1261, 1440, 1482, 1761, 1950, 3411, 3490, 3507, 3729, 4345
	fragments:	1461, 616, 352, 279, 249, 222, 191, 189, 179, 124, 88, 79, 42, 17
*Scr*FI	16 sites:	132, 172, 536, 1061, 1260, 1444, 1486, 1814, 2122, 2157, 2504, 2625, 2638, 2856, 3552, 3903
	fragments:	696, 592, 525, 364, 351, 347, 328, 308, 218, 199, 184, 121, 42, 40, 35, 13
*Sdu*I	10 sites:	281, 476, 490, 592, 1179, 1470, 2296, 2794, 3955, 4040
	fragments:	1161, 826, 604, 587, 498, 291, 195, 102, 85, 14
*Sec*I	8 sites:	115, 129, 413, 434, 548, 766, 1205, 1289, 3316
	fragments:	1843, 1188, 633, 399, 202, 78, 14, 6
*Taq*I	7 sites:	25, 340, 653, 1128, 1269, 2576, 4020
	fragments:	1444, 1307, 475, 368, 315, 313, 141
*Taq*II	6 sites:	654, 2387, 3726, 3885, 4038, 4081
	fragments:	1733, 1339, 936, 159, 153, 43
*Tth*111II	5 sites:	18, 1933, 3060, 3093, 3099
	fragments:	1915, 1282, 1127, 33, 6
*Xho*II	8 sites:	376, 1668, 3117, 3128, 3214, 3226, 3994, 4011
	fragments:	1449, 1292, 768, 728, 86, 17, 12, 11
*Xmn*I	2 sites:	2036, 3968
	fragments:	2431, 1932

Table 10. Restriction fragment lengths for φX174 DNA (5386 bp, circular)

No sites for

*Acc*III	*Alw*I	*Alw*NI	*Apa*I	*Asu*II	*Avr*II	*Bal*I
*Bam*HI	*Ban*I	*Bcl*I	*Bgl*I	*Bgl*II	*Bsa*I	*Bsp*MII
*Bst*EII	*Bst*XI	*Cfr*10I	*Cla*I	*Eco*47III	*Eco*52I	*Eco*57I
*Eco*81I	*Eco*NI	*Eco*O109I	*Eco*RI	*Eco*RV	*Esp*I	*Hind*III
*Kpn*I	*Nae*I	*Nco*I	*Nde*I	*Nhe*I	*Not*I	*Nsi*I
*Pma*CI	*Ppu*MI	*Pvu*I	*Pvu*II	*Rsr*II	*Sac*I	*Sal*I
*Sau*3AI	*Sca*I	*Sfi*I	*Sma*I	*Sna*I	*Sna*BI	*Spe*I
*Sph*I	*Sty*I	*Tth*111I	*Xba*I	*Xho*II		

One site for (coordinates in brackets)

*Aat*I (4489)	*Aat*II (2787)	*Apa*LI (4780)	*Ava*I (163)
*Ava*II (5205)	*Dra*III (5189)	*Fsp*I (158)	*Nci*I (2802)
*Pst*I (1)	*Sac*II (2863)	*Ssp*I (1010)	*Xho*I (163)

*Acc*I	2 sites:	1195, 3547
	fragments:	3034, 2352
*Acc*II	14 sites:	224, 754, 773, 1823, 1979, 2149, 2263, 2366, 2862, 3557, 4427, 4506, 5224, 5351
	fragments:	1050, 870, 718, 695, 530, 496, 259, 170, 156, 127, 114, 103, 79, 19
*Acy*I	7 sites:	719, 1021, 1135, 2784, 2978, 3365, 5227
	fragments:	1862, 1649, 878, 387, 302, 194, 114
*Afl*II	2 sites:	2915, 4414
	fragments:	3887, 1499
*Afl*III	2 sites:	222, 2147
	fragments:	3461, 1925
*Alu*I	24 sites:	162, 171, 447, 1454, 1538, 1593, 2446, 2471, 2504, 3166, 3312, 3390, 3637, 3841, 3883, 4137, 4395, 4753, 4862, 4981, 5068, 5100, 5121, 5211
	fragments:	1007, 853, 662, 358, 337, 276, 258, 254, 247, 204, 146, 119, 109, 90, 87, 84, 78, 55, 42, 33, 32, 25, 21, 9
*Ase*I	2 sites:	711, 4308
	fragments:	3597, 1789
*Ban*I	3 sites:	1020, 2479, 2977
	fragments:	3429, 1459, 498
*Bbv*I	14 sites:	158, 547, 624, 892, 1488, 1592, 1639, 3294, 3297, 3798, 4215, 4398, 4677, 5049
	fragments:	1655, 598, 501, 495, 417, 389, 372, 279, 268, 183, 104, 77, 47, 3
*Bbv*II	3 sites:	2679, 4367, 4762
	fragments:	3303, 1688, 395
*Bsm*I	4 sites:	2577, 3032, 3470, 4536
	fragments:	3427, 1066, 455, 438
*Bsp*HI	3 sites:	388, 1254, 2271
	fragments:	3503, 1017, 866
*Bsp*MI	3 sites:	3597, 4056, 5384
	fragments:	3599, 1328, 459
*Bst*XI	3 sites:	1785, 2747, 4992
	fragments:	2245, 2179, 962

Electrophoresis

Table 10. Continued

*Cvi*JI	69 sites:	160, 169, 242, 434, 445, 455, 557, 668, 737, 894, 972, 978, 1055, 1172, 1438, 1452, 1536, 1591, 1739, 1775, 2102, 2444, 2469, 2502, 2909, 3011, 3102, 3128, 3158, 3164, 3197, 3310, 3388, 3635, 3644, 3744, 3839, 3881, 3894, 3972, 4043, 4064, 4104, 4135, 4206, 4214, 4393, 4397, 4487, 4512, 4550, 4595, 4694, 4751, 4758, 4860, 4876, 4948, 4969, 4979, 4994, 5012, 5066, 5098, 5119, 5139, 5209, 5278, 5297
	fragments:	407, 342, 327, 266, 249, 247, 192, 179, 157, 148, 117, 113, 111, 102, 102, 102, 100, 99, 95, 91, 90, 84, 78, 78, 78, 77, 73, 72, 71, 71, 70, 69, 69, 57, 55, 54, 45, 42, 40, 38, 36, 33, 33, 32, 31, 30, 26, 25, 25, 21, 21, 21, 20, 19, 18, 16, 15, 14, 13, 11, 10, 10, 9, 9, 8, 7, 6, 6, 4
*Dde*I	14 sites:	351, 1363, 1528, 1534, 1837, 2023, 2041, 2343, 2381, 3308, 3701, 3711, 4253, 4739
	fragments:	1012, 998, 927, 542, 486, 393, 303, 186, 165, 38, 18, 10, 6
*Dra*I	2 sites:	330, 1409
	fragments:	4307, 1079
*Dsa*I	3 sites:	1415, 2859, 5182
	fragments:	1444, 1619, 2323
*Eae*I	2 sites:	4206, 4758
	fragments:	4834, 552
*Eco*RII	2 sites:	883, 3502
	fragments:	2767, 2619
*Fin*I	2 sites:	1214, 4339
	fragments:	3125, 2261
*Fnu*4HI	31 sites:	160, 549, 626, 740, 894, 1035, 1321, 1490, 1594, 1641, 1821, 1974, 1977, 2340, 2598, 2860, 3014, 3131, 3260, 3296, 3299, 3800, 4217, 4389, 4400, 4679, 4848, 4972, 5051, 5142, 5312
	fragments:	501, 417, 389, 363, 286, 279, 262, 258, 234, 180, 172, 170, 169, 169, 154, 154, 153, 141, 129, 124, 117, 114, 104, 91, 79, 77, 47, 36, 11, 3, 3
*Fok*I	8 sites:	228, 648, 1261, 1419, 1996, 4265, 4463, 5152
	fragments:	2269, 689, 613, 577, 462, 420, 198, 158
*Gdi*II	2 sites:	4205, 4757
	fragments:	4897, 507
*Gsu*I	3 sites:	2465, 2905, 4482
	fragments:	3369, 1577, 440
*Hae*I	6 sites:	436, 1174, 1777, 4489, 4878, 4950
	fragments:	2712, 872, 738, 603, 389, 72
*Hae*II	8 sites:	692, 877, 931, 1024, 1147, 1416, 2981, 3764
	fragments:	2314, 1565, 783, 269, 185, 123, 93, 54
*Hae*III	11 sites:	436, 670, 980, 1174, 1777, 3130, 4208, 4489, 4760, 4878, 4950
	fragments:	1353, 1078, 872, 603, 310, 281, 271, 234, 194, 118, 72
*Hga*I	14 sites:	19, 220, 718, 725, 798, 815, 1133, 1165, 1688, 3363, 3561, 4920, 5220, 5225
	fragments:	1675, 1359, 523, 498, 318, 300, 201, 198, 180, 73, 32, 17, 7, 5
*Hgi*AI	3 sites:	543, 4295, 4784
	fragments:	3752, 1145, 489
*Hha*I	18 sites:	159, 691, 775, 876, 930, 1023, 1146, 1415, 1720, 1865, 2066, 2366, 2980, 3620, 3763, 5316, 5351, 5353
	fragments:	1553, 640, 614, 532, 305, 300, 269, 201, 192, 145, 143, 123, 101, 93, 84, 54, 35, 2

Table 10. Continued

*Hinc*II	13 sites:	31, 322, 657, 954, 1295, 2352, 2514, 2593, 3363, 3708, 4203, 4815, 5025
	fragments:	1057, 770, 612, 495, 392, 345, 341, 335, 297, 291, 210, 162, 79
*Hinf*I	21 sites:	54, 205, 287, 311, 353, 393, 1119, 1185, 1898, 2016, 2265, 2678, 2878, 3378, 3444, 3544, 3684, 3732, 4149, 4702, 5129
	fragments:	726, 713, 553, 500, 427, 413, 311, 249, 200, 151, 140, 118, 100, 82, 66, 66, 48, 42, 40, 24
*Hpa*I	3 sites:	31, 1295, 5025
	fragments:	3730, 1264, 392
*Hpa*II	5 sites:	730, 1104, 2801, 3020, 3368
	fragments:	2748, 1697, 374, 348, 219
*Hph*I	9 sites:	1235, 2012, 2089, 2132, 2923, 3033, 3072, 4188, 4965
	fragments:	1656, 1116, 791, 777, 777, 110, 77, 43, 39
*Mae*I	3 sites:	3137, 3908, 5062
	fragments:	3461, 1154, 771
*Mae*II	19 sites:	260, 417, 788, 812, 828, 866, 1228, 1625, 1767, 2516, 2784, 2811, 3190, 4226, 4591, 4873, 4932, 5021, 5303
	fragments:	1036, 749, 397, 379, 371, 365, 362, 343, 282, 282, 268, 157, 142, 89, 59, 38, 27, 24, 16
*Mae*III	17 sites:	288, 399, 505, 1224, 1644, 2231, 2461, 2773, 3789, 4009, 4248, 4374, 4642, 4885, 5215, 5289, 5322
	fragments:	1016, 719, 587, 420, 352, 330, 312, 268, 243, 239, 230, 220, 126, 111, 106, 74, 33
*Mbo*II	11 sites:	480, 804, 1661, 2473, 2562, 2680, 2683, 3747, 4143, 4367, 4763
	fragments:	1103, 1064, 857, 812, 396, 396, 324, 224, 118, 89, 3
*Mlu*I	2 sites:	222, 2147
	fragments:	3461, 1925
*Mme*I	5 sites:	225, 2691, 3237, 5197, 5376
	fragments:	2466, 1960, 546, 235, 179
*Mnl*I	34 sites:	232, 436, 555, 594, 903, 990, 1217, 1280, 1436, 1523, 1671, 1680, 2042, 2300, 2344, 2700, 2921, 2998, 3016, 3126, 3346, 3702, 3934, 3957, 3970, 4041, 4062, 4410, 4485, 4489, 4510, 4611, 4683, 5276
	fragments:	593, 362, 356, 356, 348, 342, 309, 232, 227, 221, 220, 204, 158, 156, 148, 119, 110, 101, 87, 87, 77, 74, 72, 71, 63, 44, 39, 23, 21, 21, 18, 13, 9, 4
*Nar*I	2 sites:	1021, 2978
	fragments:	3429, 1957
*Nla*III	22 sites:	20, 305, 393, 684, 1041, 1259, 1320, 1640, 1653, 1718, 1845, 2169, 2276, 2398, 2744, 2830, 2985, 3072, 3829, 3957, 4558, 4777
	fragments:	757, 629, 601, 357, 346, 324, 320, 291, 285, 219, 218, 155, 128, 127, 122, 107, 88, 87, 86, 65, 61, 13
*Nla*IV	6 sites:	981, 1022, 2122, 2481, 2691, 2979
	fragments:	3388, 1100, 359, 288, 210, 41
*Nru*I	2 sites:	2263, 4427
	fragments:	3222, 2164
*Nsp*BII	5 sites:	739, 2339, 2859, 3259, 3735
	fragments:	2390, 1600, 520, 476, 400
*Pfl*MI	2 sites:	3967, 4102
	fragments:	5251, 135

Electrophoresis

Table 10. Continued

*Ple*I	10 sites:	53, 286, 352, 392, 1118, 1184, 2677, 3443, 3543, 3683
	fragments:	1756, 1493, 766, 726, 233, 140, 100, 66, 66, 40
*Rsa*I	11 sites:	416, 573, 770, 908, 1433, 1905, 2152, 2797, 3189, 4749, 4838
	fragments:	1560, 964, 645, 525, 472, 392, 247, 197, 157, 138, 89
*Sau*96I	2 sites:	979, 5043
	fragments:	4064, 1322
*Scr*FI	3 sites:	883, 2802, 3502
	fragments:	2767, 1919, 700
*Sdu*I	3 sites:	543, 4295, 4784
	fragments:	3752, 1145, 489
*Sec*I	6 sites:	880, 1415, 1434, 2859, 4681, 5182
	fragments:	1822, 1425, 1084, 535, 501, 19
*Sfa*NI	12 sites:	358, 1420, 1466, 1733, 2638, 3075, 3339, 3431, 3521, 3795, 4129, 4956
	fragments:	1062, 905, 827, 788, 437, 334, 274, 267, 264, 92, 90, 46
*Spl*I	2 sites:	413, 2794
	fragments:	3005, 2381
*Taq*I	10 sites:	57, 111, 144, 164, 491, 722, 1126, 4040, 4181, 5356
	fragments:	2914, 1175, 404, 327, 231, 141, 87, 54, 33, 20
*Taq*II	2 sites:	3070, 4602
	fragments:	3857, 1529
*Tth*111I	11 sites:	91, 251, 2662, 2932, 2992, 3040, 3226, 3317, 3523, 3590, 4099
	fragments:	2411, 1378, 509, 270, 206, 186, 160, 91, 67, 60, 48
*Xmn*I	3 sites:	3862, 4148, 5122
	fragments:	4126, 974, 286

4. REFERENCES

1. McDonell, M.W., Simon, M.N. and Studier, F.W. (1977) *J. Mol. Biol.,* **110,** 119.

2. Carmichael, G.C. and McMaster, G.K. (1980) *Methods Enzymol.,* **65,** 380.

3. Bailey, J.M. and Davidson, N. (1976) *Anal. Biochem.,* **70,** 75.

4. Lehrach, H., Diamond, D. Wozney, J.M. and Boedtker, H. (1977) *Biochemistry,* **16,** 4743.

5. Loening, U.E. (1967) *Biochem. J.,* **102,** 251.

6. Anker. H.S. (1970) *FEBS Lett.,* 7, 293.

CHAPTER 9
HYBRIDIZATION ANALYSIS

The original techniques for filter hybridization were relatively straightforward but in recent years the methodology has become increasingly complex. Several factors have contributed to this greater sophistication, the most important being:

(i)   the development of new support materials for nucleic acid hybridization;
(ii)   a greater understanding of the factors that influence hybrid stability and hybridization rate;
(iii)   a proliferation of reagents and protocols for filter hybridization;
(iv)   the development of techniques for filter hybridization using synthetic oligonucleotides as probes.

Each of these aspects of filter hybridization will be considered in turn.

1. FILTERS FOR NUCLEIC ACID HYBRIDIZATION

Five different types of filter are commonly used for nucleic acid immobilization. These are nitrocellulose, supported nitrocellulose (i.e. nitrocellulose laid onto an inert support to provide greater tensile strength), nylon, positively charged nylon and activated papers (DBM and DPT papers, see *Figure 1*). The properties of these filters are listed in *Table 1* and recommendations regarding their use are given in *Table 2*. For most procedures, nitrocellulose and nylon are virtually interchangeable and the choice between them is usually based on personal preference.

Figure 1. Structures of the active groups of DBM and DPT papers.

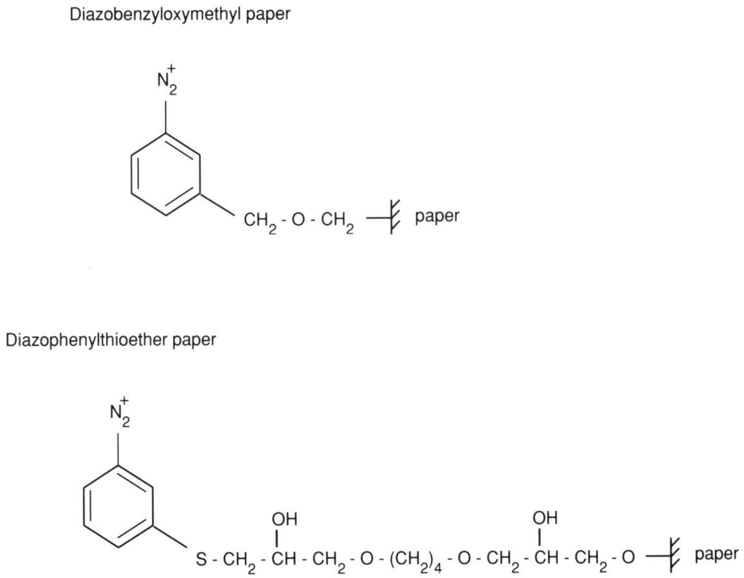

Diazobenzyloxymethyl paper

N_2^+

$CH_2 - O - CH_2$ —⌇ paper

Diazophenylthioether paper

N_2^+

$S - CH_2 - CH - CH_2 - O - (CH_2)_4 - O - CH_2 - CH - CH_2 - O$ —⌇ paper
            | |
            OH   OH

Table 1. Properties of materials used for immobilization of nucleic acids[a]

Property	Nitrocellulose	Supported nitrocellulose
Applications	ssDNA, RNA, protein	ssDNA, RNA, protein
Binding capacity (μg nucleic acid cm^{-2})	80–100	80–100
Tensile strength	Poor	Good
Mode of nucleic acid attachment	Non-covalent: results from baking for 2 h at 80°C	
Lower size limit for efficient nucleic acid retention	500 nt	500 nt
Suitability for re-probing	Poor because of fragility	Poor because of loss of signal
Suitability for electroblotting	Poor because high salt is needed for immobilization of nucleic acids Good for proteins	
Typical commercial examples	Schleicher and Schuell BA83, BA85 Amersham Hybond-C Gelman Biotrace NT PALL Biodyne A	Schliecher and Schuell Optibond Amersham Hybond-C extra

[a] Table compiled from various sources, including ref. 3 and literature pertaining to the commercial examples named.

[b] ABM (*m*-aminobenzyloxymethyl) and APT (*o*-aminophenylthioether) papers need to be converted into the active forms [DBM (diazobenzyloxymethyl) and DPT (diazophenythioether) respectively] by chemical modification immediately prior to use.

Nylon	Positively charged nylon	Activated papers
ss + dsDNA, RNA, protein	ss + dsDNA, RNA, protein	ssDNA, RNA
400–600	400–600	2–40
Good	Good	Good
Covalent: results from u.v. treatment (5 min, 312 nm)	Covalent: results from 0.4 M NaOH in the transfer buffer	Covalent: results from chemical reaction with active groups
50 nt/bp	50 nt/bp	5 nt
Good	Good	Good
Good	Good	Described for DBM paper (1, 2)
Amersham Hybond-N Stratagene Duralon-UV	Schleicher and Schuell Nytran Amersham Hybond-N+ Gelman Biotrace RP PALL Biodyne B BioRad ZetaProbe	Schleicher and Schuell ABM and APT papers[b]

Hybridization

Table 2. Recommended uses for different types of immobilization filter[a]

Application	Recommended filter type[b]	Reasons
DNA/RNA dot blots	Nylon > NC	Strength, covalent attachment allows re-probing
Colony/plaque lifts		
DNA/RNA probe	Nylon > NC	Strength, covalent attachment allows re-probing
Antibody probe	NC > nylon	Lower background
Blots		
Southern	Charged nylon > nylon > NC	Alkali transfer is a convenient method for covalent binding
Northern	Nylon > NC	Strength, covalent attachment allows re-probing
Western	NC > nylon	Lower background
Electroblotting	Nylon >>> NC	Immobilization on nylon does not require high salt
Immobilization of short nucleic acids	Activated papers	Only type suitable for molecules < 50 nt
Non-radioactive probes	NC > nylon	Lower background with some methods

[a] For sources see footnote a to *Table 1*.
[b] Abbreviation: NC, nitrocellulose or supported nitrocellulose.

2. FACTORS INFLUENCING HYBRIDIZATION

The central importance of filter hybridization in molecular biology has stimulated investigations into the factors that influence hybridization between nucleic acid molecules. The two key considerations are the thermal stability of the hybrid and the rate at which hybridization occurs.

2.1. Thermal stability

The thermal stability of a nucleic acid hybrid is expressed in terms of its melting temperature, T_m. Factors influencing the T_m of hybrids in solution are thought to hold for filter hybridization as well (*Table 3*). These factors must be taken into account when designing conditions for hybridization experiments. There have been a number of attempts to derive equations for calculating the T_m of particular hybrids; the more useful of these are listed in *Table 4*.

2.2. Hybridization rate

The hybridization rate is perhaps less important in practical terms since in most protocols hybridization is allowed to proceed for so long that factors influencing the rate become immaterial. There is also the complication that factors identified as influencing the hybridization rate in solution (detailed in *Table 5*) may not be the same as those involved in filter hybridization.

For a more detailed account of the factors influencing hybridization see ref. 3.

Table 3. Factors influencing the T_m of nucleic acid hybrids (3)

Factor	Influence on T_m
Ionic strength	T_m increases 16.6°C for each onefold increase in monovalent cations, between 0.01 and 0.40 M NaCl.
Base composition	AT base pairs are less stable than GC base pairs in aqueous solutions containing NaCl. The difference is negligible in tetramethyl ammonium chloride.
Destabilizing agents	Each 1% of formamide reduces the T_m by about 0.6°C. 6 M urea reduces the T_m by about 30°C.
Mismatched base pairs	The T_m is reduced by 1°C for each 1% of mismatching.
Duplex length	Negligible effect with probes >500 bp.

Table 4. Equations for calculation of T_m

System	Equation[a]	Reference
DNA–DNA hybrids[b]	$T_m = 81.5°C + 16.6(\log M) + 0.41(\%GC) - 0.61(\%form) - 500/L$	4
DNA–RNA hybrids[b]	$T_m = 79.8°C + 18.5(\log M) + 0.58(\%GC) + 11.8(\%GC)^2 - 0.50(\%form) - 820/L$	5
RNA–RNA hybrids[b]	$T_m = 79.8°C + 18.5(\log M) + 0.58(\%GC) + 11.8(\%GC)^2 - 0.35(\%form) - 820/L$	6
Oligonucleotide probes[c]	$T_m = 2(\text{number of AT bp}) + 4(\text{number of GC bp})$	7

[a] Abbreviations: M, molarity of monovalent cations (usually Na^+ concentration); %GC, percentage of G and C nucleotides in the DNA; %form, percentage of formamide in the hybridization solution; L, length of the duplex in base pairs.
[b] These equations hold for Na^+ concentrations between 0.01 and 0.40 M and %GC values of 30–75%.
[c] See Section 4.1 for comments on the applicability of this equation.

Hybridization

Table 5. Factors influencing the hybridization rate for nucleic acids in solution (3)

Factor	Influence on hybridization rate
Temperature	Optimal hybridization occurs at 20–25°C below T_m for DNA–DNA hybrids, 10–15°C below T_m for DNA–RNA hybrids.
Ionic strength	Optimal hybridization at 1.5 M Na^+.
Destabilizing agents	50% formamide has no effect, but lower or higher concentrations reduce the hybridization rate.
Mismatched base pairs	Each 10% of mismatching reduces the hybridization rate by a factor of two.
Duplex length	Hybridization rate is directly proportional to duplex length, increasing as the length increases.
Viscosity	Increased viscosity increases the hybridization rate when using filters, but not in solution. 10% dextran sulfate increases the rate by a factor of 10.
Probe complexity	Repetitive sequences increase the hybridization rate.
Base composition	Little effect.
pH	Little effect between pH 5.0 and pH 9.0.

3. REAGENTS AND HYBRIDIZATION CONDITIONS

Although all cloning manuals provide details of reagents and conditions for carrying out filter hybridization experiments, many practising molecular biologists simply use tried and tested protocols that have worked well for them in the past. One aspect of hybridization analysis that is subject to much debate is the relative merits of different blocking agents for reducing non-specific probe attachment to the filter surface. Several different blocking agents have been described (*Table 6*) for use individually or in combination. Recommendations for their use are given in *Table 7* but these should not be considered sacrosanct. One consideration that seems not to have been addressed is the applicability of different blocking agents to different filter types.

Optimal hybridization requires high salt which is usually provided by SSC or SSPE (*Table 8*). These solutions form the basis of prehybridization, hybridization and washing solutions, and are also used in blotting solutions for most types of transfer onto filters (though high salt is in fact needed only for transfer onto nitrocellulose). Prehybridization and hybridization are usually carried out in the same solution, either aqueous or formamide-based. The recipes that have proved most useful in the Editor's laboratory are given in *Table 9*. Recommendations for prehybridization/hybridization solutions and hybridization temperatures, again based on the Editor's experience, are given in *Table 10*. Hybridization with DNA and RNA probes should be allowed to proceed for 15 h in aqueous solution or 24 h in formamide for genomic DNA blots, or for 8 h in either solution for plasmid blots and colony/plaque lifts with a multicopy vector. With oligonucleotide probes, hybridization should proceed for at least 15 h. However, remember that homologous hybridization occurs more rapidly than heterologous hybridization so extended times increase the chance of non-specific signals.

Washing protocols are best determined empirically with reference to the T_m of the hybrid. For a DNA or RNA probe, a high stringency wash should be at a salt and temperature combination that corresponds to about 10–15°C below T_m, though standard conditions of 0.1% SSC/SSPE and 68°C are often used. When using an oligonucleotide probe, the initial wash should be at the hybridization temperature and subsequent washes at temperatures approaching the T_m. Washes above the calculated T_m may be necessary under some circumstances. Guidelines for washing protocols are given in *Table 11*.

Table 6. Blocking agents for filter hybridization

Blocking agent	Storage and use	Reference
100× Denhardts 2% (w/v) BSA Fraction V 20% (w/v) Ficoll 400 2% (w/v) polyvinylpyrollidone in water.	Stored at –20°C. Used at 5× or 10× (see *Tables 7, 9* and *10*).	8
BLOTTO 5% (w/v) non-fat dried milk 0.02% (w/v) sodium azide in water.	Stored at 4°C. Used at a final concentration of 4%.	9
Denatured salmon sperm DNA (Type II, Na salt) A 10 mg ml^{-1} solution is prepared in 0.1 M NaCl, then extracted with phenol followed by phenol–CHCl$_3$. The DNA is sheared by passage 15 times through a 17 G needle, ethanol precipitated and redissolved at 10 mg ml^{-1} in water. Finally, the solution is boiled for 10 min to denature the DNA.	Stored at -20°C. Heated at 95°C for 5 min before use at 100 μg ml^{-1}.	3
Heparin (porcine Grade II)[a] 50 mg ml^{-1} in 4× SSC or 4× SSPE.	Stored at 4°C. Used at 500 μg ml^{-1} with dextran sulfate or 50 μg ml^{-1} without.	10
Yeast tRNA 10 mg ml^{-1} in water.	Stored at 4°C. Used at 100 μg ml^{-1}.	3
Homopolymer DNA 1 mg ml^{-1} poly(A) or poly(C) in water.	Stored at 4°C. Used at 10 μg ml^{-1} for appropriate targets: poly(A) for AT-rich DNA, poly(C) for GC-rich DNA.	3

[a]See *Table 8* for preparation of SSC and SSPE stock solutions.

Hybridization

Table 7. Recommendations for blocking agent choice[a]

Blocking agent	Applications		
	Recommended	**Also suitable**	**Not recommended**
5× Denhardts	All applications	–	–
10× Denhardts	Colony/plaque lifts	–	All others
BLOTTO	Colony/plaque lifts	DNA dot blots	Southerns for a single-copy gene, Northerns, RNA dot blots[b]
Heparin	Southerns, DNA dot blots	All others	–
Salmon sperm DNA	Southerns, DNA dot blots	All others	–
Yeast tRNA	Northerns, RNA dot blots	Southerns, DNA dot blots	Colony/plaque lifts
Homopolymer DNA	Probing AT- or GC-rich DNA	–	All others

[a] Table based on the Editor's experience, as well as various sources including refs 3 and 11.
[b] BLOTTO contains a contaminating RNase activity.

Table 8. High salt solutions used in hybridization analysis

Stock solution	Composition
20× SSC	3.0 M NaCl 0.3 M trisodium citrate (filter before use)
20× SSPE	3.6 M NaCl 0.2 M NaH_2PO_4 0.02 M Na_2.EDTA (pH 7.7)

Table 9. Typical prehybridization/hybridization solutions

Solution type	Composition
Aqueous	5× SSC 5× Denhardts 1% (w/v) SDS 10% (w/v) dextran sulfate (Mol. wt 500 000) 0.3% (w/v) tetrasodium pyrophosphate 100 μg ml^{-1} denatured salmon sperm DNA
Formamide	5× SSPE[a] 5× Denhardts 50% (v/v) formamide 1% (w/v) SDS 50 mM sodium phosphate (pH 6.8) 10% (w/v) dextran sulfate (Mol. wt 500 000) 100 μg ml^{-1} denatured salmon sperm DNA

[a] SSPE is preferable to SSC for formamide-based solutions since it provides a greater buffering capacity.

Table 10. Prehybridization/hybridization solutions and hybridization temperatures for different applications[a]

Application	Prehybridization/ hybridization solution[b]	Hybridization temperature[c]
Southerns, DNA dot blots		
DNA probe	aqueous	68°C
RNA probe	formamide	42°C[d]
Northerns, RNA dot blots	formamide	37–42°C
Colony/plaque lifts		
DNA probe	aqueous, 10× Denhardts, no SDS	68°C
RNA probe	formamide, 10× Denhardts, no SDS	42°C[d]
Oligonucleotide probes	aqueous, no dextran sulfate[e]	T_m −5 to 10°C[f]

[a] These recommendations refer to nitrocellulose or nylon filters. For specialized filters, including positively charged nylon, the conditions described by the supplier should be followed.

[b] See *Table 9* for recipes.

[c] For standard DNA and RNA probes, the hybridization temperature should be set at 20–25°C below the T_m for the hybrid. The temperatures given in this table provide a guide only and should be modified if the T_m has been calculated experimentally or mathematically.

[d] For some RNA probes the hybridization temperature will need to be raised, possibly to 65°C (even in the presence of 50% formamide), to prevent background problems.

[e] Alternatively use 6× SET, 10× Denhardts, 0.1% SDS for prehybridization, hybridization and/or washing with oligonucleotide probes. 20× SET = 3 M NaCl, 0.4 M Tris–HCl (pH 7.8), 20 mM Na$_2$.EDTA.

[f] See Section 4 for additional comments on hybridization temperatures for oligonucleotide probes.

Table 11. Typical washing protocols for hybridization analysis[a]

Type of probe	Washing protocol
DNA or RNA[b]	2× 5 min, 2× SSC + 0.1% SDS, room temperature
	2× 5 min, 0.2× SSC + 0.1% SDS, room temperature (low stringency)
	2× 15 min, 0.2× SSC + 0.1% SDS, 42°C (moderate stringency)
	2× 15 min, 0.1× SSC + 0.1% SDS, 68°C (high stringency)[c]
Oligonucleotide[d,e]	4× 10 min, 6× SSC + 0.1% SDS, hybridization temperature
	4× 10 min, 6× SSC + 0.1% SDS, hybridization temperature + 3°C
	1× 2 min, 6× SSC + 0.1% SDS, T_m to $T_m + 6$°C

[a] See footnote a to *Table 10*.
[b] For RNA probes, include an additional wash for 30 min in 2× SSC, 25 µg ml^{-1} RNase A, 10 units RNase T1 to reduce background. This wash can be performed between the low and high stringency washes, or immediately before autoradiography.
[c] These conditions are generally used as a standard high stringency wash for DNA and RNA probes, but see the comments in the text.
[d] See footnote e to *Table 10* for the possibility of washing in 6× SET, 0.1% SDS.
[e] See Section 4 for additional comments on washing conditions for oligonucleotide probes.

4. SYNTHETIC OLIGONUCLEOTIDE PROBES

The development of automated procedures for oligonucleotide synthesis has opened up new possibilities for hybridization analysis in molecular biology. In general terms, oligonucleotides are employed in two ways:
(i)   a single oligonucleotide is used to detect a gene or other sequence that has already been fully characterized at the nucleotide level;
(ii) a pool of related oligonucleotides is used to detect a gene whose nucleotide sequence is not known but which has been predicted from the amino acid sequence of the translation product.

4.1. Single oligonucleotide probes

Oligonucleotides are end-labeled prior to their use as hybridization probes (see Chapter 3, Section 4), which requires that the nanomolarity of the stock solution be calculated (*Table 12*). In hybridization solutions an oligonucleotide probe is used at a concentration of 0.125 ng ml^{-1}.

The most critical factor in practical terms when using an oligonucleotide as a hybridization probe is calculation of the T_m. The equation given in *Table 4* is suitable for oligonucleotides up to 20 nt (high AT content) or 40 nt (high GC content). *Table 13* defines the T_m's for oligonucleotides of different lengths and base compositions. Some cloning manuals provide rough guides for hybridization and washing temperatures based only on oligonucleotide length but experience dictates that these rough guides should be treated with caution. In practice it is better to calculate the specific T_m for an oligonucleotide and then design the hybridization and washing conditions along the lines described in *Tables 10* and *11*.

Table 12. Data for calculation of oligonucleotide concentration

Oligonucleotide length (nt)	1 OD unit = (nmol oligo ml^{-1})[a]	1 nmol oligo = (μg)[b]
10	10.000	3.30
11	9.091	3.63
12	8.333	3.96
13	7.692	4.29
14	7.143	4.62
15	6.667	4.95
16	6.250	5.28
17	5.882	5.61
18	5.556	5.94
19	5.263	6.27
20	5.000	6.60
21	4.762	6.93
22	4.545	7.26
23	4.348	7.59
24	4.167	7.92
25	4.000	8.25
26	3.846	8.58
27	3.701	8.91
28	3.571	9.24
29	3.448	9.57
30	3.333	9.90
31	3.226	10.23
32	3.125	10.56
33	3.030	10.89
34	2.941	11.22
35	2.857	11.55
36	2.778	11.88
37	2.703	12.21
38	2.632	12.54
39	2.564	12.87
40	2.500	13.20
41	2.439	13.53
42	2.381	13.86
43	2.326	14.19
44	2.273	14.52
45	2.222	14.85
46	2.174	15.18
47	2.128	15.51
48	2.083	15.84
49	2.041	16.17
50	2.000	16.50

[a] Based on the approximation that 1 OD unit = 33 μg oligonucleotide ml^{-1}.
[b] Based on the approximation that the molecular mass of a single deoxyribonucleotide = 330 kd.

Table 13. Melting temperatures, (T_m, in °C) for oligonucleotides of different lengths and base compositions, calculated according to ref. 7.

Oligonucleotide length (nt)	Number of GC nucleotides																				
	0	1	2	3	4	5	6	7	8	9	10	11	12	13	14	15	16	17	18	19	20
10	20	22	24	26	28	30	32	34	36	38	40	—	—	—	—	—	—	—	—	—	—
11	22	24	26	28	30	32	34	36	38	40	42	44	—	—	—	—	—	—	—	—	—
12	24	26	28	30	32	34	36	38	40	42	44	46	48	—	—	—	—	—	—	—	—
13	26	28	30	32	34	36	38	40	42	44	46	48	50	52	—	—	—	—	—	—	—
14	28	30	32	34	36	38	40	42	44	46	48	50	52	54	56	—	—	—	—	—	—
15	30	32	34	36	38	40	42	44	46	48	50	52	54	56	58	60	—	—	—	—	—
16	32	34	36	38	40	42	44	46	48	50	52	54	56	58	60	62	64	—	—	—	—
17	34	36	38	40	42	44	46	48	50	52	54	56	58	60	62	64	66	68	—	—	—
18	36	38	40	42	44	46	48	50	52	54	56	58	60	62	64	66	68	70	72	—	—
19	38	40	42	44	46	48	50	52	54	56	58	60	62	64	66	68	70	72	74	76	—
20	40	42	44	46	48	50	52	54	56	58	60	62	64	66	68	70	72	74	76	78	80
21	42	44	46	48	50	52	54	56	58	60	62	64	66	68	70	72	74	76	78	80	
22	44	46	48	50	52	54	56	58	60	62	64	66	68	70	72	74	76	78	80		
23	46	48	50	52	54	56	58	60	62	64	66	68	70	72	74	76	78	80			
24	48	50	52	54	56	58	60	62	64	66	68	70	72	74	76	78	80				
25	50	52	54	56	58	60	62	64	66	68	70	72	74	76	78	80					
26	52	54	56	58	60	62	64	66	68	70	72	74	76	78	80						
27	54	56	58	60	62	64	66	68	70	72	74	76	78	80							
28	56	58	60	62	64	66	68	70	72	74	76	78	80								
29	58	60	62	64	66	68	70	72	74	76	78	80									
30	60	62	64	66	68	70	72	74	76	78	80										
31	62	64	66	68	70	72	74	76	78	80											
32	64	66	68	70	72	74	76	78	80												
33	66	68	70	72	74	76	78	80													
34	68	70	72	74	76	78	80														
35	70	72	74	76	78	80															
36	72	74	76	78	80																
37	74	76	78	80																	
38	76	78	80																		
39	78	80																			
40	80																				

The calculation is not accurate for molecules in this part of the chart

4.2. Mixed oligonucleotide probes

The degeneracy of the genetic code should be taken into account when choosing an amino acid sequence from which to predict a mixed oligonucleotide probe (*Table 14*). A region containing Met and Trp codons is usually ideal, since only one triplet specifies each amino acid. However, such a region is not always available. Nevertheless, the degeneracy of the probe can be reduced by using inosine at the third position of four-codon choices (e.g. GCI for Ala). This is because inosine base pairs with similar fidelity to each standard deoxyribonucleotide. Inosine can also be used in Ile codons (AUI) but should be avoided for two-codon choices.

'Guessmers', in which codon choice is dictated by the codon bias of the genome being probed, are becoming increasingly popular. Use the codon bias data in Chapter 7, Section 4 to design guessmers but remember that the overall codon bias for a genome may mask more subtle differences between codon usages of different genes.

Mixed oligonucleotides should be used in hybridization solutions at a concentration of 0.125 ng ml^{-1} for each unique sequence, up to a maximum of 20 ng ml^{-1}. It is difficult to predict hybridization conditions for mixed oligonucleotide probes since the T_m's of the individual sequences in the pool may vary by as much as $30°C$. Examples of successful hybridization conditions (such as those listed in ref. 11) are useful but do not provide firm guidelines to follow. Initial experiments should use hybridization and washing conditions based around the lowest T_m of the members of the pool, though if the perfectly matched oligonucleotide has a T_m in the lowest part of the range there will be little chance that its specific hybridization will be retained when all non-specifically hybridized probe has been washed off. Gradually increase the T_m in subsequent experiments and look for the hybridization signals that are retained under the most stringent conditions, but take account of negative controls. With such studies it is always advisable to run two series of experiments in parallel, using two different oligonucleotide pools predicted from different regions of the gene being screened for.

Table 14. Codon choice for different amino acids

Number of codons	Amino acid	Codons[a,b]	Number of codons	Amino acid	Codons[a,b]
1	Met	AUG	3	Ile	AU$^U_{C~A}$
	Trp	UGG	4	Ala	GCN
2	Asn	AA^U_C		Gly	GGN
	Asp	GA^U_C		Pro	CCN
	Cys	UG^U_C		Thr	ACN
	Glu	GA^A_G		Val	GUN
	Gln	CA^A_G	6	Arg	CGN, AG^A_G
	His	CA^U_C		Leu	CUN, UU^A_G
	Phe	UU^U_C		Ser	UCN, AG^U_C
	Tyr	UA^U_C			

[a] According to the 'universal' genetic code. See Chapter 7, Section 3 for genetic code variations.
[b] N = any nucleotide.

Hybridization

5. REFERENCES

1. Bittner, M., Kupferer, P. and Morris, C.F. (1980) *Anal. Biochem.*, **102**, 459.

2. Stellway, E.J. and Dahlberg, A.E. (1980) *Nucleic Acids Res.*, **8**, 299.

3. Dyson, N.J. (1991) In *Essential Molecular Biology: A Practical Approach.* T.A. Brown (ed.) Vol. 2. Oxford University Press, Oxford, in press.

4. Meinkoth, J. and Wahl, G. (1984) *Anal. Biochem.*, **138**, 267.

5. Casey, J. and Davidson, N. (1977) *Nucleic Acids Res.*, **4**, 1539.

6. Bodkin, D.K. and Knudson, D.L. (1985) *J. Virol. Methods*, **10**, 45.

7. Wallace, R.B., Shaffer, J., Murphy, R.F., Bonner, J., Hirose, T. and Itakura, K. (1979) *Nucleic Acids Res.*, **6**, 3543.

8. Denhardt, D.T. (1966) *Biochem. Biophys. Res. Commun.*, **23**, 641.

9. Johnson, D.A., Gautsch, J.W., Sportsman, J.R. and Elder, J.H. (1984) *Gene Anal. Tech.*, **1**, 3.

10. Singh, L. and Jones, K.W. (1984) *Nucleic Acids Res.*, **12**, 5627.

11. Sambrook, J., Fritsch, E.F. and Maniatis, T. (1989) *Molecular Cloning. A Laboratory Manual* (2nd edn). Cold Spring Harbor Laboratory Press, New York.

CHAPTER 10
CENTRIFUGATION

1. CALCULATION OF RELATIVE CENTRIFUGAL FIELD

The relative centrifugal field (RCF) generated by a particular rotor at a given angular velocity is calculated by:

$$RCF = (1.11 \times 10^{-5}) \; (rev \; min^{-1})^2 r$$

where RCF is the relative centrifugal field as a multiple of the earth's gravitational field ('$\times g$', with g being 980 cm sec^{-2}), rev min^{-1} is the angular velocity of the rotor, and r is the radial distance in cm. Remember that r is not uniform within a centrifuge tube, not even for vertical rotors, and that for a given centrifugation system different parts of the sample will be subject to different RCFs.

Figure 1 is a nomogram that allows approximate calculations of RCF to be made. Directions for its use are given in *Figure 1* legend. A more precise calculation can be made from the data in *Table 1*.

2. DENSITY GRADIENT CENTRIFUGATION

Table 2 gives details of common density gradient media. Guidelines for preparing density gradient solutions are provided in *Tables 3* and *4*.

3. CARE OF ROTORS

Rotors are commonly made of aluminum or titanium although some rotors are made of stainless steel. All these metals are subject to corrosion by a variety of reagents which are listed in *Table 5*. To prolong rotor life, rotors must be thoroughly cleaned after use and then dried before storage.

Centrifugation

Table 1. RCFs for different combinations of rotor speed and radial distance

Rotor speed (rev min^{-1})[a]	Multiplication factor[b]	Radial distance (cm)							
		1	2	3	4	5	6	7	8
1000	11.1	11	22	33	44	56	67	78	89
2000	44.4	44	89	133	177	222	266	311	355
3000	99.9	100	200	300	400	500	599	699	799
4000	177.6	178	355	533	710	888	1066	1243	1421
5000	277.5	278	555	833	1110	1388	1665	1943	2220
6000	399.6	400	799	1199	1598	1998	2398	2797	3197
7000	543.9	544	1088	1632	2176	2720	3264	3807	4351
8000	710.4	710	1421	2131	2842	3552	4262	4973	5683
9000	899.1	899	1798	2697	3596	4496	5395	6294	7193
10000	1100.0	1100	2200	3300	4400	5500	6600	7770	8880
11000	1343.1	1343	2686	4029	5372	6716	8059	9402	10745
12000	1598.4	1598	3197	4795	6394	7992	9590	11189	12787
13000	1875.9	1876	3752	5628	7504	9380	11255	13131	15007
14000	2175.6	2176	4351	6527	8702	10878	13054	15229	17405
15000	2497.5	2498	4995	7493	9990	12488	14985	17483	19980
16000	2841.6	2842	5683	8525	11366	14208	17050	19891	22733
17000	3207.9	3208	6416	9624	12832	16040	19247	22455	25663
18000	3596.4	3596	7193	10789	14386	17982	21578	25175	28771
19000	4007.1	4007	8014	12021	16028	20036	24043	28050	32057
20000	4440.0	4440	8880	13320	17760	22200	26640	31080	35520
25000	6937.5	6938	13875	20813	27750	34688	41625	48563	55500
30000	9990.0	9990	19980	29970	39960	49950	59940	69930	79920
35000	13597.5	13598	27195	40793	54390	67988	81585	95183	108780
40000	17760.0	17760	35520	53280	71040	88800	106560	124320	142080
45000	22477.5	22478	44955	67432	89910	112388	134865	157343	179820
50000	27750	27750	55500	83250	111000	138750	166500	194250	222000
60000	39960	39960	79920	119880	159840	199800	239760	279720	319680
70000	54390	54390	108780	163170	217560	271950	326340	380730	435120
80000	71040	71040	142080	213120	284160	355200	426240	497280	568320
90000	89910	89910	179820	269730	359640	449550	539460	629370	719280
100000	111000	111000	222000	333000	444000	555000	666000	777000	888000

[a] RCFs are not additive: for instance the RCF for a particular radial distance at 12 000 rev min^{-1} does not equal the RCF for 10 000 rev min^{-1} plus the RCF for 2000 rev min^{-1}.

[b] The RCF for a particular rotor speed is the product of the radial distance times the multiplication factor.

9	10	11	12	13	14	15	16	17	18	19	20
100	110	122	133	144	155	167	178	189	200	211	222
400	444	488	533	577	622	666	710	755	799	844	888
899	999	1099	1199	1299	1399	1499	1598	1698	1798	1898	1998
1598	1776	1954	2131	2309	2486	2664	2842	3019	3197	3374	3552
2498	2775	3053	3330	3608	3885	4163	4440	4718	4995	5273	5550
3596	3996	4396	4795	5195	5594	5994	6394	6793	7193	7592	7992
4895	5439	5983	6527	7070	7615	8159	8702	9246	9790	10334	10878
6394	7104	7814	8525	9235	9946	10656	11366	12077	12787	13498	14208
8092	8991	9890	10789	11688	12587	13487	14386	15285	16184	17083	17982
9990	11100	12210	13320	14430	15540	16650	17760	18870	19980	21090	22200
12088	13431	14774	16117	17460	18803	20147	21490	22833	24176	25519	26862
14386	15984	17582	19181	20779	22378	23976	25574	27173	28771	30370	31968
16883	18759	20635	22511	24387	26263	28138	30014	31890	33766	35642	37518
19580	21756	23932	26107	28283	30458	32634	34810	36985	39161	41336	43512
22478	24975	27473	29970	32468	34965	37463	39960	42458	44955	47453	49950
25574	28416	31258	34099	36941	39782	42624	45466	48307	51149	53990	56832
28871	32079	35287	38495	41703	44911	48119	51326	54534	57742	60950	64158
32368	35964	39560	43157	46753	50350	53946	57542	61139	64735	68332	71928
36064	40071	44078	48085	52092	56099	60107	64114	68121	72128	76135	80142
39960	44400	48840	53280	57720	62160	66600	71040	75480	79920	84360	88800
62438	69375	76313	83250	90188	97125	104062	111000	117938	124875	131813	138750
89910	99900	109890	119880	129870	139860	149850	159840	169830	179820	189810	199800
122378	135975	149573	163170	176768	190365	203963	217560	231158	244755	258353	271950
159840	177600	195360	213120	230880	248640	266400	284160	301920	319680	337440	355200
202298	224775	247253	269730	292208	314685	337163	359640	382118	404595	427073	449550
249750	277500	305250	333000	360750	388500	416250	444000	471750	499500	527250	555000
359640	399600	439560	479520	519480	559440	599400	636360	679320	719280	759240	799200
489510	543900	598290	652680	707070	761460	815850	870240	924630	979020	–	–
639360	710400	781440	852480	923520	994560	–	–	–	–	–	–
809190	899100	989010	–	–	–	–	–	–	–	–	–
999000	–	–	–	–	–	–	–	–	–	–	–

Figure 1. Nomogram for approximate calculations of RCF values (8). To calculate the RCF value at a given point along a centrifuge tube or bottle, first measure the radial distance (in mm) from the center of the centrifuge spindle to that point. Now draw a straight line connecting the radial distance (on the left of the nomogram) with the centrifuge rotor speed (on the right). The RCF value can now be read off the center column.

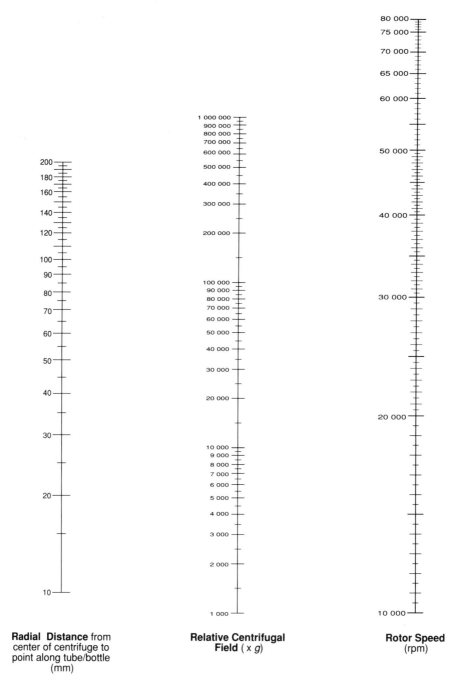

Radial Distance from center of centrifuge to point along tube/bottle (mm)

Relative Centrifugal Field (x *g*)

Rotor Speed (rpm)

Table 2. Density gradient media: physical and chemical data

	Cesium chloride	Cesium trifluoroacetate	Sucrose	Percoll
Formula	CsCl	CF₃COONa	$C_{12}H_{22}O_{11}$	Colloidal silica coated with polyvinylpyrrolidone, average diameter 21–22 nm
Molecular weight	168.36	245.93	342.30	–
Solubility (g ml⁻¹ in water at 20°C)	1.2	2.5	2.0	–
Maximum molarity (water at 20°C)	7.4	~10	5.8	–
Maximum density (g ml⁻¹ in water at 20°C)	1.91	~2.6	1.32	1.30
Risk	–	20/22–36/38	–	–
Safety	–	26–28	–	–
References	–	1, 2	–	3

Table 3. Concentration of cesium chloride required for different solution densities (2, 3, 5)

Percentage (w/w)	Concentration (mg ml⁻¹, 25°C)	Molarity (M)	Density[a] (g cm⁻³, 25°C)
1	10.05	0.056	1.0047
2	20.25	0.119	1.0125
3	30.61	0.182	1.0204
4	41.14	0.244	1.0284
5	51.83	0.308	1.0365
6	62.68	0.373	1.0447
7	73.72	0.438	1.0531
8	84.92	0.504	1.0615
9	96.30	0.572	1.0700
10	107.88	0.641	1.0788
11	119.6	0.710	1.0877
12	131.6	0.782	1.0967
13	143.8	0.854	1.1059
14	156.1	0.927	1.1151
15	168.7	1.002	1.1245
16	181.4	1.077	1.1340
17	194.4	1.155	1.1437
18	207.6	1.233	1.1536
19	221.1	1.313	1.1637
20	234.8	1.395	1.1739
21	248.7	1.477	1.1843
22	262.9	1.561	1.1948
23	277.3	1.647	1.2055

Table 3. Continued

Percentage (w/w)	Concentration (mg ml^{-1}, 25°C)	Molarity (M)	Density[a] (g cm^{-3}, 25°C)
24	291.9	1.734	1.2164
25	306.9	1.823	1.2275
26	322.1	1.913	1.2387
27	337.6	2.005	1.2502
28	353.3	2.098	1.2619
29	369.4	2.194	1.2738
30	385.7	2.291	1.2858
31	402.4	2.390	1.2980
32	419.5	2.492	1.3110
33	436.9	2.595	1.3240
34	454.2	2.698	1.3360
35	472.4	2.806	1.3496
36	490.7	2.914	1.3630
37	509.5	3.026	1.3770
38	528.6	3.140	1.3910
39	548.3	3.257	1.4060
40	567.8	3.372	1.4196
41	588.4	3.495	1.4350
42	609.0	3.617	1.4500
43	630.0	3.742	1.4650
44	651.6	3.870	1.4810
45	673.6	4.001	1.4969
46	696.0	4.134	1.5130
47	718.6	4.268	1.5290
48	742.1	4.408	1.5460
49	766.4	4.552	1.5640
50	791.3	4.700	1.5825
51	816.5	4.849	1.6010
52	841.9	5.000	1.6190
53	868.1	5.156	1.6380
54	895.3	5.317	1.6580
55	922.8	5.481	1.6778
56	951.4	5.651	1.6990
57	980.4	5.823	1.7200
58	1009.8	5.998	1.7410
59	1040.2	6.178	1.7630
60	1070.8	6.360	1.7846
61	1102.9	6.550	1.8080
62	1135.8	6.746	1.8310
63	1167.3	6.945	1.8560
64	1203.2	7.146	1.8800
65	1238.4	7.355	1.9052

[a] The concentration of CsCl solution corresponding to a particular density between 1.29 and 1.80 at 25°C is given by:

percentage (w/w) of CsCl = 137.48 – 138.11 $(1/\rho_{25})$

where ρ_{25} is the required density.

300

Table 4. Concentration of sucrose required for different solution densities (6, 7)

Percentage (w/w)	Concentration (mg ml^{-1}, 20°C)	Molarity (M)	Density[a] (g cm^{-3}, 20°C)
1	10.02	0.029	1.002
2	20.12	0.059	1.006
4	40.55	0.119	1.014
6	61.31	0.179	1.022
8	82.40	0.241	1.030
10	103.81	0.303	1.038
12	125.6	0.367	1.046
14	147.7	0.431	1.055
16	170.2	0.497	1.063
18	193.0	0.564	1.072
20	216.2	0.632	1.081
22	239.8	0.701	1.090
24	263.8	0.771	1.099
26	288.1	0.842	1.108
28	312.9	0.914	1.118
30	338.1	0.988	1.127
32	363.7	1.063	1.137
34	389.8	1.139	1.146
36	416.2	1.216	1.156
38	443.2	1.295	1.166
40	470.6	1.375	1.176
42	498.4	1.456	1.187
44	526.8	1.539	1.197
46	555.6	1.623	1.208
48	584.9	1.709	1.219
50	614.8	1.796	1.230
52	645.1	1.885	1.241
54	676.0	1.975	1.252
56	707.4	2.067	1.263
58	739.4	2.160	1.275
60	771.9	2.255	1.286
62	804.9	2.351	1.298
64	838.6	2.450	1.310
66	872.8	2.550	1.322

[a] Density calculations for sucrose solutions are complicated (7).

Centrifugation

Table 5. Chemical resistances of rotor components[a]

Chemical[b]	Resistance[c]		
	Aluminum	Stainless steel	Titanium
Acetic acid	S	S	S
Acetone	S	S	S
Ammonium hydroxide	?	S	S
Ammonium sulfate	N	?	S
Amyl alcohol	S	?	S
Benzene	S	S	S
Boric acid	S	S	S
n-butyl alcohol	S	?	S
Cesium salts	N	?	S
Calcium chloride	N	S	S
Chloroform	?	S	S
Chromic acid	U	U	S
Citric acid	S	S	S
Dimethylformamide	S	?	S
Ethanol	S	S	S
Ethylene glycol	S	?	S
Hydrochloric acid	U	U	S
Isopropyl alcohol	?	?	S
Nitric acid	N	S	N
Phenol	S	S	N
Phosphoric acid	?	S	U
Potassium chloride	N	S	S
Potassium hydroxide	U	S	N
Rubidium salts	N	?	S
Sodium chloride	N	S	S
Sodium hydroxide	U	S	N
Sulfuric acid	U	U	U
Toluene	S	?	S
Trichloroethane	?	?	S
Urea	S	?	S
Zinc chloride	N	N	S

[a] Data taken from various sources, including product data supplied by Beckman.

[b] Data refer to concentrated solutions.

[c] Abbreviations: S, satisfactory; U, unsatisfactory; N, not satisfactory due to possible long-term corrosion or other effects, though these may be avoided by scrupulous rotor care; ?, status not published.

4. REFERENCES

1. Andersson, K. and Hjorth, R. (1985) *Plasmid*, **13**, 78.
2. Zarlenga, D.S. and Gamble, H.R. (1987) *Anal. Biochem.*, **162**, 569.
3. Pertoff, H., Laurent, T.C., Laas, T. and Kagedal, L. (1978) *Anal. Biochem.*, **88**, 271.
4. Bruner, R. and Vinograd, J. (1965) *Biochim. Biophys. Acta*, **108**, 18.
5. Ifft, J.B., Martin, W.R. and Kinzie, K. (1970) *Biopolymers*, **9**, 597.
6. de Duve, C., Berthet, J. and Beaufay, H. (1959) *Prog. Biophys. Biophys. Chem.*, **9**, 326.
7. Dawson, R.M.C., Elliott, D.C., Elliott, W.H. and Jones, K.M. (1986) *Data for Biochemical Research.* (3rd edn). Clarendon Press, London.
8. Dole, V.P. and Cootzias, G.C. (1951) *Science*, **113**, 552.

CHAPTER 11
SAFETY

1. PROTECTION FROM CHEMICAL HAZARDS

1.1. Risk and safety classification systems

All reputable suppliers of laboratory chemicals provide risk and safety information with those products that present a potential hazard. The label will carry a symbol indicating the hazard category (*Table 1*) and usually this will be backed up by code numbers indicating the specific risks presented by the chemical and the safety precautions that should be observed. The most frequently used code systems are the UN Hazard Classification (*Table 2*) and the European Commission Risk and Safety Phrases (EC system) (*Table 3*). The EC system has been adopted by a number of international suppliers (e.g. BDH, Serva and Fluka) and is familiar in most countries of the world, including the USA. The EC phrases are given in full as they are used elsewhere in the *Molecular Biology Labfax* to describe risk and safety information for various chemicals and reagents. The EC system is not, however, in universal use and several companies have code systems of their own, a notable example being Sigma-Aldrich (1). Several suppliers now publish detailed risk and safety data for their products (e.g. refs 2 and 3). General information concerning chemical hazards is also available (4–15).

Table 1. Hazard symbols

Symbol	Definition	Precaution
	Explosive	Avoid shock, friction, sparks and heat
	Oxidizing	Keep away from combustible material
	Toxic	Avoid contact
	Irritating	Do not breathe vapours and avoid contact with skin and eyes
	Harmful	Avoid contact, including inhalation

Safety

Table 1. Continued

Symbol	Definition	Precaution
	Highly flammable	Various depending on sub-category[a]
	Corrosive	Do not breathe vapours and avoid contact
	Radioactive	Avoid contact, use suitable shielding

[a] Typical sub-categories: spontaneously flammable substances — avoid contact with air; highly flammable gases — avoid contact with air, keep away from sources of ignition; substances sensitive to moisture — avoid contact with water; flammable liquids (flash point <55°C) — keep away from sources of ignition.

Table 2. UN Chemical Hazard Classification

Hazard class	Description
Class 1	Explosive
Class 2	Gases
Class 3.1	Flammable liquids: flash point below −18°C
Class 3.2	Flammable liquids: flash point −18°C to +23°C
Class 3.3	Flammable liquids: flash point +23°C to +61°C
Class 4.1	Flammable solids
Class 4.2	Spontaneously combustible
Class 4.3	Dangerous when wet
Class 5.1	Oxidizing agent
Class 5.2	Organic peroxides
Class 6.1	Poisonous
Class 7	Radioactive
Class 8	Corrosive
Class 9	Miscellaneous dangerous substances
NR	Non-regulated

Table 3. European Commission Risk and Safety Phrases[a]

Phrase number	Description
Risk phrases	
R1	Explosive when dry
R2	Risk of explosion by shock, friction, fire or other sources of ignition
R3	Extreme risk of explosion by shock, friction, fire or other sources of ignition
R4	Forms very sensitive explosive metallic compounds
R5	Heating may cause an explosion
R6	Explosive with or without contact with air
R7	May cause fire
R8	Contact with combustible material may cause fire
R9	Explosive when mixed with combustible material
R10	Flammable
R11	Highly flammable
R12	Extremely flammable
R13	Extremely flammable liquefied gas
R14	Reacts violently with water
R15	Contact with water liberates highly flammable gases
R16	Explosive when mixed with oxidizing substances
R17	Spontaneously flammable in air
R18	In use may form flammable/explosive vapour–air mixture
R19	May form explosive peroxides
R20	Harmful by inhalation
R21	Harmful in contact with skin
R22	Harmful if swallowed
R23	Toxic by inhalation
R24	Toxic in contact with skin
R25	Toxic if swallowed
R26	Very toxic by inhalation
R27	Very toxic in contact with skin
R28	Very toxic if swallowed
R29	Contact with water liberates toxic gas
R30	Can become highly flammable in use
R31	Contact with acid liberates toxic gas
R32	Contact with acid liberates very toxic gas
R33	Danger of cumulative effects
R34	Causes burns
R35	Causes severe burns
R36	Irritating to eyes
R37	Irritating to respiratory system
R38	Irritating to skin
R39	Danger of very serious irreversible effects
R40	Possible risk of irreversible effects
R41	Risk of serious irreversible effects
R42	May cause sensitization by inhalation
R43	May cause sensitization by skin contact
R44	Risk of explosion if heated under confinement
R45	May cause cancer

Table 3. Continued

Phrase number	Description
R46	May cause heritable genetic changes
R47	May cause birth effects
R48	Danger of serious damage to health by prolonged exposure
R14/15	Reacts violently with water, liberating highly flammable gases
R15/29	Contact with water liberates toxic, highly flammable gas
R20/21	Harmful by inhalation and in contact with skin
R20/21/22	Harmful by inhalation, in contact with skin and if swallowed
R20/22	Harmful by inhalation and if swallowed
R21/22	Harmful in contact with skin and if swallowed
R23/24	Toxic by inhalation and in contact with skin
R24/25	Toxic in contact with skin and if swallowed
R26/27	Very toxic by inhalation and in contact with skin
R26/27/28	Very toxic by inhalation, in contact with skin and if swallowed
R26/28	Very toxic by inhalation and if swallowed
R27/28	Very toxic in contact with skin and if swallowed
R36/37	Irritating to eyes and respiratory system
R36/37/38	Irritating to eyes, respiratory system and skin
R36/38	Irritating to skin and eyes
R37/38	Irritating to respiratory system and skin
R42/43	May cause sensitization by inhalation and skin contact

Safety phrases

S1	Keep locked up
S2	Keep out of reach of children
S3	Keep in a cool place
S4	Keep away from living quarters
S5	Keep contents under . . . (appropriate liquid to be specified by the manufacturer)
S6	Keep under . . . (inert gas to be specified by the manufacturer)
S7	Keep container tightly closed
S8	Keep container dry
S9	Keep container in a well ventilated place
S12	Do not keep the container sealed
S13	Keep away from food, drink and animal feeding stuffs
S14	Keep away from . . . (incompatible materials to be indicated by the manufacturer)
S15	Keep away from heat
S16	Keep away from sources of ignition — No Smoking
S17	Keep away from combustible material
S18	Handle and open container with care
S20	When using do not eat or drink
S21	When using do not smoke
S22	Do not breathe dust
S23	Do not breathe gas/fumes/vapour/spray (appropriate wording to be specified by the manufacturer
S24	Avoid contact with skin

Table 3. Continued

Phrase number	Description
S25	Avoid contact with eyes
S26	In case of contact with eyes, rinse immediately with plenty of water and seek medical advice
S27	Take off immediately all contaminated clothing
S28	After contact with skin, wash immediately with plenty of . . . (to be specified by the manufacturer)
S29	Do not empty into drains
S30	Never add water to this product
S33	Take precautionary measures against static discharges
S34	Avoid shock and friction
S35	This material and its container must be disposed of in a safe way
S36	Wear suitable protective clothing
S37	Wear suitable gloves
S38	In case of insufficient ventilation, wear suitable respiratory equipment
S39	Wear eye/face protection
S40	To clean the floor and all objects contaminated by this material use . . . (to be specified by the manufacturer)
S41	In case of fire and/or explosion do not breathe fumes
S42	During fumigation/spraying wear suitable respiratory equipment (appropriate wording to be specified)
S43	In case of fire, use . . . (indicate in the space the precise type of fire-fighting equipment. If water increases the risk, add — Never use water)
S44	If you feel unwell, seek medical advice (show the label where possible)
S45	In case of accident or if you feel unwell, seek medical advice immediately (show the label where possible)
S46	If swallowed seek medical advice immediately and show this container or label
S47	Keep at temperature not exceeding . . . °C (to be specified by the manufacturer)
S48	Keep wetted with . . . (appropriate material to be specified by the manufacturer)
S49	Keep only in the original container
S50	Do not mix with . . . (to be specified by the manufacturer)
S51	Use only in well ventilated areas
S52	Not recommended for interior use on large surface area
S53	Avoid exposure — obtain special instructions before use
S1/2	Keep locked up and out of reach of children
S3/7/9	Keep container tightly closed in a well ventilated place
S3/9	Keep in a cool, well ventilated place
S3/9/14	Keep in a cool, well ventilated place away from . . . (incompatible materials to be indicated by the manufacturer)
S3/9/14/49	Keep only in the original container in a cool, well ventilated place away from . . . (incompatible materials to be indicated by the manufacturer)
S3/9/49	Keep only in the original container in a cool, well ventilated place
S3/14	Keep in a cool place away from . . . (incompatible materials to be indicated by the manufacturer)

Table 3. Continued

Phrase number	Description
S7/8	Keep container tightly closed and dry
S7/9	Keep container tightly closed and in a well ventilated place
S20/21	When using do not eat, drink or smoke
S24/25	Avoid contact with skin and eyes
S36/37	Wear suitable protective clothing and gloves
S36/37/38	Wear suitable protective clothing, gloves and eye/face protection
S36/39	Wear suitable protective clothing and eye/face protection
S37/39	Wear suitable gloves and eye/face protection
S47/49	Keep only in the original container at temperature not exceeding . . . °C (to be specified by the manufacturer)

[a] EC Statutory Instrument 1984 No 1244: Information approved for the Classification, Packaging and Labelling of Dangerous Substances for Supply and Conveyance by Road, Part IV (risk phrases), Part V (safety phrases).

1.2. Chemical resistance of plastic and rubber compounds

Potentially disastrous consequences can arise from the use of a container made of an inappropriate material. Plastics are susceptible to attack by various chemical and physical agents and laboratory workers should be aware of the limitations of the material they are working with (*Table 4*). Even more important is an appreciation of the protection provided by the different types of 'rubber' glove that are available (*Table 5*).

Table 4. Resistance of plastics to chemical and physical agents (5, 16)

Agent	Resistance[a]				
	PTFE[b]	PC[b]	PP[b]	HD-PE[b]	LD-PE[b]
General chemicals					
Acids, inorganic	+++	++	+++	+++	+++
Acids, organic	+++	++	+++	+++	+++
Alcohols	+++	++	+++	+++	+++
Aldehydes	+++	+	++	++	++
Alkalis	+++	n.r.	+++	+++	+++
Amines	+++	n.r.	++	++	++
Esters	+++	+	++	++	++
Ethers	+++	n.r.	+++	+++	+++
Glycols	+++	++	+++	+++	+++
Specific chemicals					
Acetic acid	+++	+++	+++	+++	+++
Acetone	+++	n.r.	+++	+++	+++
Boric acid	+++	+++	+++	+++	+++
Butanol	+++	++	+++	+++	+++

Table 4. Continued

Agent	Resistance[a]				
	PTFE[b]	PC[b]	PP[b]	HD-PE[b]	LD-PE[b]
Chloroform	+++	n.r.	++	++	+
Citric acid	+++	+++	+++	+++	+++
Ethanol	+++	+++	+++	+++	+++
Ethyl ether	+++	n.r.	n.r.	n.r.	n.r.
Formaldehyde	+++	+++	++	+++	+++
Methanol	+++	++	+++	+++	+++
Perchloric acid	++	n.r.	++	++	++
Phenol	+++	+++	++	++	++
Sodium hydroxide	+++	n.r.	+++	+++	+++
Toluene	+++	+	++	++	+
Trichloracetic acid	+++	+	+++	++	++
Urea, conc.	+++	n.r.	+++	+++	+++
Xylene	+++	n.r.	+	++	++
Physical agents					
Autoclaving	+++	+++[c]	+++	++	n.r.
Dry sterilization	+++	++[c]	++	++	+
Upper limit for heat treatment (°C)					
Short exposure	300	130	140	105	90
Continuous	260	110	120	95	80

[a] +++ = excellent resistance, ++ = adequate or good resistance, + = poor resistance, n.r. = not resistant.
[b] Abbreviations: PTFE, polytetrafluoroethylene; PC, polycarbonate; PP, polypropylene; HD-PE, high density polyethylene; LD-PE, low density polyethylene.
[c] Polycarbonate suffers mechanical damage with repeated heat treatment.

Table 5. Protection provided by glove materials (5, 16)

Chemical	Provides protection?						
	NR[a]	PE[a]	Neo[a]	Nit[a]	N-PVC[a]	HG-PVC[a]	PVA[a]
General chemicals							
Alcohols	yes	yes	yes	yes	yes	yes	yes
Aldehydes	no	no	no	yes	yes	yes	?
Chlorinated solvents	no	no	no	yes	no	yes	yes
Dilute acids	yes	no	yes	yes	yes	yes	yes
Dilute alkalis	yes	no	yes	yes	yes	yes	yes
Ethers	no	no	yes	yes	no	yes	?
Specific chemicals							
Acetic acid	yes	no	yes	yes	yes	yes	yes
Acetone	yes	no	yes	yes	yes	yes	no
Ammonium hydroxide	yes	no	yes	yes	yes	yes	yes

Safety

Table 5. Continued

Chemical	Provides protection?						
	NR[a]	PE[a]	Neo[a]	Nit[a]	N-PVC[a]	HG-PVC[a]	PVA[a]
Benzene	no	no	no	yes	no	yes	?
Butanol	yes	yes	yes	yes	yes	yes	yes
Chloroform	no	no	no	yes	no	yes	yes
Chromic acid (<50%)	yes	no	no	no	yes	yes	no
Diethyl ether	no	no	yes	yes	no	yes	?
Ethanol	yes	yes	yes	yes	yes	yes	yes
Formaldehyde	no	no	no	yes	yes	yes	?
Formic acid	yes	no	yes	yes	yes	yes	yes
Glycerol	yes	yes	yes	yes	yes	yes	yes
Hydrochloric acid (<40%)	yes	no	yes	yes	yes	yes	yes
Nitric acid (<50%)	no	no	no	no	yes	no	no
Potassium hydroxide	yes	no	yes	yes	yes	yes	yes
Isopropanol	yes	yes	yes	yes	yes	yes	yes
Sodium hydroxide	yes	no	yes	yes	yes	yes	yes
Sulfuric acid (<50%)	yes	no	yes	no	yes	yes	yes
Phenol	yes	no	yes	yes	yes	yes	?
Toluene	no	no	no	yes	no	yes	?
Xylene	no	no	no	yes	no	yes	?

[a] Abbreviations: NR, natural rubber; PE, polyethylene; Neo, neoprene; Nit, nitrile; N-PVC, normal polyvinylchloride; HG-PVC, high grade polyvinylchloride; PVA, polyvinylalcohol.

2. PROTECTION FROM MICROBIOLOGICAL HAZARDS

Although most molecular biology experiments are carried out with *Escherichia coli*, which is defined as Risk Group I (low individual and community risk), precautions must be taken in handling cultures of this microorganism. These precautions should take account of the most common routes of infection (*Table 6*) and should be defined so that personnel are aware of the function of each aspect of good microbiological practice (*Table 7*). Detailed information is provided in refs 17–21. Special procedures for handling genetically manipulated microorganisms are laid down by national and local authorities (e.g. refs 22 and 23).

Table 6. Commonest routes of infection for laboratory personnel (17–19)

Entry via	Source
Mouth	Spills, splashes via fingers, pencils, etc.
Lungs	Aerosols, caused by flaming loops carelessly, incorrect use of syringes, centrifuges, homogenizers, etc. and also generated in pouring operations
Skin	Accidental stabbing with syringes, pipettes, broken glassware, etc.
Eyes	Splashes and aerosols

MOLECULAR BIOLOGY LABFAX

Table 7. Basic precautions to prevent laboratory infections (17–19)

Primary barriers (to prevent dispersal of the microorganism)

Use pipetting devices and not mouth pipetting.

Nothing to be placed in the mouth.

Avoid hypodermic syringes.

Avoid glass Pasteur pipettes.

Replace cracked or chipped glassware.

Use sealed centrifuge tubes.

Use bacteriological loops that are completely closed and less than 3 mm in diameter.

Check homogenizers regularly for aerosol generation.

Ensure that a convenient supply of a suitable disinfectant is available.

Disinfect work surfaces regularly.

Locate discard jars and bins near each work-station and replace daily.

Secondary barriers (to protect the worker should the primary barriers be breached)

Wear proper overalls.

Remove protective clothing when leaving the laboratory.

Wash hands after handling infectious material and before leaving the laboratory.

Cover cuts, abrasions, etc.

Report illnesses to medical or occupational health personnel.

Do not handle microorganisms if under the influence of alcohol, drugs or certain types of medication.

Tertiary screens (to prevent escape into the community if the primary and/or secondary barriers are breached)

Handle microorganisms only in a properly designed laboratory.

Do not open windows if this breaches the laboratory design.

Do not allow unauthorized personnel into the laboratory.

Safety

3. PROTECTION FROM RADIOCHEMICAL HAZARDS

National and local regulations must be followed when handling and disposing of radioactive materials (e.g. refs 24 and 25). General guidelines are provided in a number of publications (26–29). In the laboratory the most important precaution with radiochemical hazards is the use of suitable shielding to minimize personal exposure to radioactive emissions (*Tables 8* and *9*).

Table 8. Shielding required to attenuate β-particles of different energies (30, 31)[a]

Energy of β-particles (MeV)	Thickness (mm) to reduce intensity by one-half		
	Perspex	Glass	Lead
0.1	0.0125	0.005	0.0011
1.0	0.38	0.192	0.042
1.71 (^{32}P)[b]	0.85	0.42	0.095
2.0	1.1	0.52	0.115
5.0	4.2	1.6	0.35

[a] Data reproduced with permission from Oxford University Press.
[b] The maximum range of ^{32}P β-particles is as follows: air, 6 m; water, 8.4 mm; Perspex, 6.4 mm; glass, 3.8 mm; lead 0.45 mm.

Table 9. Shielding required to attenuate γ-rays of different energies (30)[a]

Energy of γ-rays (MeV)	Thickness (cm) to reduce intensity of a broad beam of γ-rays by a factor of 10		
	Aluminum	Iron	Lead
0.5	20.3	6.1	1.8
1.0	25.4	8.2	3.8
2.0	32	11.0	5.9
3.0	37	12.0	6.4
4.0	40	12.7	6.3
5.0	44	13.0	6.1

[a] Data reproduced with permission from Oxford University Press.

4. REFERENCES

1. *Sigma-Aldrich Safety Data.* (1990) Aldrich Chemical Company Ltd, Gillingham.

2. *BDH Hazard Data Sheets.*(1989) BDH, Poole.

3. Lenga, R.E. (ed.) (1988) *The Sigma-Aldrich Library of Chemical Safety Data.* Sigma-Aldrich Corp, Milwaukee.

4. Sax, N.I. and Lewis, R.J. (1988) *Dangerous Properties of Industrial Materials.* (*7th edn*). Van Nostrand Reinhold, New York, Vols 1–3.

5. Bretherick, L. (ed.) (1986) *Hazards in the Chemical Laboratory.* (*4th edn*). Royal Society of Chemistry, London.

6. National Research Council (1981) *Prudent Practices for Handling Hazardous Chemicals in Laboratories.* National Academy Press, Washington DC.

7. National Research Council (1981) *Prudent Practices for Disposal of Chemicals from Laboratories.* National Academy Press, Washington DC.

8. Sax, N.I. and Lewis, R.J. (1986) *Rapid Guide to Hazardous Chemicals in the Workplace.* Van Nostrand Reinhold, New York.

9. Steere, N.V. (ed.) (1971) *CRC Handbook of Laboratory Safety.* (*2nd edn*). CRC Press, Boca Raton.

10. Lefevre, M.J. (1980) *First Aid Manual for Chemical Accidents.* Dowden, Hutchinson and Ross, Stroudsburg.

11. Pitt, M.J. and Pitt, E. (1985) *Handbook of Laboratory Waste Disposal.* Wiley, New York.

12. Young, J.A. (ed.) (1987) *Improving Safety in the Chemical Laboratory.* Wiley, New York.

13. Pipitone, D.A. (1984) *Safe Storage of Laboratory Chemicals.* Wiley, New York.

14. Collings, A.J. and Luxon, S.G. (eds) (1982) *Safe Use of Solvents.* Academic Press, Orlando.

15. Anon (1975) *Toxic and Hazardous Industrial Chemicals Safety Manual.* International Technical Information Institute, Tokyo.

16. Perbal, B. (1988) *A Practical Guide to Molecular Cloning.* (*2nd edn*). Wiley, New York.

17. Collins, C.H. (ed.) (1988) *Safety in Clinical and Biomedical Laboratories.* Chapman and Hall, London.

18. Collins, C.H. (ed.) (1988) *Laboratory-Acquired Infections.* Butterworths, London.

19. Collins, C.H. and Lyne, P.M. (1985) *Microbiological Methods.* (*5th edn*). Butterworths, London.

20. Health and Safety Executive (1988) *Control of Substances Hazardous to Health.* HMSO, London.

21. Imperial College of Science and Technology (1974) *Precautions Against Biological Hazards.* Imperial College, London.

22. Health and Safety Executive, Advisory Committee on Genetic Manipulation (1988) *Note 7: Guidelines for the Categorisation of Genetic Manipulation Experiments.* HMSO, London.

23. Anon (1984) *Federal Register,* **49** (Part VI, No 227), 46266.

24. Health and Safety Executive (1985) *The Ionising Radiations Regulations.* HMSO, London.

25. Health and Safety Commission (1985) *The Protection of Persons Against Ionising Radiation Arising From Any Work Activity.* HMSO, London.

26. Coggle, J.E. (1983) *The Biological Effects of Radiation.* (*2nd edn*). Taylor and Francis, New York.

27. United Nations Scientific Committee on the Effects of Atomic Radiation (1977) *Sources and Effects of Ionising Radiation.* UN Publication E77.ix.1.

28. International Commission on Radioactive Protection (1977) *Recommendations of the International Commission on Radiation Protection.* ICRP Publications 26 and 27. Pergamon Press, Oxford.

29. Anon (1970) *Radiological Health Handbook.* US Department of Health, Education and Welfare, Washington DC.

30. Dawson, R.M.C., Elliott, D.C., Elliott, W.H. and Jones, K.M. (1986) *Data for Biochemical Research.* (*3rd edn*). Clarendon Press, Oxford.

31. Zoon, R.A. (1987) *Methods Enzymol.,* **152**, 25.

Safety

INDEX

ABM paper 282
ABTS 74, 75
ACES 38, 39
acetic acid 28
acetone 88
acids 25, 28
acridine orange 70, 71
acrylamide 262, 263, 264
actinomycin D 31, 34
activated papers 281–284
ADA 38, 39
adenine 55, 56
adenosine 55, 57
ADP 56, 57
agarose gel electrophoresis
  denaturing 260–261
  detection of radiochemicals in
    gels 90
  dye mobilities 257–258
  effects of restriction buffer on
    resolution 115
  gelling temperature 258
  gel strength 258
  loading buffers 259
  melting temperature 258
  resolution 255–258
  running buffers 259
agarose top medium 20
alanine 29, 30
albumin 67
alkaline agarose gels 260
alkaline phosphatase
  enzyme 172
  substrates 73
alkaline protease 50–51
allyl alcohol 61
α-amylase 48–49
α-chymotrypsin 50–51
alpha DNA polymerase 146
α-fucosidase 48–49
α-glucosidase 48–49
α-mannosidase 50–51
amanitin 31, 34
amethopterin 32
amino acids 25, 29–30, 244
m-aminobenzyloxymethyl
  paper 282
aminopeptidase M 50–51
o-aminophenylthioether
  paper 282
aminopterin 31, 34
ammonium acetate 68

ammonium chloride 68
ammonium hydroxide 28
ammonium nitrate 68
ammonium peroxysulfate 68,
  262, 264
ammonium persulfate 68, 262,
  264
ammonium sulfate 68
AMP 56, 57
ampicillin 22, 31, 34
ampicillin resistance 221
AMPS 68, 262, 264
AMV reverse transcriptase 147
amyloglucosidase 48–49
ancrod 50–51
anionic detergents 44
anisole 88
anthracene 86, 87
antibiotics
  chemical and biological
    data 25, 31–37
  E. coli selection,
    concentrations for 22
antibody
  immunodetection 172
antibody probes 284
antipain 63, 65
antisense RNA synthesis 150,
  151, 152
APMSF 63, 65
aprotinin 63
APT paper 282
arginine 29, 30
asparagine 29, 30
aspartic acid 29, 30
ATP 56, 57
autoradiography 88–90
avidin 67
2′,2′-azino-di(3-ethylbenz-
  thiazoline sulfonic
  acid) 74, 75

bacteriophage λ — see also
  λ cloning vectors,
  individual vectors
  buffers 21
  codon usage 251
  DNA restriction fragment
    lengths
    265–270
  E. coli host strains 14
  gene functions 16–17

genetic map 18
genome physical data 242
packaging extracts 15
promoters 221
storage media 20, 21
Bal31 nuclease 154
basal body DNA 235, 241
bases 25, 28
BBL agar/broth 19
BBL top agar 20
BBOT 86, 87
B. cereus RNase 54
BCIP 73, 74
Becquerel 79, 80
BES 38, 39
bestatin 63, 65
β-amylase 48–49
β-fructosidase 48–49
β-galactosidase
  assay 21, 73
  enzyme 48–49
  substrates 73
β-glucosidase 48–49
β-glucuronidase
  enzyme 48–49
  substrates 73
3-β-indoleacrylic acid 221
β-mercaptoethanol 61
β-N-acetyl-D-glucosamine
  48–49
β-particle shielding 312
Bicine 38, 39
bis-acrylamide 262, 263, 264
bisbenzimide 70, 71
bis-MSB 86, 87
blenoxan 31
bleomycin 31, 34
blocking agents 287
BLOTTO 287, 288
blunt-ending of dsDNA 155,
  156, 157, 164, 166
Bluogal 73, 75
boric acid 40
bovine serum albumin 67
Brij detergents 44, 46
Brilliant Blue 70, 72
bromelain 50–51
5-bromo-4-chloro-3-indolyl-β-
  D-galactopyranoside 73,
  75 — see also X-gal
5-bromo-4-chloro-3-indolyl-β-
  D-glucuronic acid 73, 75

5-bromo-4-chloro-3-indolyl
    phosphate   73, 74
bromocresol green   70, 71, 260
5-bromo-3-indolyl-β-D-
    galactopyranoside   73, 75
bromophenol blue
    chemical data   70, 71
    mobility in agarose
      gels   257–258
    mobility in polyacrylamide
      gels   262
5-bromouracil   31, 34
BSA   67, 115
buffers   26, 38–41
n-butanol   61
butyl alcohol   61
butyl-PBD   86, 87

^{14}C
    detection of labeled
      nucleotides   89–90
    energy curve   80
    half-life   80
    radioactive emissions   80
c2RB   194, 198, 200
calcium chloride   68
calcium hypochlorite   68
cAMP   58, 59
CAPS   38, 39
carbenicillin   31, 35
carboxypeptidase A   50–51
carboxypeptidase B   50–51
carboxypeptidase P   50–51
carboxypeptidase Y   50–51
cathepsin C   50–51
cationic detergents   44
cDNA synthesis
    DNA labeling   85
    enzymes   140, 141, 147,
      148, 155, 156, 157, 159
CDP   56, 57
cefotaxime   31, 35
cellulase   48–49
centrifugation
    density gradients   295, 299–
      301
    RCF calculations   295, 296–
      298
    rotor care   295, 302
Cerenkov counting   88
cesium chloride   299–300
cesium trifluoroacetate   299
cetylpyridinium bromide   45
cetyltrimethylammonium
    bromide   45
CHAPS   44, 46
CHAPSO   44, 46
Charon vectors   222, 224
chelating agents   26, 42
chemical hazards   25, 303–310
chemical resistance
    glove materials   309–310
    plastics   308–309

rotor materials   302
CHES   38, 39
c1776 agar/broth   19
chitinase   48–49
chloramphenicol   23, 31, 35
chloramphenicol
    resistance   221
chloroform   61
4-chloro-1-naphthol   74, 75
chloroplast genome sizes   241
chromatin studies   160, 161,
    179
chymostatin   63, 65
chymotrypsin A$_4$ 50–51
citric acid   40, 41
Claforan   31
clone library sizes   235, 238
cloning vectors — *see also*
    individual vectors
    E. coli host strains   14
    M13 vectors   231
    λ vectors   221–230
    plasmid vectors   193–220
clostridiopeptidase B   50–51
clostripain   50–51
CMP   56, 57
4CN   74, 75
codon usages   245–252
coenzyme A   58, 59
colchicine   31, 35
collagenase   50–51
colony hybridization
    analysis   90, 284, 286,
      288, 289
complexanes   26, 42
Coomassie Blue   70, 72
cordycepin   31, 35
cordycepin 5'-triphosphate   57,
    58
counts per minute   79
CPB   44, 46
c.p.m.   79
CTAB   44, 46
CTP   56, 57
culture media
    E. coli media   19
    indicator media   21
    storage media   20
    top agar media   20
Curie   79, 80
cyclic AMP   58, 59
cycloheximide   31, 35
cycloserine   32, 35
cysteine   29, 30
cytidine   55, 57
cytosine   55, 56

DAB   74, 75
2'-dADP   56, 57
dam methylase   8, 13–14, 15,
    124–125
2'-dAMP   56, 57
DAPI   70, 72

2'-dATP   56, 57
3'-dATP   57, 58
DBM paper   281–284
2'-dCDP   56, 57
dcm methylase   9, 13–14, 15,
    124–125
2'-dCMP   56, 57
2'-dCTP   56, 57
3'-dCTP   57, 58
ddNTPs   57, 58
7-deaza-dGTP   58, 59
7-deaza-dITP   58, 59
decay tables for biological
    radionuclides   82–83
denaturants   26, 43
Denhardts solution   287, 288,
    289
density gradient centrifugation
    cesium chloride   299–300
    Percoll   299
    sucrose   301
2'-deoxyadenosine   55, 57
3'-deoxyadenosine   31
2'-deoxycytidine   55, 57
2'-deoxyguanosine   55, 57
deoxyribonuclease I   160
2'-deoxythymidine   55, 57
DEPC   67
detergents   26, 44–47
dextran sulfate   62, 286, 289
2'-dGDP   56, 57
2'-dGMP   56, 57
2'-dGTP   56, 57
3'-dGTP   57, 58
3',3-diaminobenzidine
    tetrahydrochloride   74, 75
diazobenzyloxymethyl
    paper   281
diazophenylthioether
    paper   281
dideoxynucleotides   57, 58
diethyl pyrocarbonate   67
dihydrofolate reductase   221
1,3-dimethoxybenzene   88
1,3-dimethoxyethane   88
dimethylarsinic acid   41
dimethyldichlorosilane   61
dimethylformamide   61
dimethyl sulfate   61
dimethyl sulfoxide   61
2,4-dinitrophenol   32, 35
1,4-dioxan   88
di-potassium hydrogen
    phosphate   40
disintegrations per minute   79
di-sodium hydrogen
    phosphate   41
dispase   52–53
dithioerythritol   43
dithiothreitol   43, 115
DMF   61
DMPOPOP   86, 87
DMS   61

Index

fluoride 64, 66
φX174 DNA restriction
    fragment lengths 277–280
phoA promoter 15
phosphodiesterases I and
    II 168
phosphoramidon 64, 66
phosphoric acid 28
piperidine 61
PIPES 38, 39
pJB8 195, 198, 208
pKC30 195, 198, 209
pKK vectors 195, 198, 209–
    210
pKT vectors 195, 198, 210
placental RNase inhibitor 67
plaque hybridization
    analysis 90, 284, 286,
    288, 289
plasmid amplification 22
plasmid cloning vectors
    E. coli host strains 14
    genetic markers 194–197,
    221
    maps 200–220
    promoters 194–197, 221
    replicons 194–197
    restriction sites 198–199
    sizes 194–197
plasmin 52–53
plastics, chemical
    resistance 308–309
P$_L$ promoter 221
pMC1871 195, 198, 211
PMSF 64, 66
pNH vectors 195, 198, 211
p-nitrophenyl-β-D-
    galactopyranoside 73, 75
p-nitrophenyl-β-D-
    glucuronide 73, 75
p-nitrophenylphosphate 73, 74
pNO1523 195, 199, 211
PNPG 73, 75
PNPGlu 73, 75
PNPP 73, 74
poly(A), blocking agent 287,
    288
polyacrylamide gel
    electrophoresis
    denaturing 261
    dye mobilities 262
    loading buffers 260, 261
    polymerization
        chemistry 264
    radiochemical detection 90
    recipes 262
    resolution 262
    running buffers 259, 261
poly(A) polymerase 177
poly(A) tail removal 159
poly(A) tail synthesis 177
poly(C), blocking agent 287,
    288

polyethylene glycol 62
polyethyleneimine 62
polymerase chain reaction 144
polymers 26, 62
Polymin P 62
polymyxin B1 32, 37
polynucleotide
    phosphorylase 176
polyvinylpyrrolidone 62
POPOP 86, 87
potassium acetate 68
potassium chloride 69
potassium dihydrogen
    phosphate 40
potassium glutamate, in
    restriction buffers 115
di-potassium hydrogen
    phosphate 40
potassium hydrogen
    phthalate 40, 41
potassium hydroxide 28
potassium iodide 69
potassium nitrate 69
potassium phosphates 40
potassium sodium tartrate 69
potassium sulfate 69
PPO 86, 87
primer extension labeling 85,
    141
pRIT vectors 195, 199, 212
proflavin 33, 37
proline 29, 30
Pronase 52–53
propan-2-ol 61
2-propen-1-ol 61
protease inhibitors 27, 63–66
proteases 26, 50–53
protein A 67
Proteinase K 52–53
protein binding studies,
    DNA 160, 162, 164, 173
proteins, general data 27, 67 —
    see also individual proteins
P$_R$ promoter 221
pSELECT-1 195, 199, 213
pseudouridine 55, 57
pSL vectors 195, 199, 212–213
pSP vectors 196, 199, 214–216
pSP6/T7-19 197
pSPORT-1 196, 199, 216
pSPT vectors 196, 199, 216
pT3/T7 vectors 196, 199, 217
pT7/T3 vectors 196, 199, 218
pT7-1, 2 196, 199, 218
pT712, 13 196, 199, 218
pTZ vectors 196, 199, 219
pUC vectors 196, 199, 219–
    220
pullulanase 50–51
puromycin 33, 37
PVP 62
pWE vectors 196, 199, 220
pYEJ001 196, 199, 220

pyroglutamate aminopeptidase
    52–53

rad 79
rad equivalent man 79
radiation absorbed dose 79
radiochemical hazards 312
radiochemicals 79–91
radiochemical shielding 312
radionuclides 80–81
random priming 85
RBE 79
RCF 295, 296–298
RecA protein 183
rec genes 11, 15
relative biological
    effectiveness 79
relative centrifugal field 295,
    296–298
rem 79
resorufin glucuronide 73
restriction endonucleases
    improving efficiency with large
        amounts of DNA 182
    inactivation
        temperatures 116–124
    methylation sensitivities
        124–136, 138
    ssDNA activities 116–124
    star activities 116–124
    Type I enzymes 114
    Type II enzymes
        alphabetical listing 93–100
        isoschizomers 101–114
        recognition sites 101–114
    Type III enzymes 114
restriction fragment lengths
    λ DNA 265–270
    M13mp18 DNA 270–273
    pBR322 DNA 273–276
    φX174 DNA 277–280
restriction mapping 154, 166,
    173
reverse transcriptase
    AMV 147
    M-MuLV 148
riboflavin 262, 263
riboflavin 5'-phosphate 262,
    263
ribonuclease inhibitors 27, 67
ribonucleases — *see also*
    individual RNases
    RNA sequencing 26, 54
ribozyme 139
rifampicin 22, 33, 37
risk and safety classification
    systems 303–308
risk and safety phrases 25,
    305–308
RNA cap structure analogs 58,
    60
RNA editing 243
RNA labeling

MOLECULAR BIOLOGY LABFAX

Index

TMB   74, 75
TM buffer   21
TMG buffer   21
toluene   88
topoisomerases   170–180 — *see also* individual enzymes
TPCK   64, 66
TPE buffer   259
tracking dyes   27, 70–72
transcript mapping   155, 156, 157, 178
trc promoter   221
trp promoter   221
trichloracetic acid   43
Tricine   38, 39
trifluoromethylumbelliferyl glucuronide   73
trimming dsDNA   154
Tris   38, 39
tri-sodium citrate   40
Triton detergents   45, 47
tRNA anticodon modification   171
tRNA misacylation   171
trypsin   52–53
tryptophan   29, 30

tunicamycin   33, 37
Tween detergents   45, 47
tyrosine   29, 30

UDP   56, 57
UMP   56, 57
UN chemical hazard classification system   304
uracil   55, 56
uracil-DNA preparation   15
urea   43
uric acid   55, 56
uridine   55, 57
UTP   56, 57

V8 protease
valine   29, 30
vanadyl ribonucleoside complex   67

water top agar   20
Western blots/ hybridization   90, 284

xanthine   55, 56
X-gal
  agar medium   21

chemical data   73
  structure   75
X-glu   73, 75
X-phos   73, 74
X-ray film   88–90
xylene   88
xylene cyanol
  chemical data   70, 72
  mobility in agarose gels   257–258
  mobility in polyacrylamide gels   262

YAC vector, genomic library sizes   235, 238
yeast tRNA, blocking agent   287, 288
YT agar/broth   19
2YT — *see* dYT

zinc chloride   69
zinc sulfate   69
Zwittergent detergents   44, 46
zwitterionic buffers   26, 38–39
zwitterionic detergents   44